Botanicallly interesting areas mentioned in the text

1. Hebrides, machair
2. Bettylhill
3. Durness
4. Ichnadamph
5. Rassal Aswood
6. Flowe Lands
7. Cairngorms
8. Rothiemurchus
9. Black Wood of Rannoch
10. Culbin Sands
11. Ben Lawers
12. Glen Clova
13. Lake District(especially Helvellyn)
14. Gait Barrows
15. Hutton Roof
16. Ingleborough
17. Malham Cove and Tarn
18. Upper Teesdale
19. North York Moors
20. Spurn Head
21. Derbyshire Dales
22. Snowdonia
23. Cwm Idwal
24. Newborough Warren
25. Great Orme
26.. Brecon Beacons
27. The Gower
28. Kenfig Dunes
29. Wye Valley
30. Wyre Forest
31. Mendips
32. Brean Down
33. The Levels
34. Exmoor
35. Dartmoor(Wistman's Wood)
36. Berry Head
37. Axmouth-Lyme Undercliffs
38. The Lizard
39. Pewsey Downs
40. Chesil Beach
41. Isle of Purbeck, Studland
42. New Forest
43. Surrey Heathlands
44. Dungeness
45. The Weald
46. Orford Ness
47. The Broads
48. Fenland
49. Norfolk Marshes
50. Breckland
51.The Burren
52. Shannon Callows
53. The Gereagh
54. Killarney Oakwoods
55. Magilligan Foreland
56. Lough Owel and Deeavaragh
57. Mountains of Mourne
58. Ben Bulben
59. Macgillycuddy's Reeks
60. Wicklow Mountains.

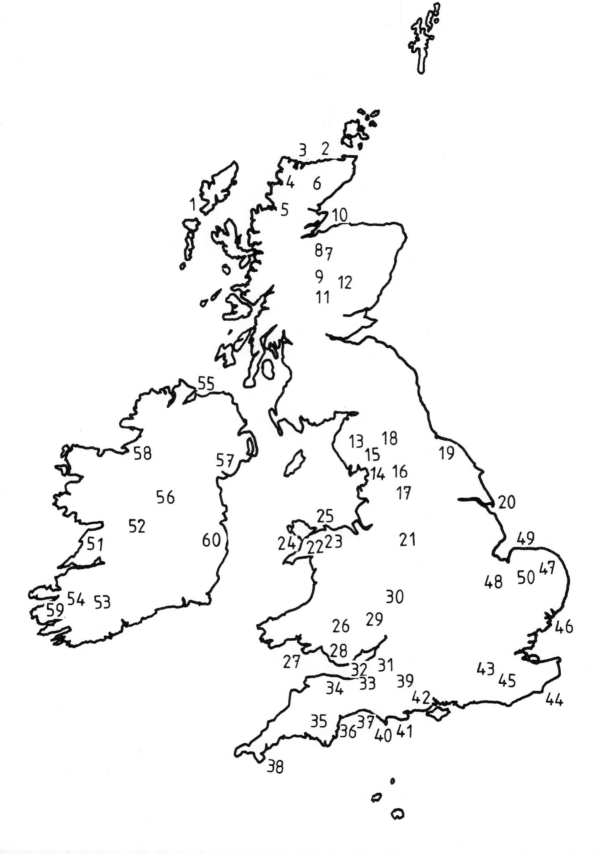

HISTORICAL ECOLOGY OF THE BRITISH FLORA

HISTORICAL ECOLOGY OF THE BRITISH FLORA

Martin Ingrouille

Birkbeck College
University of London
UK

CHAPMAN & HALL

London · Glasgow · Weinheim · New York · Tokyo · Melbourne · Madras

Published by Chapman & Hall, 2–6 Boundary Row, London SE1 8HN, UK

Chapman & Hall, 2–6 Boundary Row, London SE1 8HN, UK

Blackie Academic & Professional, Wester Cleddens Road, Bishopbriggs, Glasgow G64 2NZ, UK

Chapman & Hall GmbH, Pappelallee 3, 69469 Weinheim, Germany

Chapman & Hall USA, One Penn Plaza, 41st Floor, New York NY 10119, USA

Chapman & Hall Japan, ITP-Japan, Kyowa Building, 3F, 2-2-1 Hirakawacho, Chiyoda-ku, Tokyo 102, Japan

Chapman & Hall Australia, Thomas Nelson Australia, 102 Dodds Street, South Melbourne, Victoria 3205, Australia

Chapman & Hall India, R. Seshadri, 32 Second Main Road, CIT East, Madras 600 035, India

First edition 1995

© 1995 Martin Ingrouille

Typeset in 10/12 pt Palatino by Cambrian Typesetters, Frimley, Surrey

Printed in Great Britain by Clays Ltd, St Ives plc, Bungay, Suffolk

ISBN 0 412 56150 6

The publisher makes no representation, express or implied, with regard to the accuracy of the information contained in this book and cannot accept any legal responsibility or liability for any errors or omissions that may be made.

A catalogue record for this book is available from the British Library

Library of Congress Catalog Card Number: 94–74700

♾ Printed on permanent acid-free text paper, manufactured in accordance with ANSI/NISO Z39.48-1992 and ANSI/NISO Z39.48-1984 (Permanence of Paper).

CONTENTS

PREFACE

The native British flora is today relatively impoverished. Today the British Isles has a flora of only about 1500 species of native flowering plants. France and Spain, each geographically only about twice the area, have 3–4 times as many species each. The comparison is more marked when considering the endemic species, those specialities of each geographical region which grow nowhere else. If only normal sexual species are considered, then there are only about 13 endemic species in the British Isles while 1000 species are endemic to Spain.

However, the poverty of the British flora is not a unique phenomenon. The whole of north-western Europe, an area including northern France and much of Germany and Scandinavia, has only about 2000 native species. This region has a shared history in which its present flora is almost entirely immigrant. Not once, but several times, glacial ice and cold deserts have advanced south, wiping the slate clean. Temperatures were at times so low that very few plant species could survive within the geographical area of the British Isles. They have had to recolonize from the south. The last time this happened was only 14 000 years ago. In soil-development terms this is a very short time. In some ways our vegetation and soil are still adapting to the last major climatic change.

In the British Isles there may be another reason for our relatively poor flora. The sea forms a barrier to plant migration. At the end of the glacial period, ice melted in vast quantities and the sea level rose. Britain was isolated from the continent as the North Sea flooded, about 7800 years ago. Some import-ant species on the continent, such as *Picea abies* (Norway spruce), did not get into Britain in time. However, we must not over-emphasize the importance of Britain being an island. A comparison of floras on either side of the English Channel shows that there are species present in England and not in northern France as well as vice versa. Many of the species present in northern France but absent from England are weeds adapted to French agriculture. Others may be limited not by the sea but by the climate.

Nevertheless, the example of Ireland, which was isolated much earlier than the rest of the British Isles, does show the effect of isolation because it does have a much poorer flora and fauna. It has only about 1000 native species of flowering plants and it lacks the mole, the common shrew, wild cat, roe deer and brown hare as native species, and never got the beaver, aurochs, and elk which were once present in Britain.

There might be a third reason for our impoverished flora; the narrow range of habitats present within these islands. For example, we do not have the great range of altitude of the Pyrenees and Alps and we lack the climatic range of France from its moist Atlantic coast to the seasonally dry Mediterranean region. These climatic and topo-graphical variations have allowed a range of vegetations to develop. In Britain 5000 years ago our natural landscape was dominated by a rather species-poor wildwood, even up the high slopes of our hills. Much of the diversity of our vegetation has arisen since then, mostly as a result of human activities. Today many distinct plant communities have been recognized within the British Isles each with a distribution related to soil type and climate and crucially the kind of management or interference they have experienced.

In this book I have tried to tell the whole history of the British and Irish vegetation from its earliest beginnings. Although its present flora is relatively poor, no other part of the world of equal size to the British Isles has such a rich described fossil flora, from the earliest land plants to plants of the Ice Age. Almost the whole story of the evolution of land plants can be told using British fossils as examples. Hardly any other part of the world has undergone such remarkable climate and vegetation changes; tropical forests, hot sandy deserts, broad-leaved deciduous woodlands and arctic tundra are all part of this history. Then more recently there is a rich archaeological history of how humans have used, managed and changed the vegetation. Evidence of the changing flora and vegetation comes from the shape of the landscape itself and the fossils preserved in the rocks and soils. One of the most remarkable achievements of science is the way in which palaeontologists and geomorphologists have learned how to read the landscape and its fossils. There are many difficulties, which they have tried to overcome. The British landscape and its fossils tell a fascinating story.

I have chosen to tell the history of the British and Irish vegetation in time sequence as if it was a film unrolling before us. I do this in the full knowledge that this is a ridiculously ambitious thing to attempt. Where other books describe past floras they choose favourable ages and sites from around the world for the preservation of fossils; they describe individual plants, fossils or reconstructions; or they limit themselves to the Quaternary where there is relatively abundant fossil evidence and at least the fossils are probably of plant species very like those of today. Sir Harry Godwin in his ground-breaking book, called *The History of*

the British Flora, but almost entirely on the Quaternary history of the British flora, suggested that 'an ounce of historical geological fact was worth a ton of historical speculation' (Godwin, 1975).

Godwin's work was subtitled '*A Factual Basis for Phytogeography*'. It was concerned with the changing distribution of individual species. In this volume I have tried to do something different; to tell the story of changing vegetations, the flora in its landscape. Fossil assemblages are the only hard evidence of past vegetations but on their own they are not enough. They need to be interpreted. The ecology of living species and the present vegetation helps us to understand the very distorted and limited picture that fossils provide of the past. This kind of speculation is a dangerous activity but I think it helps to bring the past alive. There are many diversions on the way into the ecology and evolution of species and vegetations.

The botanist who is trying to discover the story of our vegetation is like a film restorer trying to piece together an epic movie from the fragments of a film. Each film-frame shows a collection of fossils. At first there are very few frames and many of them are damaged. Most are taken in close-up when we need to see the wide-angle. The skills of a detective are needed because the botanist does not even know how many scenes the story has, or sometimes which order to put them in. As time passes there are more and more frames, but there are many strange changes of scene and new characters appear and disappear with what looks like great rapidity. In parts it looks as if the film has been exposed twice or two stories are going on at once. Our fossil collections are mixed from different times or from quite different habitats. Only towards the final few minutes of the film are there enough film-frames so that we begin to get a continuous sequence, but even then it is still a very difficult movie to watch. It jumps about and is obscured. It is a silent movie until the last few moments when we have the first eyewitness report from Julius Caesar.

The book tells the story of a changing flora and vegetation over millions of years. Throughout that long period the same natural processes have been acting to change the flora and vegetation. Early on, the time scale is very long, which makes visible to us changes due to the very slow processes of continental drift and the evolution of new plant forms. We can see the changes because, in the same way as a time-lapse film can show us a process which is otherwise invisible to us because film-frames have been taken at long intervals, fossil assemblages have been preserved at long intervals of time. Gradually, the story rolls forward, and fossil assemblages are preserved more frequently. Now changes due to other processes, such as the erosion or uplifting of land masses and climate change, especially from the alternation of glacial and interglacial stages and changes in sea level which accompany them, become more visible and seem to be more important. Then, with a more complete fossil

sequence, changes from processes that occur over a shorter period, like the maturing of soils, the migration of plants and epidemics of disease, come to the fore. At each age different processes seem to be important, but throughout the whole length of the story all have been occurring. It is only right at the end that, like an explosion, the disturbance caused by humankind is so profound that it masks or distorts all the changes from all other natural processes.

A sequence of ages has been named (Tables 1.2, 1.4, 2.4). There are geological periods of time subdivided into epochs and so on, but there can be considerable difficulty in placing a fossil assemblage in the correct part of the time sequence. The whole sequence is never present in any one locality. However, we are very lucky in the British Isles in having almost every part of the sequence represented, but there are some ages missing and others do not contain plant fossils. For some periods and in some places a more continuous sequence may be recorded, but only for a relatively short period. Even

Table The geological time scale (after Anderton *et al.*, 1979)

Era	Period	Epoch		Age when started (millions years BP)
Precambrian				c.4700
Palaeozoic	Cambrian			c.570
	Ordovician			c.492
	Silurian			c.435
	Devonian			c.412
	Carboniferous			c.345
	Permian			c.286
Mesozoic	Triassic			c.235
	Jurassic			c.194
	Cretaceous			c.135
Cenozoic	Tertiary	Palaeocene		65
		Eocene		53.5
		Oligocene		37.5
		Miocene		22.5
		Pliocene		5
		Quaternary	Pleistocene	1.8
			Holocene	0.01

then the process is intermittent, the movie flickers, as sedimentary basins appear, only to fill up and disappear. The older rock strata are generally lower than more recent ones, but this pattern may be greatly disturbed by folding and uplifting of the rocks. Some strata have been eroded away and lost completely. Old fossils may be washed out of rocks and redeposited in young strata. There are several different methods for ageing fossil deposits (section 1.1.4). They all have their limitations and experts may differ on the age of a fossil deposit or even its relative position in the fossil sequence. A sequence of fossil sites has been adopted which seems logical to me.

Species of our present flora are named according to the *New Flora of the British Isles* by Stace (1991). The vernacular or common names, as well as the scientific name of species, have been used. Where a vernacular name has been used in the plural it generally refers to a kind of plant and not to one particular species or genus, so that 'heathers' means all heather and heath-like plants in the Ericaceae and related families, and not just *Calluna vulgaris* (heather). The naming of fossil species poses another difficulty. Usually either a name referring to a kind of plant is used or a Latin scientific name. It may be unwise to use the names of living taxa for fossils before the Quaternary period but this has been done in the description of Tertiary fossils because it helps the reader to picture Tertiary vegetations.

The fossil flora has to be interpreted in the light of what we know about the behaviour of living plants in living vegetations: the way they grow, reproduce, compete; the way their distribution is limited by soils and climate. This is a fairly dubious activity because of the strangeness of many extinct species, but if we want to imagine what it was like in the past, and I for one do, in the absence of a time machine, there really is no alternative. Things get easier from the Tertiary onwards, when more and more plants are

recognizably like those of today, but even so, different kinds of plants which we might not expect to find growing together today, are found in the same fossil assemblage, as if they were growing together in the past. It is important to remember throughout this book that I am indulging in this dubious activity of reconstructing a past vegetation in the image of an existing vegetation just by the words I use; 'marsh', 'forest', 'savannah' and 'swamp' all describe existing vegetations. I use these in the absence of any other readily understandable words, but a Devonian marsh or a Carboniferous swamp would seem very strange to us. Remember, the past was a very strange place indeed.

There is emphasis throughout the book on the adaptations of individual plant species. Much of the information for this has come from numerous authors in the *Journal of Ecology*, especially the accounts for a biological flora of the British Isles. I have also been greatly aided by *Comparative Plant Ecology: A Functional Approach to Common British Species* by Grime, Hodgson and Hunt (1988). Another invaluable aid has been the series *British Plant Communities* edited by Rodwell, three of which have been available at time of writing (Rodwell, 1991, 1992a, 1992b). This classification of British plant communities has been adopted throughout the book and, wherever possible, the code letter used for particular communities in *British Plant Communities* has been included in parenthesis; so, for example, *Calluna vulgaris–Festuca ovina* heath has the code H1. In addition, I acknowledge the help of the two volumes of *A Nature Conservation Review* edited by Ratcliffe (1977), which has proved an invaluable guide to botanically interesting sites in Britain. The *Atlas of the British Flora* (Perring and Walters, 1982), Cleal (1988) and, of course, Godwin (1975) have proved very useful for fossil sites.

This book is divided into three chapters. In the first, the rich fossil flora of the British Isles is described. This is the story of how plants adapted for life on land, their changing forms, the evolution of plants as shown by British fossils. By the Tertiary period many of the species are recognizably like living species today; the stage is set and the cast list has been more or less established. There follows the play; the story of a changing vegetation responding to the alternation of cold and warm periods. The play is different, although the list of characters remains much the same, each time. However, over the very long period covered by this chapter, there was time enough for evolutionary change and the origin of new species.

In the second chapter the establishment of the natural vegetation of the British Isles after the last glacial period, relatively uninfluenced by human activity, is described. This is mainly the history of the spread and establishment of the wildwood, but there is also a description of natural communities which existed outside the woodland. These 'natural' vegetations are represented in today's vegetation only by the modified remnants of those Early Holocene communities, but they include most of the interesting areas for British and Irish botanists.

In the final chapter, the establishment of our present vegetation, the flora in today's landscape, is told. There is an account of the use of the flora by prehistoric man and the gradual conversion of the natural vegetation into a countryside managed by human activity. Then the changing patterns of agriculture and landscape use in the historic period and how they affected the flora are described. Finally, recent changes in our flora, from the loss of habitats and pollution to the introduction of exotic species, are described.

ACKNOWLEDGEMENTS

I would like to thank colleagues, and especially several anonymous referees, who have read parts of, or the whole, text and offered many helpful suggestions. Most importantly, this one is for my parents, for a lifetime of support and encouragement.

Many of the photographs are by Jon Wilson. Those of fossils belong to W. Chaloner. Figures have been made freely available by many publishers and authors and used either directly or modified in diverse ways.

1.1 HARD EVIDENCE FROM THE PAST

Fossils are the primary source of evidence for reconstructing past environments. In many cases the fossil assemblage itself is also used to date the rock strata in which it is found. Plant fossils are of two main kinds: macro-fossils, generally fragments or casts of parts of plants; and microfossils, pollen and spores.

1.1.1 MACROFOSSILS

Fossils are the clues of the botanical detective, the palaeobotanist, but interpreting them is very difficult. Different parts of plants are preserved to a different extent. Soft tissues rot too quickly for fossils to be formed. Fossils may be large and visible to the naked eye (macrofossils) or only visible under a micro-scope (microfossils). Macrofossils are of various sorts: pieces of twig or timber, leaves, reproductive structures and also the frag-ments of animals – the teeth and bones of large animals or pieces of insects.

Permineralized macrofossils are those in which the soft tissue has been infiltrated by minerals such as silicates. The minerals have then precipitated out to provide a rock matrix that supports the organic tissues. Per-mineralized fossils are particularly useful for discovering the internal structure of plants. A classic example of permineralized fossils are those of the Rhynie chert described in section 1.2.5. Because of the vicinity of an active volcano, a marsh at Rhynie was periodically flooded by boiling, silica-rich water. The water killed and sterilized the plants and the subsequent infiltration with silica preserved beautiful fossils in the chert rock. Some of the fossils have a vertical orientation which suggests that they were preserved *in situ* (in growth position).

By cutting and polishing the petrified fossil the palaeontologist can obtain a thin section of the fossil-bearing rock. Observed under a microscope, almost miraculously, it shows cellular detail like any section of a living plant. The acetate peel technique, which is more commonly used these days, relies on the fact that the organic material of the plant is preserved within the rock matrix. A cut and polished surface is dipped in acid to dissolve the rock but not the organic material so that the organic material stands up in relief. When the surface is flooded with acetone and a cellulose acetate sheet is placed over it, the sheet partially dissolves and then hardens around the fossil. The cellulose acetate sheet, plus the fossil, can then be peeled off and mounted on a glass slide for microscopic examination.

Other fossils include the coalified com-pressions of plants. They result from the collapse of the plant material and the chemical altering of plant residues under pressure. Sometimes on these compressions there is a thin film of the remains of the crushed plant or its waxy outer coat, the cuticle. The cuticles of plants are particularly resistant to decay. They can be removed and examined under a microscope. Pollen and spores also have a resistant outer coat, called the exine, made up of a resistant substance called sporopollenin. Maceration of the fossiliferous rock by applications of a strong oxidizing reagent and then a strong basic solution releases the spores or fragments of cuticle for examination.

A third type of fossil is the cast. A mould or impression is formed around the soft tissues of a plant. When the plant tissues rot away the space becomes filled with sediment, providing a cast. A fourth kind of fossil is the direct preservation of the original silicified or calcareous coats of some organisms, notably corals and diatoms.

The fossil plant is rarely preserved as a complete specimen, more likely it survives only as a number of separate plant parts, or fragments mixed with those of other species. Each form or kind of fragment is given a different taxonomic name (a form genus). Some structures, such as root structures, are so similar, even though they are found in a number of different species, that they are given a single fossil name. Only as information accumulates, with the discovery of specimens showing a physical connection of diverse parts, can a whole picture of the plant be reconstructed. The separate parts of one particular fossil coal swamp tree were given the following names: *Stigmaria* for the basal horizontal stem, *Knorria* for the bark impression, *Lepidophloios* for the impression of leaf scars, *Lepidophylloides* for the leaves, *Lepidostrobus* for the male cone, *Lycospora* for the microspore, *Lepidostrobophyllum* and *Lepidocarpon* for the female cone scale and *Cytosporites* for the dispersed megaspore (Thomas and Spicer, 1987). The problem does not end there because then three-dimensional shapes and orientations have to be estimated to reconstruct the plant.

Another difficulty is that we know from living plants that a single individual changes in size, and even in shape, as it grows and matures. Even within a plant, at any one time, there may be different shaped leaves on different parts of the plant. Each kind of leaf might be given a name as if it is a different fossil species. Also, living species contain many varying, genetically different, individuals within each population. There is no reason to think that extinct plants were different. In addition, a species may have different geographical races. Only because we can see the full range of variation of a living species is that variation seen as part of a coherent pattern. In the fossil record only isolated representatives of part of a spectrum of variation may have been preserved, and these variants may be given different names. Only as more and more fossils are discovered is a true picture of ancient ecosystems emerging. Important in this process of scientific detection, of piecing together the past, is taphonomy, the study of the processes of fossil formation.

1.1.2 POLLEN AND SPORES

The spores and pollen grains of plants have been the most important kind of fossils for recording the changing vegetation. Pollen and spores have the advantage of being very resistant to corrosion. Sometimes the kind of plant that produced each kind of pollen grain is known, and so a plant assemblage can be reconstructed from a fossil pollen assemblage. However, in the distant past, it may be difficult to connect the pollen to any particular kind of plant, so that although we have the fingerprint we do not have the portrait. Then the pollen assemblage gives no picture of what the vegetation was actually like. However, for the fossil pollen assemblages of the past few hundred thousand years, hundreds of different species can be recognized by the size and shape of the pollen grain, especially from the outer coat, which is sculpted in various ways (Figure 1.1). Unfortunately in some groups, the most important being the grasses, the pollen grains are nearly identical for all species.

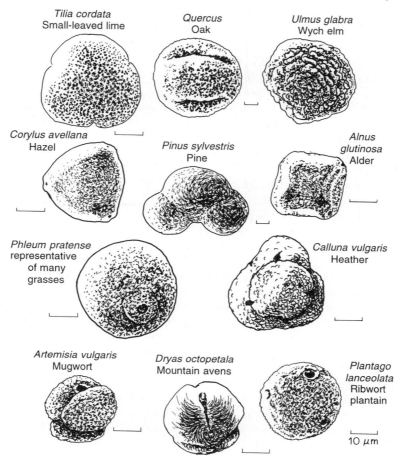

Figure 1.1 Pollen grains of some native British species important for reconstructing past vegetations. (Drawn from photographs in Moore, Webb and Collinson, 1991.)

Pollen grains are obtained from a soil, lake-sediment or peat sample by a variety of different methods (Moore, Webb and Collinson, 1991) involving the digestion of other materials, filtration and centrifugation. Peat can be partly digested by 10% potassium hydroxide solution. Treatment with 10% hydrochloric acid removes carbonates, and concentrated hydrofluoric acid treatment removes silica. The last stage is normally acetolysis, with a 9 : 1 mixture of acetic anhydride and concentrated sulphuric acid, which removes polysaccharides such as cellulose but leaves the highly corrosion-resistant pollen coat made of sporopollenin.

Pollen types are counted to assess the relative contribution of different species to the vegetation. The spectrum of species represented in fossil pollen assemblages can be used to determine likely environmental conditions. The assumption is made that change in a fossil pollen assemblage is due to a change in the climate or soils. However, caution has to be exercised in the reconstruction of past environments. The assemblage can change for several reasons not related to climate or soils. Microfossils are no different from macrofossils in one respect: species differ in how well they preserve. Some species of plants produce spores and pollen which preserve very readily and others do not. For example, the best conditions for the preservation of pollen are where oxygen is excluded and calcium is absent. As a result wetland species or peatland species are well recorded in the fossil record compared to those of dry, calcareous grassland.

Pollen and spores are tiny (5–200 µm) and dispersed in the air or water. Also, some kinds are more easily transported than others and so they may be deposited well away from the plants by which they were produced. There is the additional problem that the individual plants of some species produce abundant quantities of spores and pollen and others very little, so that the relative abundance of plants is difficult to assess from the abundance of spores or pollen grains; a correction factor of pollen productivity has to be applied to convert pollen abundance to plant abundance (Andersen, 1970). More pollen is produced by wind-pollinated species such as *Pinus sylvestris* (Scots pine) than by insect-pollinated species such as *Tilia cordata* (small-leaved lime) (Table 1.1).

1.1.3 RECONSTRUCTING PAST VEGETATION

A major problem in the interpretation of the evolution of floras from fossils is the patchiness of the fossil record, both ecologically and geographically. Plant fossils are found in sedimentary rocks, such as shale and mudstone, and, as a result, the types of fossils preserved are likely to be biased towards those kinds of plants which grew in the close vicinity of a sedimentary basin, lakes and swamps, or near the coast. Marsh plants are abundantly represented. Plants growing in dry, upland habitats, where the rocks were eroding, are likely to be sparse or absent in the fossil flora. Only plant fragments that were small and tough enough to survive

Table 1.1 Relative pollen deposition rates beneath a forest canopy (source: Andersen, 1970)

		Pollen productivity
Fagus sylvatica	Beech	1
Pinus sylvestris	Scots pine	4
Quercus spp.	Oak	4
Betula spp.	Birch	4
Alnus glutinosa	Alder	4
Carpinus betulus	Hornbeam	3
Ulmus spp.	Elm	2
Tilia spp.	Lime	0.5

being washed in streams and rivers down to the sedimentary basin are likely to have been preserved.

For the relatively recent past the pollen record gives us the best idea we have of palaeovegetation. By comparison to the climatic limits of species in the present-day vegetation, past climates can be estimated, assuming that climate is the main determinant of species distributions and past behaviour is similar to present behaviour. Neither of these preconditions can be taken for granted. However, by using a combination of species a fairly accurate picture of past climate can be obtained.

The evidence for past changes in the environment is strengthened when it comes from many other sources (Lowe and Walker, 1984). Many other kinds of organisms provide evidence of change. They include fungi, algae, diatoms, rhizopods, cladocera, ostracods, chironomids and molluscs. Beetles have been especially useful because they seem to respond very rapidly to climate change by migrating (Coope, 1977). Some species of beetle are good climate indicators: they seem to be unchanged from the past and have very narrow climatic limits today. They may provide very good maximum and minimum thermometers of past climates. The macrofossils of the megafauna, large animals, can be very revealing about the nature of the vegetation. The size and form of large herbivores, and especially the nature of their teeth, provide clues about grazing suffered by the vegetation.

There are also physical and chemical methods of discovering past climates. Palaeotemperatures have been inferred in a number of ways. In particular, the oxygen isotope ratio $^{18}O/^{16}O$ in precipitation (snow or rain) depends on climate. The ratio of oxygen isotopes, in polar ice cores or in the carbonate sediments of lakes and seas, from the bodies of planktonic foraminiferans or from molluscs, records changes in temperature. The ratio varies between 1 : 495 and 1 : 515 with a mean of 1 : 500. During evaporation there is preferential evaporation of lighter $H_2^{16}O$ molecules. This is particularly marked at high latitudes where cold air is increasingly unable to support the heavy $H_2^{18}O$ molecules. Thus the oxygen isotope ratio in precipitation is indirectly a record of the sea surface temperature. In cold phases large quantities of $H_2^{16}O$ are trapped in glacial ice, thereby enriching the sea with $H_2^{18}O$. A 1°C reduction in temperature gives 0.02% enrichment.

1.1.4 DATING THE FOSSILS

Given all the considerable problems of interpreting the fragmentary fossil record, it is remarkable how much has been discovered (Lowe and Walker, 1984). However, the sequence of events is still uncertain in many respects. The largest remaining problem is the accurate dating of fossil deposits so that they can be compared across regions and ages (Jones and Keen, 1993). The older rock strata are generally lower, buried below more recent ones, but this layer-cake may be greatly disturbed by folding and uplifting of the rocks. The fossils themselves, especially the range of species present, are commonly used to help identify the relative age of the rock stratum. However, this can lead to some perilously circular reasoning concerning the development of floras; if fossils are used to date the rocks, the rocks cannot be used to put the fossils in sequence to show the development of the vegetation. There is the extra difficulty of comparing floras from different areas. Two adjacent areas may differ in climate, or in some edaphic factor such as the wetness of the soil, and so have very different floras, so different that they might seem to be from quite different ages.

There are many different methods for giving an absolute age involving 'isotopic clocks', unstable forms of chemical isotopes that radioactively decay to more stable forms at a known rate called a 'half-life'. Radiocarbon dating is the most important and has been used very effectively to date biological remains of a few thousand years old (Pilcher, 1991). It relies on the fact that radiocarbon, ^{14}C, is in relatively constant proportion to 'normal' ^{12}C in the atmosphere. The ratio of these two atoms is maintained by the barrage of cosmic radiation which converts atmospheric nitrogen to radioactive ^{14}C which is almost immediately oxidized to $^{14}CO_2$. $^{14}CO_2$ is incorporated into living material in the same proportion as in the atmosphere, but when an organism dies the proportion of ^{14}C declines with age because it decays to 'normal' nitrogen by the emission of a β-particle. The half-life is 5730 years (an earlier estimate of 5570 years is used by convention). After 10 half-lives, about 50 000 radiocarbon years, there is little ^{14}C left in the dead organic matter, and little detectable radiation, so dating becomes inaccurate.

There are several potential sources of error in radiocarbon dating. Laboratories will differ in the date they measure for the same material. Dates are normally reported ± an age range. For example, the date for the earliest evidence of neolithic settlement in the British Isles at Ballynagilly in County Tyrone is 5745 ± 45 BP (ApSimon, 1976). This means that there is 1 chance in 20 that the true radiocarbon age lies outside twice the range reported, outside the range 5935–5655 BP. A very important potential source of error comes from the accidental or unavoidable mixing of samples of different ages. Dead roots of younger plants can be incorporated in older deeper layers. Contamination by even a small percentage of modern carbon will greatly reduce the age measured; 0.1% modern carbon in a sample 48 000 years old will give it an age of 45 300 BP (Pilcher, 1991). Radiocarbon dates also systematically underestimate calendar date. They have been calibrated with the use of dendrochronology, counting tree rings, back to about 8000 BP. At this age radiocarbon date underestimates calendar age by about a thousand years. The calibration to calendar age is not straightforward because the amount of radiocarbon in the atmosphere has varied over the ages as a consequence of variations in solar activity and the amount of cosmic radiation reaching the Earth's atmosphere. A decrease in atmospheric radiocarbon is recorded from the nineteenth century onwards because of the burning of fossil fuels, but this trend was reversed post-1945 because of atomic bomb testing. The perturbations of the world climate during the glacial period may have had a dramatic effect by changing the solubility of carbon dioxide in the oceans.

Another dating method relies on the proportion of radioactive potassium, ^{40}K, and one of its decay products, argon, ^{40}Ar, trapped in volcanic rocks as they cool. This method is rather coarse but has been used to date rocks 4 billion years old. It is generally only valuable for dating fossils where volcanic rocks can be found interleaved with fossiliferous rocks. Another method measures ^{40}K indirectly by the proportion of $^{39}Ar/^{40}Ar$ (Smart and Frances, 1991). This has a greater sensitivity and can date minerals of an age of a few hundred thousand years. Other radiometric measures can be applied to bone, stalagmites and stalactites, carbonate material from lakes, corals, mollusc shells and peat. They rely on the radioactive decay of soluble uranium salts to other radioactive intermediates, such as protactinium or

thorium. The ratio changes with age because different isotopes have different half-lives. In particular the ratio $^{230}Th/^{234}U$ has proved the most useful, giving ages back to about 350 000 years and partly bridging the gap between radiocarbon and potassium/argon methods (Smart and Frances, 1991).

These isotopic dating methods provide an absolute chronology of true age. There is a wide range of alternative methods. Thermoluminescence has been used to date archaeological deposits (pottery and burnt stone) but also works effectively for the quartz and feldspar once exposed to sunlight in wind-blown sediments (loess) within the past few hundred thousand years (Smart and Frances, 1991). Electron spin resonance (ESR) can be used to date molluscs, corals and tooth enamel (Smart and Frances, 1991). Counting laminations in lake sediments, varves due to regular algal blooms, has provided evidence of the time period of a deposit. One method uses the reversals in the Earth's magnetic field to correlate different ocean sediment cores. Systematic changes in amino acids after death, the change from protein L-amino acids to non-protein mirror-image D-amino acids, provide a chemical signal for obtaining the relative age of bone and shell.

Accurate dating becomes more and more difficult with greater age. For this reason old dates, greater than 10 half-lives, must be treated with caution as very rough approximations. With our present inadequate knowledge, different experts do disagree about the relative age of some sites, sometimes by millions of years.

1.2 THE SILURIAN AND LOWER DEVONIAN

The story of the British vegetation, as it is preserved today, starts about 440 million years ago in the Upper Silurian rocks of a few quarries in Wales and the Welsh borderland. Britain lay in the tropics just south of the equator. A shallow sea lay in a depression between a north European and a North

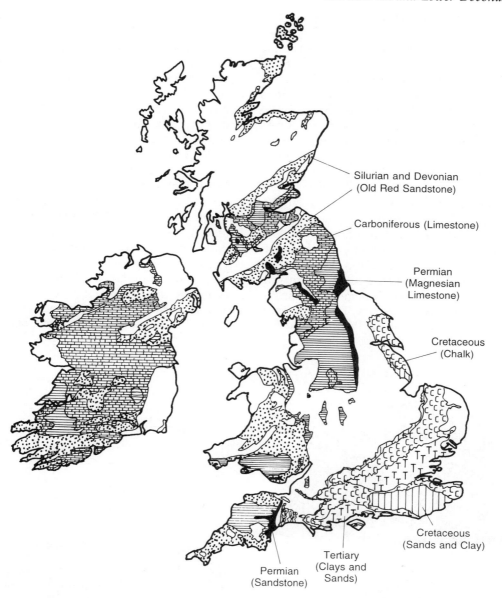

Silurian and Devonian
(Old Red Sandstone)

Carboniferous (Limestone)

Permian
(Magnesian
Limestone)

Cretaceous
(Chalk)

Cretaceous
(Sands and Clay)

Tertiary
(Clays and
Sands)

Permian
(Sandstone)

Figure 1.2 The geographical distribution of rock strata of different ages. (Adapted from British Museum (Natural History), 1975.)

American continent which had collided and were fusing together. Gradually the Welsh Basin, which contained the sea, filled up with silt from the land around, and with carbonates from corals and other marine fauna from a primeval sea. Contained in these sediments are some of the earliest of all fossils of complex, multicellular plants (Figure 1.2).

1.2.1 THE FIRST MULTICELLULAR PLANTS

The fossils are strange and unlike anything living today. In the Pen-y-Glog Quarry in Clwyd, or later sites like Perton Lane in Hereford and Worcester, there are very peculiar fossils, which may have been fungus or alga (Figure 1.3a, c–e) (Bassett, 1984). Some consist of a set of parallel tubes, with, perhaps, the tubes on the outside bending out at right angles to create a kind of hairy outer layer (Edwards, 1982; Burgess and Edwards, 1988). There is also a fossil that consisted of little spherical bodies, about 0.5 cm in diameter, with a central medulla of woven tubes and an outer cortex of radially arranged tubes (Figure 1.3b). Like a ball, it may have rolled across the sea floor with the tide. There were other unusual plants: in the Llangammarch Wells Quarry, Powys, there is one called *Powysia* and in Rockhall Quarry,

Hereford and Worcester, one called *Inopinatella*.

The sea was teaming with life, but imagine a landscape where the only variations from the dull colours of rock and dust were stripes of slime; green, purple and red algae in damp hollows or on the banks of rivers and streams and at the margins of lakes and seas. Insignificant slime, but, nevertheless, it, along with aquatic algae, was changing the world. The oxygen level 440 million years ago was only 10% of that today and, without an ozone layer to protect the Earth, high levels of damaging ultraviolet (UV) light reached the land. By photosynthesizing, fixing carbon dioxide into carbohydrates and releasing oxygen, the first plants were helping to build a protective ozone layer (Cloud, 1976).

Some of the strange Silurian and Early Devonian forms have morphological features that may indicate that they were becoming

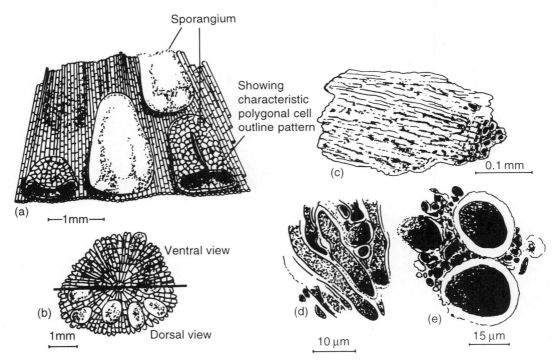

Figure 1.3 Silurian plants: (a) reconstruction of *Parka*; (b) reconstruction of *Pachytheca*; (c)–(e) *Protoaxites*. (From Niklas, 1976; Edwards, 1982; Burgess and Edwards, 1988.)

adapted to life on land (Gensel and Andrews, 1987). At a range of quarries in Powys, Dyfed and Salop (Cwm Craig Ddu, Capel Horeb, Perton Lane, Targrove Quarry, Freshwater East) as well as *Protoaxites* and *Pachytheca* there are fossil forms with definite terrestrial adaptations. At the Capel Horeb Quarry in Powys and Targrove Quarry in Salop there is a range of forms, including a thallose plant, called *Nematothallus*, which may have encrusted rocks. It had a thick, upper, reticulate-patterned protective cuticle.

A later thalloid plant from the Lower Devonian was *Parka*, first described from Parkhill near Newburgh in Fife, Scotland, and also found in several places in Scotland, England, Germany and North America (Niklas, 1976). *Parka decipiens*, was a flat, lobed thallus from 5–70 mm long, like a kind of liverwort. It was covered with groups of spores, the groups covered by a thin layer of cells, and looking like squashed raspberries. It is a characteristic of all land plants to produce spores in tetrads, in groups of four, following a special kind of cell division called meiosis. In *Parka* the spores are not in tetrads. Perhaps these are not spores but vegetative propagules, like the gemmae of living liverworts, or else *Parka* was an alga and not a liverwort.

1.2.2 ADAPTATIONS FOR LIFE ON LAND

A particularly important fossil, because it is the earliest plant to show several adaptations for life on land, is *Cooksonia* (Figure 1.4) (Chaloner, 1970; Edwards and Fanning, 1985). It had rounded or kidney-shaped sporangia, organs in which hundreds of air-dispersed spores were produced. The presence of these trilete spores, which have a characteristic shape from being produced in tetrads, is the earliest firm evidence of land plants in many fossil rocks. The earliest *Cooksonia* has been discovered in rocks older than 420 million years, from the Devilsbit Mountain of Co. Tipperary in Ireland. Already several species existed, differing in the shape of their sporangia (Figure 1.4).

Cooksonia had a very simple branching pattern. The short stems forked at a wide angle and each stem terminated with a rounded nob of a sporangium. Plants probably grew adjacent to stream channels or on top of sand bars, rapidly maturing as the habitat dried, so that there was a flush of sporangia. The outer layers of the stem and sporangia had cells with thick cell walls, sclerenchyma, which supported the plant and may have conferred some protection from drought. A cuticle and pigmentation protected against high UV light.

The evolution of *Cooksonia* was a portent of a remarkable change in the world. For millions of years nothing much had changed on land. Now the plants were making the terrestrial environment potentially habitable for animals (Jeram, Selden and Edwards, 1990). For example, at Ludford Lane near Ludlow, in Salop, 414 million years ago, along with *Cooksonia* there were two kinds of centipedes and a trigonotarbid arachnid which must have been preying on smaller arthropod detritus feeders. If the whole history of the Earth took place in one day, the first cells had evolved before 8 a.m. but the land was not properly colonized until 10 p.m. Then at the beginning of the Devonian period, 412 million years ago, after all the waiting, like a kettle suddenly coming to the boil, a full terrestrial vegetation and an accompanying arthropod fauna appeared in just a few million years.

1.2.3 THE OLD RED CONTINENT

Britain with Europe, and North America were part of the same continent of Euramerica or 'the Old Red Continent' (Figure 1.5), most of it covered by sea. The red colour of the rocks, the Old Red Sandstone, and the evidence of other soil processes which were acting in the Devonian, like the development of calcrete, calcareous soil horizons, or

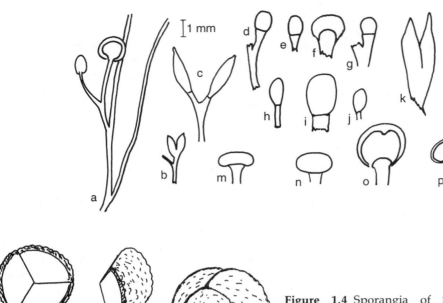

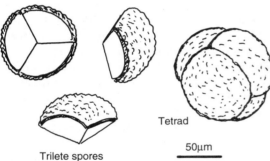

Figure 1.4 Sporangia of Silurian and Early Devonian plants: (a) *Renalia*-type specimen; (b)–(c) *Salopella marcensis*; (d)–(g) *Cooksonia hemisphaerica*; (h)–(j) *Tarrantia salopensis*; (k)–(l) *Torticaulis transwalliensis*; (m) *Cooksonia pertoni*; (n) *C. cambrensis*; (o)–(p) *C. calidonica*. (From Edwards and Fanning, 1985; Edwards and Davies, 1990; Fanning, Edwards and Richardson, 1990.)

carbonate beds, were results of a hot and dry climate. Similar soils develop today in dry tropical latitudes where the mean annual temperature is 16–20°C and annual rainfall is 100–500 mm, but where there is some seasonal rainfall. In these conditions early land plants may have been confined for most of the year to water margins and marshy areas.

The Devonian period, from 412–354 million years ago, is named from the Old Red Sandstone found in Devon (Arber and Goode, 1915) but fossil plant deposits of the Lower Devonian in Britain are found in two main geographical areas: in the Welsh borderland and South Wales (Croft and Lang, 1946) and in the Midland Valley of Scotland. In the south, rivers meandered across subsiding coastal plains, depositing their alluvium of

mud and sand in deltas, on the coastal shelf of the Devonian Sea. Freshwater sediments interdigitate with marine ones. Underwater volcanoes erupted to produce pillow lavas of basalt. In northern Britain, sediments were laid down in a freshwater lake in the basin of the Midland Valley of Scotland. It was ringed by volcanoes which spewed ash and lava. As the Devonian progressed volcanic activity became more intense.

1.2.4 EARLY DIVERSIFICATION

In these conditions a diverse flora was evolving (Chaloner and MacDonald, 1980). *Cooksonia* was present throughout the Late Silurian and Early Devonian, but now it was accompanied by more and more diverse types (Figure 1.4) (Edwards, 1968, 1970, 1976;

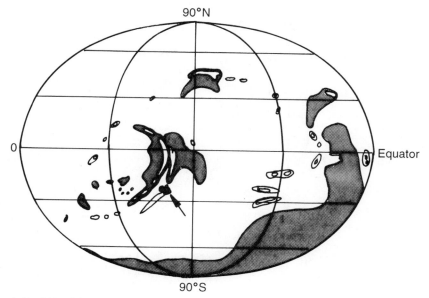

Figure 1.5 Map of the Late Silurian world (425 million years ago) showing the position of the British Isles, south of the equator. Shaded area = land and high ground. (Courtesy of C.R. Scotese, Palaeomap Project, University of Texas, Arlington.)

Edwards and Richardson, 1974; Shute and Edwards, 1989; Fanning and Edwards, 1992). *Salopella* and *Torticaulis* had an elongated sporangium. From Targrove Quarry, *Tarrantia* had spherical sporangia and *Uskiella* (Figure 1.6d) had flattened ovoid ones.

Other kinds had different branching patterns. *Renalia* had a main stem with lateral branches (Figure 1.4a). *Zosterophyllum* is often represented by a species, first described from Myreton, in Scotland, called *Z. myretonianum* (Figure 1.6e) (Lang, 1932a; Lele and Walton, 1961; Edwards, 1975; Edwards and Kenrick, 1986). The zosterophylls had a complex branching pattern called H-branching. Two branches were produced shortly after each other from a horizontal stem; one was directed upwards and one downwards into the soil or water. They may have looked like tiny species of rush. Another species was *Z. llanoveranum*, first described from the Llanover Quarry in Gwent. Unlike *Cooksonia*, which had sporangia situated individually,

Zosterophyllum, and the related *Hicklingia* from Scotland, had sporangia on small lateral shoots as well as a terminal one on the apex of the stem (Figure 1.6f). The sporangia were relatively large and kidney-shaped, and opened by a transverse slit to release the spores.

Another evolutionary line had forms with stems without a terminal sporangium. This allowed indeterminate growth, the ability to keep growing, producing lateral sporangia, as long as conditions remained favourable (Niklas and Banks, 1990). This greater plasticity of growth was favourable because it adapted species to growing in varying seasons. Some forms also had short projections, like spines, which increased their photosynthetic area (Figure 1.6a). They include *Sawdonia* and *Dawsonites* from Scotland (around Callander in Perthshire and Orkney), *Thrinkophyton* from near Carmarthen, and *Gosslingia*, *Tarella* and *Deheubarthia* from the Senni Beds of Wales, at the Llanover Quarry

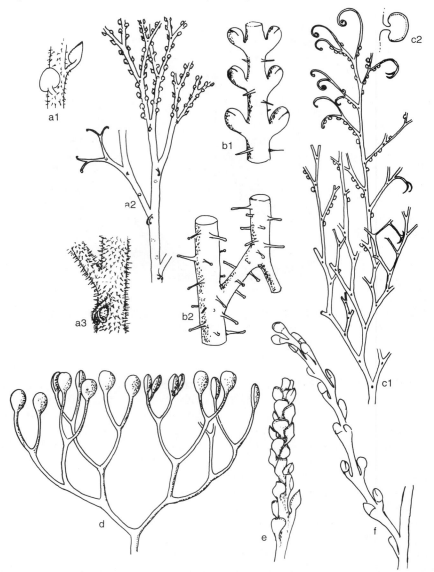

Figure 1.6 Lower Devonian plants: (a) *Deheubarthia splendens* (= *Psilophyton, Dawsonites*) (a1, sporangia; a2, branching stem; a3, lateral appendage); (b) *Sawdonia ornata* (b1, stem with sporangia; b2, branching point); (c) *Gosslingia breconensis* (c1, branching stem; c2, sporangium); (d) *Uskiella spargens*, (e) *Zosterophyllum myretonianum*; (f) *Hicklingia edwardii*. (From Edwards and Richardson, 1974; Edwards, 1976; Rayner, 1983; Edwards and Kenrick, 1986.)

near Abergavenny, and in the Brecon Beacons Quarry.

These evolutionary lines differed from *Cooksonia* in having a kind of water-conducting cell, called G-type after *Gosslingia* (Figure 1.6c). These cells are related to the water-conducting cells, called tracheids, found in almost all living vascular land plants (Edwards and Davies, 1976, 1990). Like tracheids, the spiral or annular thickening of the cell wall, which gave it strength, was decay resistant and therefore probably impregnated with lignin.

1.2.5 A DEVONIAN MARSH

Probably the most important fossil site of the Devonian period is found in a small area near the village of Rhynie in Aberdeenshire (Figure 1.7). It was a reed-like marsh which lay in the shadow of a volcano. The remarkable nature of this site is that many plants look as if they have been preserved *in situ* (Figure 1.8). It is like a Devonian Pompeii for plants (Kidston and Lang, 1917–21).

It is difficult to be sure where the Rhynie marsh plants fit into the time sequence. The primitive morphology of many of the species, only a little advanced on *Cooksonia*, is like many fossils of an Early Devonian age, but

the chert occurs in a group of rocks of Middle Devonian age. The Rhynie plants may already have been primitive at the time they were growing because they lack the sophisticated adaptations for life on land that some of the other Lower Devonian plants possessed. Possibly, the Rhynie swamp lay at high altitude and preserved a relic of Early Devonian vegetation, while new forms continued to evolve at lower altitudes. Even today, primitive clubmosses, perhaps direct descendants of some Rhynie plants, grow at high altitudes on our hills.

The Rhynie plants were all only a few centimetres tall (Figure 1.9). Three species, *Rhynia*, *Aglaophyton* and *Horneophyton*, were closely related to *Cooksonia* but were slightly more advanced. Simply branching, upright stems arose from either a horizontal rhizome or a basal corm which had fine filaments,

Figure 1.7 Rhynie muir looking west from Rhynie, Aberdeenshire. (Courtesy of W.G. Chaloner.)

Figure 1.8 Rhynie chert. (Courtesy of W.G. Chaloner.)

called rhizoids, reaching into the superficial part of the soil. All these plants had pores, called stomata, in the epidermis, so that water was drawn up the stem by evaporation from the surface of the plant. There is a central strand of tissue in the centre of the stems (Figure 1.10a). This 'stele' conducted water. It contained elongated cells with thickened cell walls. In contrast to the G-type tracheids of *Zosterophyllum* (section 1.2.4), these alternative S-type water-conducting cells, like those of a few living mosses, had a spongy thickening with only a very thin, decay-resistant inner coating (Kenrick and Crane, 1991).

Also present was a *Zosterophyllum*-like plant called *Nothia*, with its sporangia on short lateral branches. The tallest and most sophisticated plant at Rhynie was *Asteroxylon*. It was 50 cm tall. It looked rather like a large kind of clubmoss. Unlike the other Rhynie plants it had small leaves and a complex star-shaped internal anatomy (Figure 1.10b), with G-type tracheids. It also had forked roots. Perhaps the ancestors of two fundamental lineages of plants was represented at Rhynie: the mosses by *Rhynia*, *Aglaophyton* and *Horneophyton* and the vascular land plants by *Nothia* and *Asteroxylon*.

One other species present at Rhynie, called *Lyonophyton*, is interesting because it had the sex organs that are present at one stage in the life cycle of living land plants; the male antheridia in which motile sperm are produced, and the vase-shaped archegonia, in each of which a single egg is produced.

In the Rhynie soil there were fungal filaments as well as stems, roots and rhizoids. Perhaps the fungi were part of a kind of mycorrhizal association, the association between the roots of plants and a fungus, which is found in many living land plants and helps them to garner rare nutrients like phosphate from the soil. Another interesting association is shown by the presence of some of the first land animals: mites and springtails (Rolfe, 1985). The plants were making the land habitable for animals. Some plants show lesions indicating that they were also a source of food to these first land animals. They include a sap-sucking mite called *Protacarus*. Some sporangia contain trigonotarbid mites, called *Palaeocharinus*, which may have been spore-eating.

The Rhynie marsh was dominated by one or other species at different times. Lowest in the chert, and possibly representing the oldest kind of vegetation, was *Rhynia*. It seems to have colonized open areas and disappeared when the marsh was flooded. Later *Asteroxylon* and *Horneophyton* formed clonal patches in a marsh which was periodically flooded and destroyed. Eventually *Asteroxylon* failed to re-establish. *Horneophyton*, which had a basal corm allowing it to

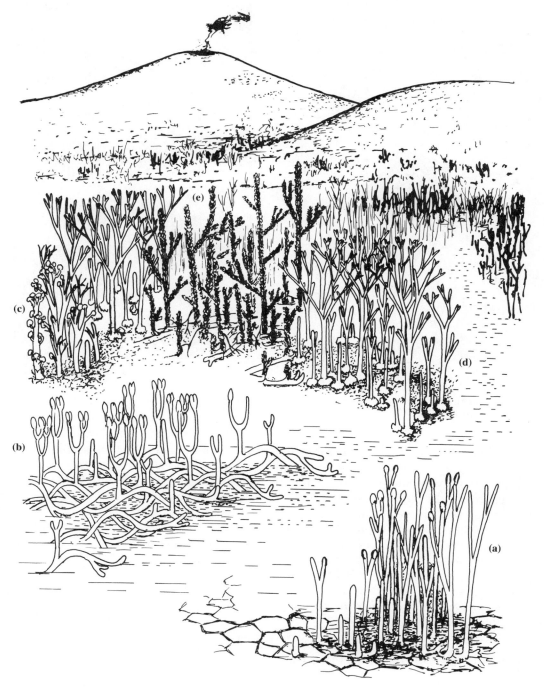

Figure 1.9 A view of the Devonian marsh at Rhynie in Aberdeenshire: (a) *Rhynia*; (b) *Aglaophyton*; (c) *Nothia*; (d) *Horneophyton*; (e) *Asteroxylon*.

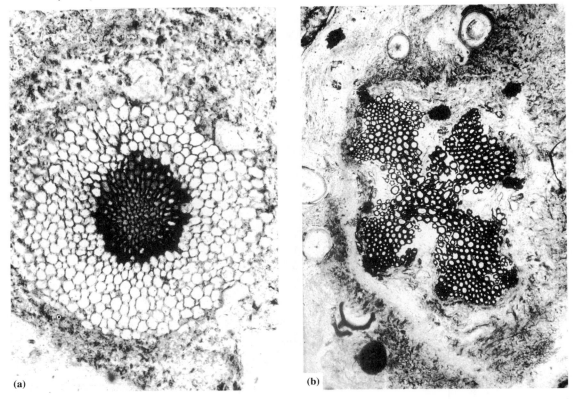

Figure 1.10 Transverse sections of the vascular strand of (a) *Aglaiophyton*, (b) *Asteroxylon*. (Courtesy of W.G. Chaloner.)

survive relatively dry periods, became the dominant species. Unlike the others, it did not grow extensively laterally, but seems to have relied on the efficient production of spores to spread to new areas. In wetter areas, the first species to recolonize the marsh each time it was flooded was *Aglaophyton*. It formed a highly competitive turf. It had a horizontal stem which bent up and down into the better-aerated upper layers of the waterlogged soil or pools of water in which it was growing.

1.3 THE MIDDLE AND UPPER DEVONIAN

The Rhynie chert is an inlier of the deposits of a large non-marine lake of the Middle Devonian which lay over north-east Scotland, the Orkneys and Shetlands, the Orcadian Basin, which was ringed by mountains. Fossiliferous rocks are found in the Caithness Flagstones and the Stromness Beds, among others (Lang, 1932b; Chaloner, 1972). The fossils show a change in the land flora. There were no large terrestrial vertebrates to graze plants back and so the land was a green tangle of plants scrambling for space. There was intense competition for light. One solution was to grow tall to stop being shaded. Soon there was the evolution of diverse kinds of trees (Figure 1.11).

As if a higher evolutionary gear had been selected, the diversification of land plants continued apace, filling and creating new niches, clothing the land in green. New fundamentally different lineages were

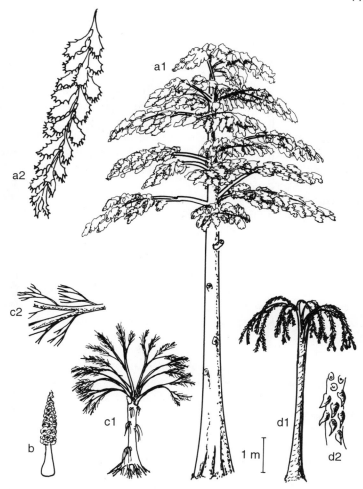

Figure 1.11 Devonian trees: (a) *Archaeopteris* (a1, tree; a2, part of foliage); (b) *Duisbergia*; (c) *Pseudosporochnus* (c1, tree with trunk clothed in roots; c2, part of branching axis); (d) *Cyclostigma* (d1, tree; d2, branch with leaves).

evolving. The cast list of terrestrial plants multiplied hugely. Spore-producing vascular plants were now represented by three main kinds which have living descendants. The lycophytes were spore producers with simple, small leaves and simple, usually dichotomous, branching. Fossil lycophytes were herbaceous or arborescent but all living lycophytes are herbaceous and include the clubmosses *Lycopodium* and *Selaginella* and the quillwort, *Isoetes*. The fossil sphenophytes were also herbs or trees. There is only one genus of living sphenophyte, the horsetails, *Equisetum*, which are all herbaceous. The sphenophytes are recognizable by their radial symmetry, with leaves and branches arranged in whorls and stems with a distinctly jointed appearance. The last group of spore producers present were the primitive ferns, which were also herbs or trees. Superficially many of these looked like living ferns but they produced spores in simple primitive sporangia with a thick sporangial wall. Only a few genera of living ferns have these primitive sporangia. They survive mainly in the tropical and subtropical forests of Madagascar, South-East Asia and the Antipodes.

1.3.1 FIGHTING FOR SPACE

Many of the terrestrial plant species that evolved in the Middle and Late Devonian became widespread in continents of the northern hemisphere. At Ballanucator Farm, central region of Scotland, there was *Sawdonia* (Figure 1.6b) and the lycophyte *Drepanophycus*, a tough, shrubby plant with scale-like leaves. In *Drepanophycus spinaeformis*, the scale-leaves had the extra sophistication of each having a vascular trace, to conduct water and the products of photosynthesis in and out. An extraordinary plant has been identified from a borehole at Harmansole in Kent. It was *Duisbergia mirabilis*, a great long cone of a plant, 2–3 m tall (Figure 1.11b). It reminds one of the giant rosette plants of tropical mountains. Perhaps it, too, was adapted to extreme daily temperature fluctuations.

One very remarkable fossil from Llanover Quarry is *Sporogonites exuberans*. It was thalloid with sporangia on thin, unbranched stalks 4–6 cm long (Andrews, 1960). This may represent the third fundamental lineage of living land plants, the liverworts, but the sporangia are more like the capsules of mosses, with a basal sterile region, the apophysis, and a central sterile columella inside the spore mass. Another intriguing plant was *Sciadophyton steinmanni*, with flattened axes, which sometimes branched, radiating out from a central area, some of which terminated with shallow cup-like structures marked by dark dots, which may be archegonia or antheridia (Remy and Remy, 1980).

Middle Devonian fossils have come from Scotland (Chaloner, 1972). *Protolepidodendron karlsteini* from the Spital Quarry, Halkirk in Caithness, may have been a shrubby plant or a small tree up to 7 m tall (Lang, 1926). It had branches 1 cm in diameter with characteristic rows of cushion scars. The related species, *P. scharianum*, may have been a shrubby plant with a creeping horizontal stem giving rise to upright stems about 30 cm high. These were ancestral to the huge lycophyte trees of the Carboniferous. Another tree was *Pseudosporochnus*, found in the Stromness Beds of Orkney and also from Burren Hill, east of Beltra Lough in Ireland (Lang, 1927) (Figure 1.11c). It was about 3 m tall. The trunk rose up from a swollen base and then divided again and again so that the final twigs were clothed in fine forked appendages.

One important evolutionary step was the origin of larger leaves called megaphylls, to catch light and out-shade competitors. They evolved by the flattening and webbing of groups of branches. *Psygmophyllum* from the Middle Devonian of Caithness had fan-shaped leaves more than 3 cm long. *Sphenopterys* has fern-like fronds. *Rellimia* (*Protopteridium*) was a shrub or small tree which branched profusely and had spirally arranged forked leaves (Walton, 1957). It is found in a fossil assemblage from the Bay of Skaill, Orkney. It comes from an important group of plants called the Progymnosperms which had evolved from plants like *Asteroxylon*, and which were, in their turn, to give rise to seed-plants. At Sloagar in Shetland there was another member of the group, called *Svalbardia* (Figure 1.12).

1.3.2 NEW WAYS OF REPRODUCING

Upper Devonian fossils are found in rocks of south-west England and Ireland. The Kiltorcan Beds in Co. Kerry had a tree called *Cyclostigma kiltorkense* (Chaloner, 1968) (Figure 1.11d). *Cyclostigma* was a progymnosperm. It grew up to 12 m high, the trunk rising up from a thick rhizome. It had forked branches hanging over like a parasol at the top and clothed in long, narrow leaves arranged in spirals and whorls. Another tree

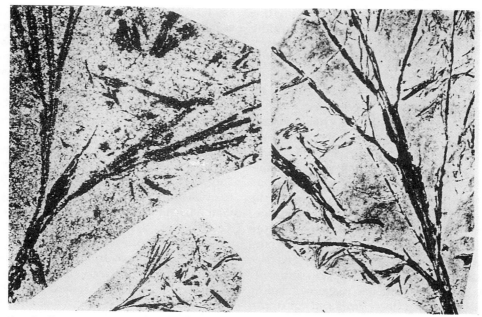

Figure 1.12 *Svalbardia polymorpha* from the Upper Devonian of Shetland. (Courtesy of W.G. Chaloner.)

has been described by some workers from these rocks. This is a species of progymnosperm called *Archaeopteris* (Figure 1.13), a very tall tree, widespread around the world from Australia to North America. The world's flora was truly global and without any great regional diversification.

At the ends of the finer twigs of *Cyclostigma* there were cones 6 cm long and 2–3.5 cm in diameter. The cones had a spiny appearance with 1 cm long scales. At the base of the scales were sporangia containing spores of one of two different sizes. The production of spores of different sizes, heterospory, was an important step towards the evolution of seeds. There were small male microspores and large female megaspores. The megaspores of some later lycophytes, such as *Lepidophloios*, were released in massive seed-like reproductive structures, aquacarps, which may have been dispersed by water (section 1.5.2).

The evolution of seeds was a very important event in the adaptation of plants to life on land. A seed is like a set of Russian dolls

Figure 1.13 *Archaeopteris hibernica*, crown of frond-like branches. (courtesy W.G. Chaloner.)

protecting the embryo seedling at its centre. The embryo arises from an egg after it has been fertilized by a male nucleus or sperm. The sperm or male nucleus is released from the male tissue in the microspore or pollen grain after it has landed near the egg. The egg is part of a female tissue, the female gametophyte. After fertilization the female gametophyte can provide a nutritive tissue, called the perisperm, surrounding the embryo. It is the innermost layer of the seed surrounding the embryo. The female gametophyte/perisperm arises inside a female spore, the megaspore, which, except in primitive plants, is not dispersed from the plant. The megaspore wall provides the next protective layer of the seed called the nucellus. Outside this is the wall of the sporangium, within which the megaspore arose. This in turn is protected by a layer or layers of integument. The whole structure is called an ovule before fertilization and a seed after fertilization. Very often the seeds are also protected by surrounding branches or bracts forming a cupule (Figure 1.14). Seeds are a means by which young embryo plants can be dispersed.

In comparison to seeds, spores are only weakly protected and are dispersed before fertilization. They have to land in moist conditions to enable the sperm to swim to the egg. Many are wasted. They fail to germinate and even then fertilization may not occur. Spores tend to be small and provide little energy for the establishment of the plant. Because they develop after fertilization, less energy is wasted in the production of seeds. The embryo is well protected by the seed coat which develops from the integument and seeds contain energy to feed the growing

seedling. This is particularly important where seedlings have to establish in harsh or competitive conditions, where they have to rely on their own energy reserves for a long time. Later in evolution seeds evolved the ability to remain dormant until conditions favourable for seedling establishment arose. In this way seed-plants could be dispersed in time as well as space.

Some of the earliest seed-plants have been discovered in Upper Devonian rocks of Devon, for example from the Baggy Beds of the Plaistow/Sloley Quarry, and Baggy Point near Barnstaple in north Devon. For example, *Xenotheca devonica* had a cup-like organ, with a toothed margin, which probably contained seeds. *Hydrasperma*, from Berwickshire, East Lothian and Co. Kerry, from the Upper Devonian/Lower Carboniferous, shows a stage in the development of an intricate seed and cupule complex (Matten, Lacey and Lucas, 1980) (Figure 1.14e). The seed coat was drawn out into an elongated funnel, called the salpinx, which caught the air-dispersed pollen, and groups of four seeds were surrounded by a cup of sterile branches.

Already at this stage in the Late Devonian two different kinds of seed can be recognized. One kind, called platyspermic seed, was flattened and had bilateral symmetry. The

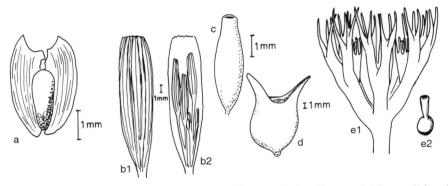

Figure 1.14 Seeds and cupules from the Devonian and Lower Carboniferous: (a) *Spermolithus devonicus* (a platyspermic seed); (b) *Calathospermum fimbriatum* (b1, cupule; b2, section through cupule); (c) *Stamnostoma huttonense* (= *Pitus*) (a radiospermic seed); (d) *Lyrasperma scotica*; (e) *Hydrasperma tenuis* (e1, a four-seeded cupule; e2, a single seed showing elongated salpinx). Bar length 1 mm. (From Chaloner, Hill and Lacey, 1976; Matten, Lacey and Lucas, 1980; Retallack and Dilcher, 1988.)

earliest platyspermic seed has been found in the rocks of Kiltorcan (Chaloner, Hill and Lacey, 1976). Plants with platyspermic seeds were to give rise eventually to the conifers. Those with radially symmetrical seeds (radiospermic) like *Hydrasperma* may have given rise to the cycads and other seed-plants (Figure 1.14). Many radiospermic seed-plants had fronds like ferns. They were trees and have been called pteridosperms or seed-ferns (pterys = fern, sperma = seed). Fossil fronds of different sorts called *Sphenopteridium*, *Sphenopteris* and *Telangium* have been named.

1.4 THE LOWER CARBONIFEROUS

The Carboniferous period started 345 million years ago. In the Lower Carboniferous, Britain lay near the equator as part of the great continental mass of the Old Red Continent. Shallow seas advanced over the land, leaving Britain a series of elongated islands separated by gulfs running east–west. In northern Britain, in the Northumberland Basin and especially the midland valley of Scotland, delta mud and sand, derived from the north, was deposited, alternating with limestone. In some areas, as in the Kilpatrick Hills north of Glasgow, basalt lavas were interleaved. As the sea retreated and advanced over the land, forests grew up and then were destroyed, to be preserved as beds of coal or oil shale.

Spore-producing plants continued to be important, dominating many lowland assemblages throughout the Carboniferous period, but a greater development of seed-plants is shown in the Lower Carboniferous. The Whiteadder River in the borders region of Scotland has one of the most diverse assemblages of permineralized seeds from the Lower Carboniferous in the world.

Trees of various sorts dominated the vegetation. In some places there were stable habitats where large trees such as the lycophyte *Lepidodendron* and the seed-fern *Stamnostoma* (= *Pitus*) were growing in wet woodlands or coastal swamps (Figure 1.15a). Tree stumps from the Lower Carboniferous are preserved *in situ* on the banks of a small stream, the Kingwater, north-west of Brampton in Cumbria. *Protopitys*, a heterosporous plant, described from Dunbartonshire and Yorkshire, had wood tissue and was probably tree-like (Walton, 1957, 1969). At Lennel Braes in Northumbria there is one important large tree species, a fossil trunk called *Pitus antiqua*. From its wood anatomy it was probably a seed-fern. The massive trunk of one specimen from East Lothian in Scotland, nearly 15 m long and with a diameter of 1.5 m at the base has been displayed in the grounds of the Natural History Museum in London.

1.4.1 FIRE AND FOREST

Rich fossil assemblages from the Lower Carboniferous are found in Berwickshire at Edrom, Langton Burn, Burnmouth, Newton Farm near Foulden, and Cove, Oxroad Bay in East Lothian, Loch Humphry Burn and Glenarbuck in the Kilpatrick Hills and Pettycur in Fife (Scott, Galtier and Clayton, 1984). Several of these assemblages are dominated by small, shrubby plants which took advantage of transient habitats available because of periodic flooding or forest fire.

The assemblage at Oxroad Bay has been studied closely. It was preserved in a volcanic landscape with frequent ash fall and mass flow of the unstable immature soils brought on by periodic heavy rainfall (Bateman and Rothwell, 1990; Bateman and Scott, 1990). Some plants had large rootstocks which provided water storage during periods of drought. Low-growing communities quickly established on new soils. Wildfire, started by volcanic activity, ran through the vegetation intermittently.

There were primitive ferns, a short stocky one called *Cladoxylon* and a prostrate one called *Stauropteris*, and in later assemblages sphenophytes called *Archaeocalamites*. There

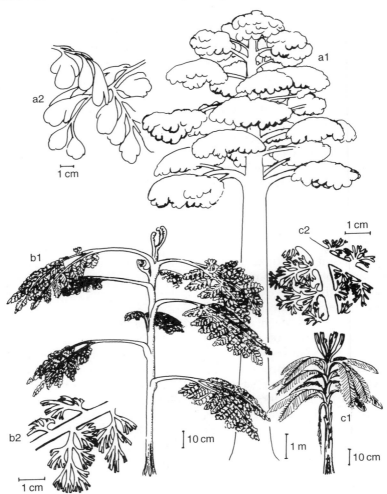

Figure 1.15 Reconstruction of Lower Carboniferous seed-ferns from Scotland: (a) *Stamnostoma huttonense* (= *Pitus*) (a1, tree; a2 part of the foliage); (b) *Lyrasperma scotica* (b1, tree; b2, part of foliage); (c) *Calathospermum fimbriatum* (c1, tree; c2, part of foliage). (From Retallack and Dilcher, 1988.)

were two species of *Oxroadia*, prostrate or semiprostrate lycophytes. Scorpions hunted below the foliage. Especially important in these transient habitats were small, shrubby seed-ferns such as *Calathospermum* and *Lyrasperma* (Figure 1.15).

Pettycur and the adjacent Kingswood End have fossil assemblages which have provided a picture of a range of Lower Carboniferous plant communities (Gordon, 1909; Rex and Scott, 1987; Scott, 1990). The fossils are preserved in a number of states; as permineralized fragments, as fusain or fossil charcoal, and as compressions in coalified deposits. There was a broad range of kinds of plants. One assemblage is represented in the Kingswood limestone. It is dominated by several species of seed-ferns and the sphenophyte called *Archaeocalamites goeppertii*. These fossils are almost always fusainized. The vegetation they formed was often subject to fire.

A different assemblage, also subject to fire, is represented in the Pettycur limestone (Figure 1.16). Ferns scrambled over the open upland habitat. The vegetation was dominated by one kind of fern with a characteristic stem anatomy, called a zygopterid fern, with the lycophytes, *Lepidodendron pettycurensis* and *Anabathra pulcherrima*. The latter are described in more detail below. The climate was generally hot, with mean annual temperatures in excess of 30°C and semiarid. The presence of growth rings in *Archaeocalamites* indicates that there was an alternation of seasons when rapid growth took place, with others, perhaps seasons of drought, when vegetative activity slowed. Alternatively, it is possible that plant growth was not seasonal but merely inhibited periodically because of the deposition of volcanic ash. In these communities volcanic activity regularly ignited the vegetation.

Some species produced two kinds of fossil, either fusainized or permineralized. Others are never fusainized. In the Kingswood limestone, the latter include fossil roots called *Ameylon*, and two species of lycophyte, called *Achlamydocarpon* and *Oxroadia*. Perhaps this lake-shore community, dominated by *Oxroadia*, grew in conditions that did not favour wildfire, or perhaps it grew at a time when volcanic activity had ceased in the close vicinity. *Ameylon* is the name given to the fossil roots of the plant, *Cordaites*, which was one of the earliest representatives of the group that gave rise to conifers. It has a number of morphological and anatomical characters found in living mangrove trees, and in some reconstructions *Cordaites* is given a remarkable mangrove-like platform of roots. It grew to about 30 m with a trunk diameter of 1 m. The tree had a few stout branches bearing long, possibly leathery, strap-like leaves up to 1 m in length and 15 cm across. Some species of *Cordaites* may have grown on drier land.

The main community represented at Pettycur grew at a time when there was diminished volcanic activity. It was a peat swamp vegetation preserved only as permineralized fossils. It has the widest range of species of lycophytes, sphenophytes, primitive ferns and seed-ferns. The different distribution of individual components of this fossil assemblage shows that different plant species dominated in different areas or at different times. For much of the time the

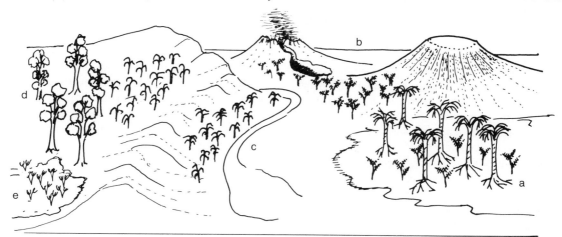

Figure 1.16 Generalized view of the distribution of reconstructed vegetation types at Pettycur: (a) lowland 'swamp' with arborescent lycophytes and the seed-fern *Heterangium*, (b) zygopterid ferns in a fire 'savannah'; (c) seed-fern woodland; (d) seasonal upland woodland with primitive conifers and seed-ferns; (e) *Oxroadia* 'marsh'. (Modified from Rex and Scott, 1987.)

community was dominated by lycophytes, including the tree *Anabathra* (*Paralycopodites*), which was like *Lepidodendron* but lacked its leaf scars and had lateral branches, and the seed-fern, *Heterangium*. *Anabathra* was essentially a colonist of potentially temporary habitats, reaching reproductive age quickly and saturating the environment with continuous production of female megaspores and male microspores (Phillips and DiMichelle, 1992). *Heterangium* had a stem up to 4 cm in diameter which branched only rarely. Its leaves were divided into two equal branches and then bore finely divided lobes.

The kind of peat swamp community represented in an early form at Pettycur later became the dominant vegetation over much of the British Isles, and even over much of Europe and North America. It has been called the *Lepidodendropsis* flora. Characteristic genera of this flora, which were also present at Pettycur, were the tree lycophytes, *Lepidodendron* and *Lepidophloios*, the sphenophytes, *Archaeocalamites* and *Sphenophyllum*, and primitive-fern or seed-fern foliage, called *Adiantites* and *Sphenopteridium*. At different Lower Carboniferous sites in Britain there are other widespread species, such as the fern *Triphyllopteris* which has been found at Puddlebrook Quarry in Gloucestershire along with the seed-fern *Diploteridium* (Rowe, 1988).

One very special Lower Carboniferous site is the East Kirkton quarry, near Bathgate, west of Edinburgh. Here, 335 million years ago, a hot, sterile lake in a volcanic area preserved both a fauna and flora in remarkable detail (Wolfe *et al.*, 1990). Black and cream mineral layers containing fossils alternate with layers of volcanic ash washed in by torrential rainstorms. Many fossils were discovered in the slates from the quarry used to build walls in the surrounding district. Giant lycophytes and primitive seed-plants found at Pettycur are here, but also, very remarkably, there is a rich fauna, with ancestors of frogs and salamanders, and

including the earliest known reptile. This was a lizard 20 cm long which has been nicknamed 'Lizzie'. The amphibians grew up to 50 cm long, and included a limbless one that slithered through plant debris. There were scorpions up to 80 cm long, a harvestman, and millipedes and eurypterids, like stream-lined lobsters, up to 50 cm long. The forest was full of the clatter and splash of these creatures and perhaps the first grunts and croakings of amphibians could be heard.

1.5 THE UPPER CARBONIFEROUS

As the Carboniferous period progressed, sediments of alternating marine and terrestrial origin were deposited in northern Britain in regular cycles, called Yoredale cycles. The alternation of sedimentary beds can be clearly seen on Ingleborough (Strahan, 1910) (Figure 1.17).

The marine sediments provided the Carboniferous limestone whose distribution is so important to the flora of the British Isles today. The terrestrial sediments were sands deposited in deltas which advanced and retreated several times. They gave rise to the sandstone of the Yoredale Beds and the Millstone grit. The reason for the cycles occurring so regularly may have been because of world-wide changes in sea level or, on a more local basis, as deltas changed course and sediments subsided.

Later the marine basins, one north and one south of a central Wales–Brabant land mass, had become choked and, although cyclic deposits still accumulated, the marine beds were now of marine mud and not limestone. Erosion and sedimentation meant that a broad band of Europe and North America was covered by vast delta swamps in which forests were constantly forming and then being drowned and coalified (Trueman, 1954). Britain lay in an equatorial position (Figure 1.18).

This was the major period of coal formation. Continuing subsidence of the Earth's crust

(a)

(b) Hiatus or unconformity — Ordovician — Silurian — Carboniferous — Millstone Grit — Yoredale Beds (interleaved limestones)

Figure 1.17 Ingleborough: (a) view of Ingleborough, exposed Carboniferous limestone in foreground; (b) section showing geological strata, vertical scale exaggerated. (Adapted from Edwards and Trotter, 1954.)

allowed the build up of coal-bearing sediments of great depth. There are 3050 m of coal stacked vertically in the Lancashire–north Staffordshire area, 2440 m in South Wales, 1060 m in the midland valley of Scotland and 760 m in Kent.

1.5.1 THE ECOLOGY OF COAL

The coal seam and its associated strata record the changing ecology of a sedimentation cycle (Raistrick and Marshall, 1939) (Figure 1.19).

As well as the coal seam there was a rootlet zone, called 'seat-earth' with *in situ* tree-trunk bases, a black non-marine shale or mudstone containing bivalves, coarsening via silt to sandstone, and a marine band, which is often missing. The cycle in part or in whole was repeated many times over. In the South Wales coalfield there are 29 coal seams and 11 marine bands. Workable coal seams

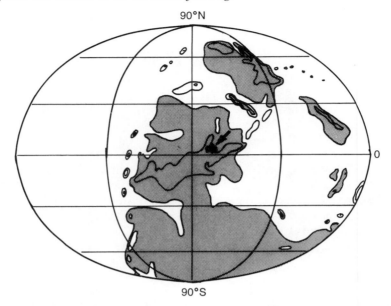

Figure 1.18 Map of the Late Carboniferous world (306 Ma), showing the position of the British Isles just north of the equator. Shaded area = dry land. (Courtesy of C.R. Scotese, Palaeomap Project, University of Texas, Arlington.)

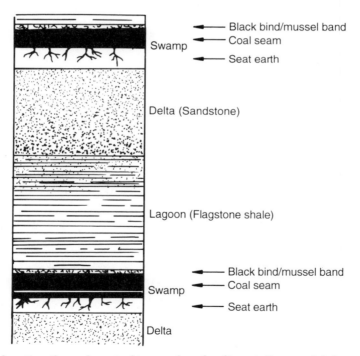

Figure 1.19 Vertical section through part of two cycles of sedimentation, containing coal seams and the long-term marine inundation of costal swamps and deltas.

are generally in the range 1–2 m in thickness but seams range from thin films to many metres thickness. The 'Thick Coal' of south Staffordshire is over 9 m thick, but it consists of 12 united seams, some separated by narrow 'dirt partings' of a few centimetres thickness. The coal peat was formed when the coal swamp was more or less static in relation to water levels. A depth of 15 m of dead vegetation provided 1 m of coal. There was an 80% reduction in thickness of the peat as it was coalified, compacted and de-volatized (Trueman, 1954). The seams are composite in nature. Bottom coal is often softer. The harder middle coal is called 'brights' and the upper coal is dull and called or 'hards'.

As well as smaller compression fossils of foliage, twigs and cones, upright tree trunks or fallen logs are often preserved as internal mud casts within the shale or mudstone, surrounded by a thin film of bright coal formed from the bark. Prostrate logs are known as 'cauldron-bottoms' or 'horse-backs' by miners. The stumps of trees, 'baum pots' or 'pot holes' are a serious hazard in mining because, after the supporting coal seam has been removed, the 'bright' coal skin of the cast can giveway allowing the stump to fall into the tunnel.

Changing water level was the most import-ant determinant of coal swamp vegetation. Two other factors were important. The coal measures contain a lot of fossil charcoal. Forest fires were frequent and severe enough even to burn the peat. After burning, the area may have degenerated into an area of stand-ing water before it was recolonized. Another important ecological factor was the absence of large herbivores (Scott and Taylor, 1983). Large amphibians swam and clambered around the swamps, but they were not adapted for grazing or digesting the vegeta-tion. Arthropods, especially insects like cock-roaches, were evolving, and bite marks have been seen in *Neuropteris* leaves, but herbivory was not yet a major factor in the vegetation.

Nevertheless, fossil droppings, coprolites, do contain plant cuticles and spores. Spores provided a rich source of food for many animals.

1.5.2 COAL SWAMP TREES

Tree lycophytes, lepidodendrids, dominated the Late Carboniferous swamp vegetation (Crookall, 1929, 1955). At least five different genera have been recognized (Bateman, DiMichele and Willard, 1992). These trees could grow to great heights, rising to over 50 m (Figure 1.20).

In *Lepidodendron* the stem was a massive pole which forked equally near the top,

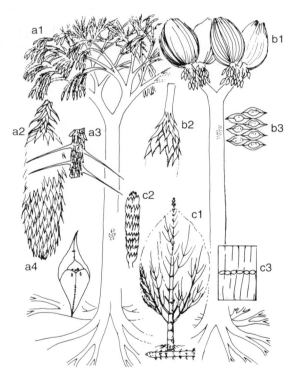

Figure 1.20 Reconstructions of coal swamp trees: (a) *Lepidodendron* (a1, tree; a2, spore cone; a3, branch and foliage; a4, leaf 'cushion'); (b) *Sigillaria* (b1, tree; b2, part of spore cone; b3 leaf scars); (c) *Calamites* (c1, part of plant showing horizontal and upright trunks; c2, cone; c3, node, inner cast).

forking again and again dichotomously very profusely towards the apex. The tree had little woody tissue but was supported by a thick, waterproof and decay-resistant bark. The whole surface was covered with scale-leaves, but on older parts of the plant these had fallen off. The trunk and branches were embossed with the cushion-like leaf-bases, which probably functioned photosynthetically. On a related genus, called *Lepidophloios*, the leaf-bases enlarged after leaf fall. One fossil, called *Knorria*, is the trunk cast of these lepidodendrids. It shows the vascular traces

reaching up into the leaves. Some species produced deciduous branches, dropping them as they gained height. *Anabathra* (*Paralycopidites*) and *Diaphorodendron* retained some lateral branches on the pole-like trunk. The laterals branched dichotomously.

The coal swamp looked like a forest of poles of young, unbranched and little-branched youthful trees (Figure 1.21) (Scott, 1979). Perhaps the most peculiar part of these trees was their massive platform-like base, which, like the base of some living species of mangroves, supported them in the swamp

Figure 1.21 A view of a Carboniferous coal swamp: (a) *Lepidodendron* swamp-forest; (b) seed-ferns on the flood plain and levee banks; (c) *Calamites* on sand banks and lake margins; (d) fringing marsh of herbaceous calamites. (Modified from Scott, 1979.)

mud. These have provided some of the most remarkable fossils, for example the fossil forest in Victoria Park in Glasgow (Figure 1.22). The fossilized base of different trees is very similar and together they are called *Stigmaria*, or stigmarian axes (Figure 1.23). *Stigmaria* consists of branched horizontal rhizomorph stems up to a metre in diameter which extended up to 12 m out from the trunk. On the younger parts of the rhizomorph there were spirally arranged lateral root-like appendages which were largely air filled. These may have been photosynthetic if supported in shallow water (Phillips and DiMichelle, 1992). In soft mud the roots spread out as a 'star-burst' pattern. As the rhizomorph aged, the root appendages were shed, leaving round root scars.

In early stages of growth the rosette of rhizomorph axes grew more rapidly than the pole-like trunk. This elongated rapidly only when the base had grown enough to provide a wide platform, as if the trunk and branches were preformed and simply extended from a massive basal bud. Many trees were relatively narrow stemmed. They grew close together and relied on mutual support for strength. In one fossil forest in Nova Scotia there are 96 vertical trunks in an area less than 37 × 5 m. The architecture of the lepidodendrids is simple and repetitive. Those without lateral branches had very limited ability to adapt to variations in light and shade of different parts of the canopy by differential growth.

Lepidodendrids were mostly short lived, dying in less than 15 years (Phillips and DiMichelle, 1992). *Lepidodendron* and *Lepidophloios* were monocarpic; after they had reproduced once, they died. *Lepidodendron* on muddy swamps and *Lepidophloios* on peaty ones were ecologically separated. *Lepidophloios* may have been especially tolerant of stagnant, standing water. *Lepidophloios* had separate male and female cones. The female cones produced, as a disseminule, a single megaspore protected by sporangium wall and integuments, called an aquacarp. It was

Figure 1.22 The fossil grove, Glasgow. (Courtesy of W.G. Chaloner.)

Figure 1.23 Compression fossil of *Stigmaria ficoides*, showing the round scars of root appendages. (Courtesy of W.G. Chaloner.)

probably dispersed aquatically before being fertilized by the tiny male microspores.

One peculiar aspect of the coal-swamp forest in comparison to swamps of today was its relative lack of epiphytes. Only herbaceous lycophytes called *Lycopodites* and *Selaginellites*, almost indistinguishable from some modern clubmosses, may have clambered over fallen trunks. The tree-fern *Psaronius* (*Pecopteris*) was sometimes a component of the swamp. The vegetation was quite unlike anything we might see today, open above but chaotic below, with pole-like young plants pushing their way up between the half fallen trunks of dead and dying trees.

Some species of *Diaphorodendron* were polycarpic but they, too, produced aquacarps, and some were more tolerant of brackish conditions than the other lepidodendrids. The seed-plant *Cordaites* may have been part of this fringing, mangrove-like vegetation.

1.5.3 PLANTS OF THE FLOOD PLAIN

A different and more diverse vegetation is represented not in the thick coals and seat-earths of the swamp deposit itself, but in a rich fossiliferous plant bed, the soft carbonaceous shale preserved today in the roof of the seam. This was a flood-plain vegetation dominated by a rich diversity of primitive ferns and seed-ferns. One seed-fern, called *Neuropteris* was especially important (Figure 1.24g–h). Seed-ferns formed a low, scrambling bush vegetation. *Neuropteris heterophylla* had beautiful, forked fronds about 1.5 m

long. Part of this scrub vegetation included thin-stemmed, clambering sphenophytes, *Sphenophyllum*. At the margins of the swamp, and also present on the banks of the meandering rivers of the flood plain and on point bars and spits in the delta (Figure 1.21), a giant sphenophyte, called *Calamites* (Figure 1.20c) grew. Like a giant horsetail up to 10 m high, it had a jointed stem, with a whorl of branches at each joint. It may have grown in pure stands and colonized areas of bare, drying mud, spreading vegetatively by its massive rhizome.

Scattered lepidodendrids, *Sigillaria*, or *Bothrodendron*, grew on the flood plain, poking up through the bush. *Sigillaria* was a lepidodendrid of sporadic occurrence, similar in many respects to *Lepidodendron* but less branched. It looked like a very peculiar palm with, at maturity, a single fork bearing two great tussocks of grass-like leaves. The leaves were arranged not in spirals but in parallel lines. It seems to have been less competitive and colonized open sites of major disturbance, such as river and stream banks. Once it was established it may have been relatively long lived, producing spores regularly. *Bothrodendron* was like *Lepidodendron*, but had small leaf scars.

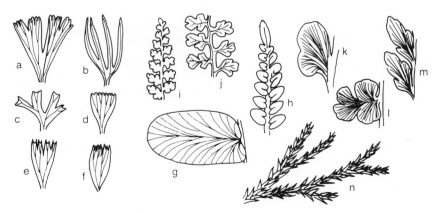

Figure 1.24 Foliage of Carboniferous plants: (a)–(f) *Sigillaria* ((a) *S.majus*; (b) *S. myriophyllum*; (c) *S. trichomatosum*; (d) *S. cuneifolium*; (e) *S. saxifragaefolium*; (f) *S. oblongifolium*); (g)–(h) *Neuropteris*; (i)–(m) *Sphenopteris* ((k) *S. dilatata*; (l) *S. striata*; (m) *S. macilenta*); (n) *Lycopodites*. (From Chaloner and Collinson, 1975.)

1.5.4 EVOLUTIONARY CHANGE

Evolution in the swamp was slow. The same species seem to have existed over millions of years. However, different species of lycophyte trees were present and dominated at different times (Smith and Butterworth, 1967; Pfefferkorn and Thomson, 1982).

Commonly occurring spores show marked changes of abundance over time and also show regional variation in abundance (Phillips, 1979). For example, one called *Cristati sporites* was abundant between Middle and Late Carboniferous times but only in South Wales. Another, called *Densosporites sphaeroangularis*, was a widespread species abundant in the Lower Coal Measures but died out in the Middle Coal Measure time. It, or something like it, reappeared, but slightly smaller, in the Upper Coal Measure time of Bristol and Somerset.

Changes in the composition of vegetation can be observed over the 45 million years of the Late Carboniferous. In the Lower Coal Measures times, *Sigillaria* was rare and the coal-swamp flora homogeneous. The flora of the Middle Coal Measure time was much more diverse. *Sigillaria* was abundant and *Calamites* was very common, with numerous different species. This may represent a period of lower rainfall, related to a period of glaciation in other parts of the world. Later, swamp vegetation spread again.

In the Upper Coal Measure times there was another change as conditions became drier again. Lepidodendrids and *Calamites*, although still common, become less plentiful, and primitive ferns, especially one called *Pecopteris* (*Psaronius*), became more common. Seed-ferns rose to abundance and *Cordaites* was very common. Of the lepidodendrids, *Sigillaria* remained most abundant. On the flood plains in the middle Late Carboniferous, on levees and deltas, the vegetation dominated at first by seed-ferns and sphenophytes changed to one in which primitive ferns were abundant. Towards the end of the Carbon-

iferous the seas withdrew from north of the Wales–Brabant island and the climate became drier, ultimately the result of global climatic change driven by the growth of polar glaciation.

Throughout the Carboniferous there was another vegetation in the drier uplands which is hardly represented in the lowland fossil assemblages (Florin, 1951). This was dominated by conifers. The earliest conifer, called *Swillingtonia*, has been identified from fragments of leafy shoots obtained by maceration of shale of middle Late Carboniferous in Yorkshire (Scott and Chaloner, 1983). Other fragments of different species, called *Lebachia* or *Walchia* (*Geinitzia*), have been found from the Late Carboniferous in sites near the margins of the uplands. However conifer pollen, *Potonieisporites*, has been found right back to the beginning of the Late Carboniferous.

The upland conifer vegetation became most significant as waters receded and conditions grew drier at the end of the Carboniferous. Coal-swamp plants became extinct, first *Lepidondendron* in Europe towards the end of the Carboniferous, with *Sigillaria*, *Sphenophyllum* and *Calamites* surviving for a little while into the Lower Permian. Seedplants, especially the shrubby seed-ferns, including *Neuropteris*, became relatively more important. The arrival of a new species of seed-fern, called *Callipteris conferta*, has been used by some workers as a marker for the base of the Permian. *Callipteris* was a kind of seed-fern called a peltasperm, which had its seeds around the margin of a stalked disc.

1.6 THE PERMIAN AND TRIASSIC

At the beginning of the Permian period, *c.* 286 million years ago, the British Isles lay in the hinterland of the newly formed supercontinent of Pangaea. Deserts became widespread. There was considerable uplift of the terrain, exposing older sediments to erosion by flash floods or wind. Permian sandstone has quartz grains of sand, rounded and frosted by wind action. Fans of coarse-grained angular fragments, called conglomerate, were deposited as scree around the base of the hills. They are preserved, for example, in the Brockram rock of the Appleby area in Cumbria, and in the Penrith sandstone crescent-shaped dunes of desert landscape are preserved. To the east and south of the Pennines there was a gently rolling plain reaching across the area of the North Sea to Germany. Running through the desert, marked by gravelly and pebbly beds, were dry wadis which ran with torrential streams after desert-pours. Further east, in a North Sea Basin, there was at times a large land-locked desert-playa lake. It reached up to the eastern edge of the Pennines and the northern edge of the London–Brabant Massif in the south. The lake grew and retreated, filling with water and then evaporating, leaving layer upon layer of 'Rotliegendes' siltstone and claystone with evaporites, which are 30% halite in the centre. The shoreline is marked by desiccation cracks (Anderton *et al.*, 1979).

The landscape shimmered in the heat. The sediments were reddened by iron-containing solutions percolating through them, coating grains with iron oxide. Reddening of the soil profile reached deep into pre-existing Carboniferous strata. This would only have been possible in the absence of humic acids from organic matter. Red Permian rocks can be seen to good effect in the cliffs of the east Devon coast north from Paignton. In the North Sea the porous Permian sandstone later provided the reservoir for North Sea gas, percolating up from devolatilizing Carboniferous coal measures below.

1.6.1 A DESERT VEGETATION

Within the desert landscape a thin and patchy vegetation grew around lakes and along river beds. One small Permian desert lake existed in the Mauchline Basin of Ayr-shire. Like the Rotliegendes lake, the Mauchline lake margin fluctuated, perhaps seasonally, as it filled and evaporated. Its sediments are fine-grained and water-lain. It has fossil beds interleaved with lavas, which have allowed it to be dated to 275 to 272 million years ago (Wagner, 1983). The plants are herbs or shrubs, including the seed-ferns, primitive ferns and sphenophytes. The conifer *Lebachia* (Figure 1.25) was also present. Elsewhere the landscape was perhaps nearly devoid of vegetation, with straggling lines of drought-adapted seed-ferns and conifers following the wadis, surviving by tapping into a deep water table, like some species of desert palms and acacias do today. There are rare vertebrate footprints in the Permian sandstone and so there must have been some plants to eat.

1.6.2 THE MARL SLATE FLORA

Later in the Upper Permian the sea flooded the North Sea Basin from the north, forming the Zechstein Sea which reached across Europe as far as Poland (Pattison, Smith and Warrington, 1973). In Britain an area from Durham to Lincolnshire was flooded (Figure 1.26). On the opposite side of Britain a Bakevillea Sea penetrated the Irish Sea Basin, including parts of Cumbria and Cheshire. Around the margins of the seas, marine life flourished. The activities of bryozoans and algae built reefs. This reef limestone and associated deposits gave rise to the Magnesian limestone of north-east England. There were at least five separate episodes of flooding. As the basins dried out thick salt deposits were left. Plant fossils are rare (Harris, 1938; Cox, 1954; Chaloner, 1962; Schweitzer, 1986). At the base of the evaporation cycle is marl slate which sometimes contains fossils. It rarely outcrops naturally in Britain but it is 9 m thick in the Kimberley railway cutting near Nottingham. The plants from there, probably from the earliest evaporation cycle, were all, to a greater or lesser extent, adapted for dry

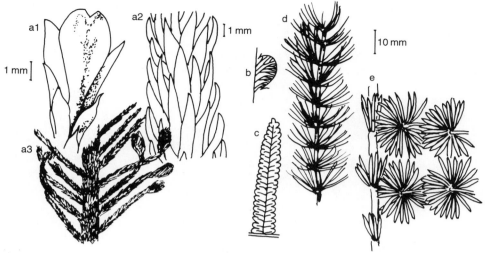

Figure 1.25 Fossil plants from the Permian: (a) *Lebachia* (a1, seed shoot; a2, foliage; a3, branched stem); (b) *Pecopteris*; (c) *Odontopteris*; (d) *Annularia*, (e) *Asterophyllites*. (Redrawn from Andrews, 1961.)

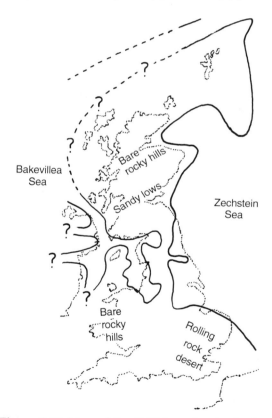

Figure 1.26 Map of the British region in the late Permian. (Adapted from Smith *et al.*, 1974.)

habitats (Stonely, 1958). They had leathery leaves and thick cuticles. They include a plant similar to *Cordaites*, the conifers *Pseudovoltzia* and *Ullmannia* (Figure 1.27) and the peltasperm, *Callipteris martinsi*.

In the marl slates of Durham, especially at Middridge Quarry east of Shildon Station, but also at Cullercoats Bay, these species are present again (Figure 1.28) with the exception of *Cordaites*. An important addition is the horsetail *Neocalamites*, a small, branched sphenophyte, which probably formed a marsh in the shallow waters of river mouths. Two important groups of seed-plants had evolved, the cycads, represented by *Pseudoctenis* and *Taenopteris*, and ginkgoes, represented by *Sphenobaieria* with its divided leaf. The seed-ferns included *Sphenopteris*. The most abundant species were three different species of conifer: *Ullmannia bronni*, *U. frumentaria* and *Pseudovoltzia*. The flora was poor and dominated by conifers. Perhaps the vegetation consisted of a thin woodland of conifers with an understorey of cycads and seed-ferns. Interestingly, the conifers have growth rings, probably indicating that rainfall was slightly seasonal. Ferns and clubmosses are absent from the fossil assemblage.

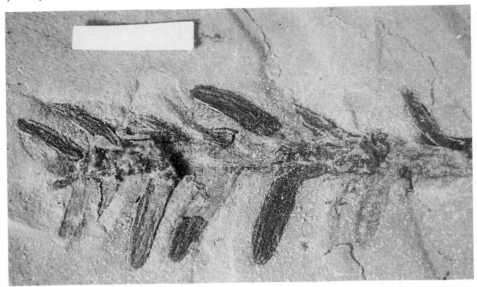

Figure 1.27 *Ullmannia* from the Upper Permian rocks of Nottinghamshire. (Courtesy of W.G. Chaloner.)

Figure 1.28 Plants of the Durham Marl slates: (a) *Plagiozamites*; (b)–(d) *Sphenopteris* ((b) *S. patens*; (c) *S. kukukiana*; (d) *S. dichotoma*); (e) *Ullmannia frumentaria*; (f) *Pseudoctenis middridgensis*; (g) *Neocalamites mansfeldicus*; (h) *Pseudovoltzia liebeana*. (Drawn from Schweitzer, 1986.)

The Eden Shale from Cumbria, especially the Hilton Plant Beds, from a low cliff on the southern bank of Hilton Beck in the Vale of Eden, have the same range of fossils, but a spore of the royal fern *Osmundacidites* has also been discovered.

1.6.3 DESERT STORMS

Red rocks show the continuation of desert climate conditions into the Triassic period which began 225 million years ago. Britain lay at a latitude of 40°N. Nodular calcareous concretions and other features are sometimes present, which may have formed in similar circumstances to the 'caliche' being formed today in the desert of Chile or Peru. Rainfall was less than 300 mm/year and may have fallen sporadically in tremendous monsoonal storms. Flash floods and torrential rain continued to erode the landscape. Screes were deposited in fans at the base of the higher land, and valleys were filled by mud flows carrying pebbles and gravel. The Bunter pebble beds of the north Midlands originated as desert-fan and valley-fill deposits formed by torrential streams running off the highlands to the south and west. Further away from the highlands silt-stone deposits were left as alluvial fans braided by streams (Hains and Horton, 1969; Anderton *et al.*, 1979).

A large part of the landscape was a flat monotonous plain. The desert plain could become a temporary playa lake following torrential rain. A transient vegetation may have taken advantage of the temporarily improved conditions. There are occasional green beds within the red Triassic sediments, showing reducing geochemical conditions, perhaps because of humic substances from the vegetation.

1.6.4 PHASES OF GREATER HUMIDITY

During the Middle Triassic the sea spread from southern Europe across Germany and the southern North Sea to flood much of the Midland plain. In the Keuper sandstone, especially the 'Waterstones' of Staffordshire, Warwickshire and Worcestershire, there is preserved a rich fauna and flora, which took advantage of the increased humidity on the remaining dry land. Ferns became more abundant and sphenophytes, clubmosses and cycads were also present (Wills, 1910). In the Building Stones Formation at Bromsgrove there are fossil horsetails like living ones, but sometimes rather larger, and another herbaceous sphenophyte with a divided leaf sheath, called *Schizoneura*. The Keuper sand-stone shows ripple marks, rain pitting and desiccation cracks as well as the casts of plants. There were many shallow pools which dried up periodically. Fishes in streams and pools included the lungfish *Ceratodus*, which could survive in a burrow as the pools dried up. There were also amphibians, reptiles, dinosaurs and scorpions. On drier land there were conifers, such as *Voltzia*, and another, *Yuccites vogesiacus*, which may have been a kind of *Cordaites*. They had tough, leathery leaves adapted to drought. A cone, *Masculostrobus willsii*, may have come from a tree related to the family of the living monkey-puzzle tree, Araucariaceae. At the base of the Kenilworth breccia there is a clay, which has been dug for brick-making. It contains fossils of two species of the conifer *Lebachia*. Else-where in the Corley Beds of Warwickshire there is *Asterophyllites* and silicified wood of *Dadoxylon* and *Cordaites*.

The generally improved conditions did not last long. The sea retreated. Conditions became harder for plants and a mainly desert landscape returned. Wind-blown red dust settled in the desert pools and salt pans, forming together the Keuper marl.

In the Late Triassic there was another marine transgression. Greyish and greenish mudstone, tea-green marls and grey marls, replace the earlier red Keuper rocks. Pangaea had started to break up, with the formation of the Tethys Ocean between a northern super-continent of Laurasia and a southern super-continent of Gondwana. Britain was still probably mainly arid but was now more subject to a monsoon with abundant rainfall. Periodically the desert bloomed with a transi-ent vegetation. Laid down in the Rhaetic Sea which covered central England, there is a rich marine fossil fauna, although plant fossils are rare, scattered and of poor quality.

The bennettite *Otozamites* has been found in a quarry near Radstock in Somerset. The bennettites were like cycads but had bisexual cones, looking a little like flowers. Their leaves are distinguishable from those of cycads by the presence of a particular kind of arrangement of cells around the stomata. Caverns in the Carboniferous limestone filled with fossil debris. The fossil site at Cnap Twt, mid-Glamorgan has conifer remains like

Voltzia. Elsewhere there are fossils of *Baiera*, a kind of ginkgo.

A very important kind of fossil first seen at this time was of the extinct conifer family Cheirolepidaceae (Alvin, 1982). Early species had overlapping leaves like a cypress. Later species, the frenelopsids, had foliage looking superficially very like the living *Tetraclinis*, in the Cupressaceae, the cypress family, a species which grows in Spain and North Africa. The cuticle was thick and the pores were protected in deep pits, perhaps an adaptation or response to physiological drought for arid or saline-maritime habitats, or even as a mangrove species. The reproductive structure was a cone called *Classo-strobus*. Associated with these is the pollen, *Classopollis*, with its characteristic equatorial ring of striations. The pollen has a complex outer coat analogous to that of flowering plants which carries a chemical signal to mediate a self-incompatibility which prevents self-fertilization (Rowley and Srivastava, 1986). In flowering plants, self-incompatibility is associated with great morphological diversity and ecological success. Perhaps the presence of a similar breeding system in the Cheirolepidaceae explains their rapid diversification (Watson, 1988).

In the area of Glamorgan and the Mendips there was an archipelago of tropical islands, fringed with shell-sand beaches. *Hirmeriella* (= *Cheirolepis*) in the Cheirolepidaceae grew on the islands. A complex of shallow marine lagoons developed. In the monsoon they became at first brackish and then freshwater as they were flushed by rainwater. As this happened they were colonized by a plant called *Naiadita*. This fossil, which is preserved, for example, at Hapsford Bridge in Somerset, is especially interesting because it was a leafy liverwort (Harris, 1939). It was a tiny plant. Its presence shows that liverworts had been evolving for millions of years alongside other land-plants but almost unseen. *Naiadita* only grew 2 cm high and had thin leaves, 1–5 mm long, arranged spirally around the stem on

which capsules and archegonia, have been detected. Along with *Naiadita* there was *Hepaticites*, a thalloid liverwort. Filamentous structures, rhizoids, tiny 'rootlets' of what may have been a moss, have also been detected.

In the Triassic, for the first time in terrestrial history, herbivores became an important factor in the development of the vegetation, because the first large herbivorous dinosaurs had evolved. Dinosaurs were to dominate the world and its vegetation for the next 130 million years. By their activities they may have helped to maintain different kinds of vegetation. One of the earliest dinosaurs to be seen in Britain was the primitive *Theco-dontosaurus*. Only 2 m long, it was bipedal, and may have reached up into shrubby seed-ferns and bennettites to strip cones and young leaves from the twigs, chopping them into fine pieces with its saw-edged spoon-shaped cheek teeth. The related *Plateosaurus* was much larger, 8 m long, and although it had a short neck, it could rise up on its hind legs to browse the canopy of ginkgoes and sapling conifers.

1.7 THE JURASSIC

Conditions did not change very radically in the Early Jurassic, which started 194 million years ago. The fossil plant flora was poor at first but evidently there was enough vegetation to supply herds of the dinosaur, *Scelido-saurus*, a four-legged plant eater about 4 m long with an armoured back.

The fossil record from the Middle Jurassic onwards improves dramatically and now the fossil flora and dinosaur fauna provide evidence of a subtropical, mainly humid climate. Much of Europe had been transformed into a shallow sea by the Middle Jurassic. The estuarine Middle Jurassic fossil beds of north-east Yorkshire, where there were three separate marine transgressions, have an especially rich flora (Figure 1.29), but a rich fossil vegetation is not confined to

Figure 1.29 Fossil hunting over Jurassic plant beds of the Yorkshire coast.

Yorkshire. Stonesfield in Oxfordshire and Eyeford in Gloucestershire have similar Middle Jurassic assemblages. Bearreraig in the Highland region also had a vegetation dominated by conifers. There may have been a kind of gallery forest along streams or rivers. Most plants grew away from the site of fossilization. Material fell into streams and rivers before floating out into the deltas where it sank.

1.7.1 THE DOMINANCE OF THE CONIFERS

Some Jurassic fossil sites have a very diverse fossil flora (Harris, 1961, 1964, 1969). Over 300 species have been identified in all. At Roseberry Topping in Cleveland, 70 different species are recorded. At Red Cliff in Yorkshire there are 90 species. The conifers are particularly rich in different species (Figure 1.30). Hill Houses Nab, North Yorkshire has

a high proportion of conifers. One conifer family, now extinct, was represented by *Czekanowskia* leaves and the seed-bearing structures called *Leptostrobus*. The Cheirolepidaceae had several species, with *Pagiophyllum maculosum* growing in uplands, and *P. kurri* (*Hirmeriella*) on leached sandy soils above the flood plain.

Many different fossils resembled living species (Stewart and Rothwell, 1993). *Elatides* was like *Cunninghamia* in the Taxodiaceae, which grows in East Asia. *Marskea* (*Taxus jurassica*) was like *Torreya* (Californian nutmeg or kaya nut) in the Taxaceae. The Cephalotaxaceae, the plum-yew family, may have been represented by *Thomasiocladus zamoides*. The Araucariaceae, the monkey-puzzle tree family, were represented by *Cycadocarpidium* seed cone and scales and foliage *Podozamites*, which were very like the living *Agathis australis* (kauri), from New

Figure 1.30 Jurassic conifers: (a) *Protocupressinoxylon purbeckensis* (a1, tree; a2, foliage (=*Cupressinocladus*); a3, male cone (= *Classostrobus*); a4, pollen (= *Classopollis*); a5, female cone before and after seeds are shed (*Hirmeriella*); a6, seed scales); (b) *Marskea thomasiana* ovule; (c) *Araucarites phillipsii* cone scale; (d) *Pagiophyllum veronense* leafy branch; (e) *Elatides thomasii* (e1, leafy branch with two seed cones; e2, branching twig); (f) *Brachyphyllum macrocarpum* twig with overlapping scale-leaves. (Drawn from Miller, 1977; Francis, 1983; Hill *et al.*, 1985.)

Zealand. Other fossil foliage has been assigned to form-genera. *Brachyphyllum* was like the foliage of living podocarps *Dacrydium colensoe* and *Podocarpus ustus* (Podocarpaceae). *Cyparissidium* was like *Dacridium cupressinum* (rimu from New Zealand). *Geinitzia* was like *Araucaria heterophylla* (Norfolk Island pine) or *Cryptomeria* (Japanese red cedar). *Elatocladus* was like *Taxus* (yew), *Taxodium distichum* (swamp cypress), *Sequoia sempervirens* (Californian redwood) or *Metasequoia glyptostroboides* (dawn redwood). *Pityocladus* had the needles of many Pinaceae.

1.7.2 LIVING FOSSILS AND EXTINCT FORMS

Jurassic assemblages include both plants very like living plants and forms which have become entirely extinct.

Ginkgoes were diverse and represented by *Baiera*, *Eretmophyllum*, *Pseudotorellia* and *Sphenobaiera* (Figure 1.31) as well as, at Scalby Ness, by the living fossil, *Ginkgo*, itself.

Another surviving group are the cycads. They are abundant in fossil assemblages at Broughton Bank and Cloughton Wyke. Cycads included *Ctenis*, *Pseudocycas*, *Paracycas* and *Nilssonia*, with its male and female reproductive organs called *Androstrobus* and *Beania*, respectively.

In contrast, the bennettites have since become extinct. The fossil assemblage at Wrack Hills near Runswick Bay is a classic site for the reproductive structures of the bennettites (Figure 1.32) (Hill *et al.*, 1985). Most Yorkshire Jurassic assemblages are coalified but at Wrack Hills the three-dimensional structure of fossils is preserved in ironstone. There were many kinds. The bisexual reproductive axis with male parts in a whorl around the female parts was analogous to a flower. Even the seeds were in a fruit-like head protected behind an external rind formed from the expanded head of interseminal scales (Watson and Sincock, 1992). Some are fossils of different parts of the same

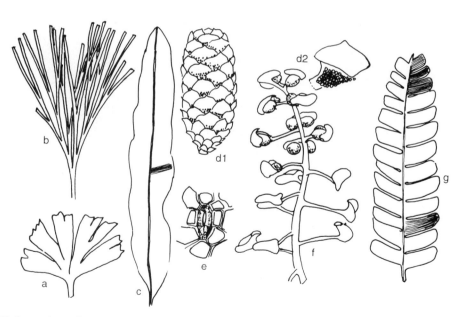

Figure 1.31 Jurassic ginkgos and cycads: (a) *Ginkgo* leaf; (b) *Sphenobaiera* leaf; (c) *Nilssonia tenuinervis* leaf; (d) *Androstrobus* (d1, pollen cone; d2, cone scale with microsporangia); (e) cycad stoma; (f) *Beania* seed cone; (g) *Nilssonia compta* leaf. (Drawn from Andrews, 1961; Harris, 1961–69; Wesley, 1973.)

Figure 1.32 Jurassic bennettites: (a) *Williamsonia spectabilis* (from Whitby); (b) reconstruction of *Williamsoniella coronata* (b1, section through 'flower-like' reproductive axis; b2, branching stem with leaves and 'flowers'; (c) reconstruction of *Williamsonia sewardiana* (from India); (d) reconstruction of *Wielandiella angustifolia* (from southern Sweden); (e) *Anomozamites*; (f) *Otozamites*; (g) stoma; (h) *Cycadeoidea*. (From Nathorst, 1911; Thomas, 1913; Sahni, 1932; Andrews, 1961; Bold, Alexopoulos and Delevoryas, 1980; Hill *et al.*, 1985.)

Figure 1.33 *Williamsonia* female reproductive structure axis surrounded by leaves, from Cayton Bay, Yorkshire. (Courtesy of W.G. Chaloner.)

plant, so *Williamsonia* is the female cone, and *Weltrichia* the male reproductive structure, of the species with leaves, *Pterophyllum* (Figure 1.33). *Bucklandia* is the stem of the plant with leaves *Zamites* (Figure 1.34). There were 14 different species of one kind of bennettite leaf called *Otozamites*. Several of these kinds of bennettites remained relatively unchanged for millions of years into the Cretaceous period.

Another extinct group are the seed-ferns. The last flourishing of seed-ferns was in the Jurassic period. An important kind was one with leaves named *Sagenopteris*, and male and female reproductive structures called *Caytonanthus* and *Caytonia*, respectively. It was described from Cayton Bay in Yorkshire (Figure 1.35) (Thomas, 1925).

Caytonia was once thought to be a potential ancestor of flowering plants because it has

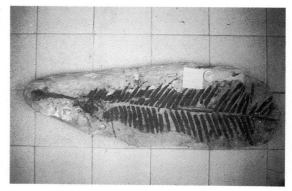

Figure 1.34 *Zamites gigas* from the Middle Jurassic of Yorkshire. (Courtesy of W.G. Chaloner.)

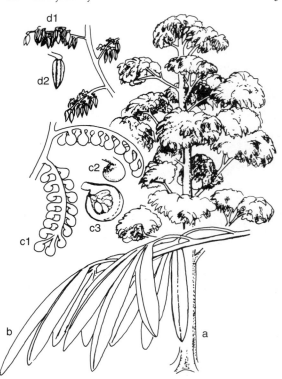

Figure 1.35 Reconstruction of *Caytonia nathorstii*: (a) tree; (b) foliage; (c) female shoot showing cupules; (d1–2) male shoot with microsporangia. (From Retallack and Dilcher, 1988.)

seeds protected within a cupule, rather like the way seeds are protected inside the pistil in flowers. The Caytoniales reached a peak of abundance in the Jurassic.

Growing with seed-ferns were primitive ferns, including some living kinds like *Marattia*, with its large, primitive sporangia (Figure 1.36). Today *Marattia* is a low-growing fern with huge fronds, which grows in temperate and subtropical regions of the southern hemisphere. Modern ferns probably evolved in the Jurassic period. Jurassic fossils are interesting in having several kinds showing intermediate evolutionary stages. These include ferns in the family Matoniaceae with their characteristic fan-like frond. The genus *Matonia* has two living species in

South-East Asia. They have sporangia intermediate between the large, robust ones of primitive ferns, which are called eusporangia, and the small, delicate ones of modern ferns, which are called leptosporangia. The fossil genus *Dictyophyllum* in the Dipteridiaceae was another important 'intermediate' fern. One species, *Dictyophyllum rugosum*, is taken as a marker for the Jurassic period. The Dipteridiaceae has only one genus today, found in Asia and Polynesia. Typical Jurassic ferns are also *Todites*/*Cladophlebis* and *Osmundopsis*, related to our living *Osmunda* (royal fern) which also has rather primitive leptosporangia. One Jurassic modern fern was the tree-fern *Dicksonia*, which, like many of the other Jurassic ferns, has living relatives in South-East Asia, tropical America and the Antipodes.

Horsetails were represented by *Schizoneura* and *Annulariopsis*. *Neocalamites* survived as a relict species.

1.7.3 THE JURASSIC VEGETATION

There is little evidence for evolutionary change in Jurassic floras (Spicer and Hill, 1979). Different fossil assemblages record different conditions of fossil deposition or different habitats rather than newly evolved species. Assemblages dominated by ferns and horsetails probably represent a coastal or marshy vegetation. Maw Wyke in north Yorkshire is a particularly rich assemblage of ferns. They were preserved *in situ* in some sites, even forming thin coal seams, some of which have been mined. They must have formed a thick marsh vegetation.

At Hayburn Wyke and Runswick Bay there are particularly rich assemblages of bennettites and cycads which may represent a drier habitat or time.

Seed-ferns may have come from a lowland and coastal vegetation. One species, called *Pachypteris* or *Thinnfeldia*, has given its name to the flora of one Jurassic fossil plant bed, the 'Thinnfeldia' Leaf Bed of Roseberry

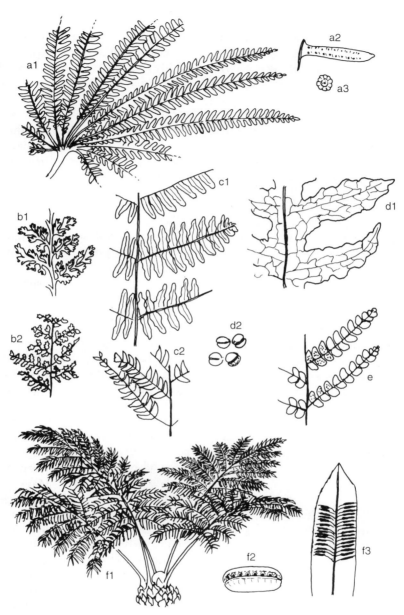

Figure 1.36 Primitive ferns, fossil and living: (a) *Matonidium* (a1, fossil frond; a2, fossil leaflet; a3, group of sporangia (sorus) of living *Matonia pectinata*); (b) *Coniopteris* (b1–2, different fossil species); (c) *Cladophlebis* (c1–2, different fossil species); (d) *Dictyophyllum* (d1, fossil frond; d2, sporangia from living specimen); (e) *Todites* (fossil frond); (f) *Marattia* (f1, living plant; f2, group of sporangia; f3, fossil frond). (Fossils drawn from Wesley, 1973; Hill *et al.*, 1985.)

Topping. *Pachypteris* was a large bush which formed mangrove thickets along the margin of river mouths. Growing with *Pachypteris* in the 'saltmarsh' community was the conifer *Brachyphyllum crucis* in the Cheirolepidaceae.

The Yorkshire Jurassic assemblages are rich but they are usually derived from elsewhere and mixed. Perhaps a better idea of the character of the Jurassic vegetation can be obtained from the Jurassic Purbeck formation of Dorset. At Chalbury Camp, Portland, and around Lulworth cove there are silicified tree stumps preserved *in situ*. Around some of the tree stumps, algae formed burrs of stromatolitic limestone after the forest was flooded. The forest was dominated by a single species of Cheirolepidaceae, *Protocupressinoxylon purbeckensis*. The wood had very variable growth rings. The climate was very irregular from year to year, sometimes favourable for growth, sometimes not. Ginkgoes, cycads and shrubby bennettites grew as a sparse understorey in conifer forest.

Other evidence about the vegetation comes from the adaptations for reproduction. There was a predominance of seed producers in the dry conditions. Seeds of conifers were small and winged. The cones of some cycads, such as *Beania* and *Androstrobus*, were relatively open for wind pollination.

1.7.4 THE AGE OF THE DINOSAURS

Another approach in reconstructing the vegetation is to look at the dinosaur fauna that grazed it (Lambert, 1990). Some conifers were evidently very tall, encouraging the evolution of huge dinosaurs, such as the sauropod *Cetiosaurus* (18 m long), and the diplodocid *Cetiosauriscus* (15 m long). With their massive pillar-like legs and long necks they could reach up into the canopy to pluck tender young buds and cones. These were the giraffes and elephants of the dinosaur fauna, but much bigger. The most giraffe-like of them to be found in Britain was *Pelosaurus* (24 m long). It is no accident that modern

kinds of conifer became abundant in the Jurassic. They had diffuse branching patterns with many shoot apices, and small scale-leaves. This limited herbivore damage and allowed them to recover from attack more readily. Taller trees were relatively safe from herbivores once mature. A few conifers, such as those related to *Araucaria*, the monkey-puzzle tree, retained spiny leaves on their trunk, which may have helped deter predation.

Most plant species had adaptations that limited their palatability. They were fibrous and resinous. The sauropod and diplodocid dinosaurs had huge digestive systems, first grinding the vegetation with stones in a stomach gizzard, and then fermenting it. The fermentation process may have been important in reducing plant toxicity. The foliage of living cycads is toxic to mammals, and living ferns are also rich in toxic substances that could have upset the dinosaur digestion. The horsetails had their own protection, their own armour-plating, a coating of minute silica shields over the whole plant.

The bennettites were more varied vegetatively than cycads. The stems were variously thick or thin, and weakly or profusely branched. The profusely branched habit and the scattering of reproductive organs over the plant, protected at the bases of leaves, may have allowed bennettites to withstand being browsed by dinosaurs better than the thick-stemmed cycads, which had exposed cones at the apex of the trunk.

Alternatively, the ginkgoes may have taken advantage of dinosaurs for seed dispersal. They produced large seeds surrounded by a fleshy aril. The seeds probably fell to the ground and may have been odoriferous to attract dinosaurs. The inner seed was protected by a hard inner coat, so that once eaten they may have survived and been dispersed in dinosaur dung.

Stegosaurs, like *Dacentrurus* and *Lexovisaurus*, or the nodosarids *Cryptodraco* and *Priodontognathus*, cropped the vegetation at a lower level than the sauropods and diplo-docids. They were only about 5 m long. Slow-moving, they relied on back armour to protect them from predators. They were the rhinoceroses of the dinosaurs. They had short necks and a horny beak for cropping the ground vegetation of ferns and horsetails, and nipping the tender cones and seeds from the low-growing cycads, bennettites and seed-ferns. The leaves of these plants were particularly leathery and fibrous, providing a very indigestible meal, but the stegosaurs had small, leaf-shaped cheek teeth which could macerate the plant tissue finely to make it more digestible.

The largest dinosaurs would have had no difficulty strolling through even relatively dense scrub, but by their activities they must have created many open areas. Herds of stegosaurs and their like must have had a very dramatic impact on the vegetation. Their grazing may have prevented the establish-ment of young conifer saplings. Open savannah-like areas are indicated by the abundance of ferns in some fossil assem-blages. Herbaceous ferns, clubmosses and horsetails may have provided the bulk of the food for young dinosaurs and those without long necks.

As well as large dinosaurs, there were also relatively small kinds. *Callovosaurus* was only 3.5 m long. It cropped the ground vegetation on all fours but could rise up on two legs to escape predators such as *Megalosaurus*. The bipedal *Dryosaurus* (3–4 m long) was fleeter yet. It was the gazelle or antelope of the dinosaur fauna. The bipedal *Echinodon* from the Late Jurassic was only 60 cm long and was different again. Probably a specialist eater, it had distinctive teeth, some long and sharp which may have been designed to chop into cones to reach the ovules and young seeds. These fleet-footed dinosaurs occupied open terrain below a thin tree canopy, relying on being able to use their speed in the open fern plains to escape attack.

Perhaps the more agile bipedal forms

clambered up, or reached into, bushes and small trees of seed-ferns, cycads and bennettites to get the tender, nutrient-rich reproductive structures.

1.7.5 SEASONAL CLIMATES

Towards the end of the Jurassic there was a marked increase in seasonality, shown by a rise in the importance of dry-season adapted trees of the Cheirolepidaceae (Francis, 1984). *Classopollis* pollen became abundant (Norris, 1969). Much of it probably came from *Cupres-sinocladus valdensis*, which dominated the forests bordering a hypersaline gulf in southern England (Francis, 1983). Ginkgoes, ferns and the cycad *Nilssonia* became rarer (Oldham, 1976). The forest had only a sparse undergrowth. Drought-adapted bennettites remained important. The period of aridity, which is associated with a decline of sea levels, lasted into the beginning of the Cretaceous period, 135 million years ago, but it may not have prevailed everywhere. An Upper Jurassic assemblage from Culgower in Sutherland in northern Scotland, has a fossil flora representing a lush river margin vegeta-tion merging with freshwater swamps and upland forest. It shares many plants with the Middle Jurassic of Yorkshire (Van Konijnen-burg-van Cittert, 1989). It may come from just before the period of aridity. There are 41 species of all sorts.

1.8 THE CRETACEOUS

Later in the Lower Cretaceous there was a general rise in humidity, but it was still seasonally dry. The London–Brabant island was uplifting and eroding to provide material to fill the Weald Basin, which was subsiding (Batten, 1974). The relative rates of these processes and intermittent marine trans-gressions from an East Anglian Sea deter-mined the nature of the terrain in the Weald Basin. Fossils were laid down in estuarine and freshwater conditions.

There was a rise in species diversity, including liverworts such as *Hepaticites*. Club-mosses, *Lycopodites* and *Selaginella*, and especially more and more different kinds of modern ferns, were present. The vegetation was still dominated by conifers. At Cliff End in East Sussex there are *in situ* stumps of *Tempskya*, a tree-fern (Figure 1.37). It had an unusual stem, which thickened by repeated branching and by the production of a thick mantle of roots.

In uplands, conifers were abundant (Figure 1.38). The pine family, the Pinaceae, were represented for the first time by species variously called *Abietites*, *Pinites* and *Pityostrobus*. One species, called *Pseudoaraucaria major*, was central to the diversification of the family. Another family, the Cupressaceae, the cypress family, which had evolved in the Jurassic, was now represented in the British area. Other families were the Araucariaceae family with *Elatides* and *Araucarites*, the Taxodiaceae and the Cheirolepidaceae. Bennettites were also important. At Luccombe Chine and Hanover Point, on the Isle of Wight, the Lower Greensand flora is dominated by conifers, but also includes bennettite cones, while seed-ferns and ginkgoes were of lesser importance.

Even when flooded, the Weald Basin was very shallow (Hughes, 1975). It was first colonized by the horsetail, *Equisetum lyellii*, which grew with its base in water (Figure 1.37) (Watson and Batten, 1990). Drier meadowland was colonized by *Lycopodites* and another horsetail species *Equisetum burchardtii*, which survived the dry season by producing a resistant tuber. Dinosaurs left their footprints in the drying mud as they trampled through the marshes. On the driest parts of the sand bars a fern called *Weichselia* colonized (Alvin, 1974). Anatomical studies on fusainized material show that it was very xeromorphic, highly adapted to dry conditions, with small, thick leaflets (pinnules), with a thick cuticle and sunken stomata to prevent desiccation and a fibrous leaf anatomy to prevent wilting. The presence of fusain, fossil charcoal, shows that the vegetation was sometimes subject to fire.

1.8.1 ADAPTATION TO DROUGHT

At about the Middle Cretaceous the climate became drier again. Evidence for increased aridity is the continued decline of the cycad, *Nilssonia*, and a rise in diverse bennettites (Ruffel and Batten, 1991; Watson and Sincock, 1992). The arid period is marked by a rise of conifer pollen, though *Classopollis* pollen was not as abundant as it had been in the Upper Jurassic, rising to only 4–5% of all pollen.

The tree *Pseudofrenelopis parceramosa* was probably responsible for much of the *Classopollis* deposited at this time. It had several adaptations allowing it to survive drought (Figure 1.37). It was a tall tree with succulent photosynthetic stems in bunches at the ends of its branches (Alvin, 1983). The leaves were small and sheathed the stem. Although most of the stem and leaf had a thick cuticle, there was a line of hairs on the margin of the leaf on which dew could precipitate at night. The dew could then be absorbed through a thin area of the cuticle at the base of the leaf. In a drought the tree became dormant, giving rise to growth rings in the wood, and whole photosynthetic stems could be shed to restrict water loss. It grew rapidly when it became humid again.

The Cretaceous is marked by a change in the fauna. There were still massive sauropod dinosaurs, represented now by *Macrurosaurus* (12 m long) and armoured nodosaurids and stegosaurids such as *Polacanthoides*, *Acanthopolis* and *Craterosaurus* (4–5 m long). However, there were also new, smaller, fleet-footed kinds, *Yaverlandia* (1 m long), *Hypsilophodon* (2.5 m long) and *Valdosaurus* (3 m long). The small size of these herbivores was perhaps favoured now because of the availability of a new supply of digestible herbage and the spread of coastal marshlands.

These dinosaurs grazed at medium to low

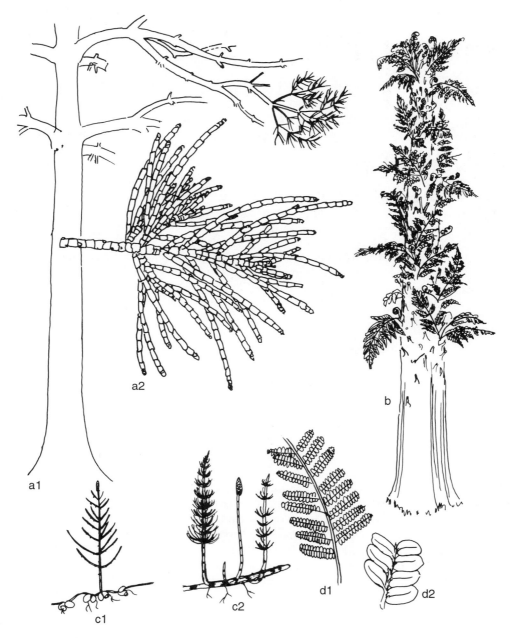

Figure 1.37 Reconstruction of Wealden plants: (a) *Pseudofrenelopsis* (*Classopollis* plant) (a1, part of tree; a2, terminal portion of branch); (b) *Tempskya*; (c) horsetails (c1, *Equisetum burchardtii*; c2, *E. lyellii*); (d) *Weichselia* (d1, frond; d2, part of pinna). (Redrawn from Andrews, 1948; Alvin, 1974, 1983; Watson and Batten, 1990.)

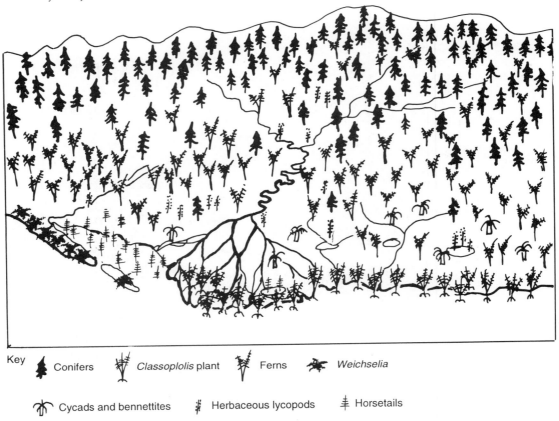

Key   Conifers   *Classoplolis* plant   Ferns   *Weichselia*

Cycads and bennettites   Herbaceous lycopods   Horsetails

Figure 1.38 The Cretaceous landscape of the Weald. (Modified from Batten, 1974.)

levels. Their activity may have prevented the establishment of the older kinds of trees, favouring low-growing herbaceous and semi-woody plants. *Hypsilophodon* and *Valdosaurus* had horny beaks and self-sharpening cheek teeth which were replaced as they wore out. Another Cretaceous dinosaur was *Vectisaurus*, a kind of iguanodon about 4 m long. It had a horny beak, and closely packed cheek teeth for chewing the vegetation. It could rise up on two legs and reach into the vegetation to grasp, with its prehensile tongue, tasty shoots and reproductive structures.

1.8.2 THE ORIGIN OF FLOWERING PLANTS

The evolution of new kinds of dinosaur was accompanied by the most important event of the Cretaceous, which was the evolution of the flowering plants, the angiosperms (Friis, Chaloner and Crane, 1987; Crane and Lidgard, 1990).

Unfortunately, flowers do not fossilize well, and so early fossil flowers are very rare. From fossil pollen and a fragmentary macro-fossil record it is likely that the flowering plants originated in the region of West Gondwana from where they spread over the supercontinents of Laurassia and Gondwana. Some of the earliest evidence for the presence of angiosperms in the northern hemisphere, from 35°N or more, comes from fossil pollen of the Lower Greensand and the Weald of south-eastern England from the Barremian epoch (mid Early Cretaceous) (Kemp, 1968). An earlier record of what may be angiosperm pollen is from the Upper Jurassic of France (Cornet and Habib, 1992).

In the Wealden record there are pollen grains of several sorts, including those of conifers and bennettites. Less than 1% of the pollen has the complex tectate exine which makes it recognizably angiosperm (Hughes, Drewry and Laing, 1979). The earliest angiosperm pollen, called *Clavatipollenites*, has a single furrow and has been likened to the pollen of the living genus *Ascarina* in the family Chloranthaceae, a small genus of shrubs and small trees living in parts of South-East Asia to New Zealand (Couper, 1958). Other early fossil angiosperm pollen grains have been likened to those of other living genera in the same family (Friis, Crane and Pedersen, 1986) and the putative earliest angiosperm leaf macrofossils also share some characteristics with those of living Chloranthaceae: with pinnate–palmate reticulate venation (Lidgard and Crane, 1990). In addition the earliest angiosperm fossil flowers from the Lower Cretaceous of eastern North America and southern Sweden are of fragments like parts of the flowers of living species in the Chloranthaceae (Crane, Friis and Pedersen, 1986). The flowers are small with short, stubby stamens and often unisexual.

The absence of angiosperm wood from the earliest part of the Cretaceous, even where fossil conifer wood is abundant, has been taken in part as evidence that the earliest angiosperms may have been small rhizomatous or scrambling perennial herbs (Taylor and Hickey, 1992). From Bedfordshire one genus of wood has been named *Woburnia*.

From slightly later in the Wealden pollen record there are diverse kinds of a different kind of angiosperm pollen grain, a three-furrowed pollen called *Tricolpites*. The same kind of pollen has been found inside the anthers of an Early Cretaceous fossil flower which is like a species of *Platanus* (plane) (Crane, Friis and Pedersen, 1986). These fossils are evidence of the early and rapid diversification of the angiosperms. In the British area angiosperms remained a relatively unimportant part of the Cretaceous vegetation (Hughes and McDougall, 1990). Only one site in the Aptian epoch, 110 million years ago, has 10% flowering plant pollen. Some fossil wood, of several different kinds, is like that of angiosperms. By the Upper Cretaceous flowers related to the saxifrages were growing in southern Sweden (Friis and Skarby, 1981) and diverse flower types have been discovered elsewhere. It was rather as if the green monochrome film of the Cretaceous landscape had been touched up here and there with just a splash of colour from a flowering shrub or herb.

1.8.3 A CHALKY GAP, THE LATE CRETACEOUS

As the Cretaceous continued, the break-up of the supercontinents continued (Figure 1.39). By the end of the Cretaceous the shape of our modern world was becoming obvious. The break-up of the continents dissipated the monsoonal climates which had dominated the world over the previous 250 million years. Towards the end of the Cretaceous, 65 million years ago, the continents were maximally dispersed. Conditions were warmer than today but, more importantly, latitudinal climatic gradients were much less marked. In the Albian epoch of the Cretaceous a major marine transgression began, so that by 90 million years the British area was covered by a sea in which chalk deposits were being laid down. Only the Highlands and Southern Uplands of Scotland, and perhaps the mountains of North Wales and the Wicklow Mountains in Ireland, remained as islands in the chalk sea, and even these may have been briefly inundated. There are no fossil plants preserved in the British area for the Upper Cretaceous.

There are several possible reasons for this. Most of the chalk was several hundreds of miles from the nearest land. Terrestrial fossils

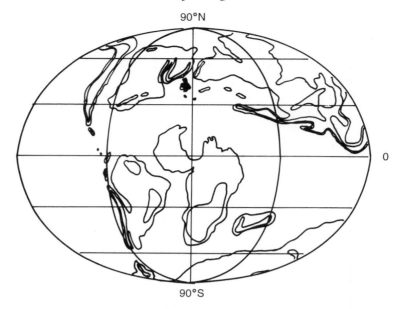

Figure 1.39 Map of the Cretaceous world (94 Ma), showing the position of the British Isles. (Courtesy of C.R. Scotese, Palaeomap Project, University of Texas, Arlington.)

were not transported into them. Chalk laid down closer to the Scottish islands was eroded away almost completely after the Cretaceous. It is generally agreed that the climate was non-seasonal and hot, but it may either have been desert or tropical forest. If the Scottish islands were desert, they would have provided little plant debris to be fossilized and no seasonal river flood to carry the debris out to sea. Phosphatic nodules are found in the chalk, which are comparable to those produced in seas subject to arid conditions today. However, even if a thick tropical forest covered the British islands, as has been suggested, the oxidizing conditions of the chalk sea would have rotted any plant material very rapidly. Very few dinosaur fossils have been discovered, perhaps from the odd, bloated corpse which floated far out to sea before it sank.

Towards the end of the Cretaceous the chalk sea receded as there was uplift of the land. Then, at the end of the Cretaceous, 65 million years ago, there was a cataclysm.

Whatever the reason, either a collision with a massive asteroid, or greatly increased volcanic activity, the cataclysm had a profound effect. It changed the climate, like a prehistoric analogue of a nuclear winter, even if for only a short period. All dinosaurs, tropical reef invertebrates and calcareous planktonic organisms were eliminated. It was as if the curtain had been brought down, calling an intermission upon a whole fossil history. When the curtain rose again everything was different – and in Technicolor, as flowers spread rampant across the landscape.

1.9 RECOGNIZING THE PAST

The Tertiary period, which started 65 million years ago, has been divided into a number of epochs. The first was the Palaeocene.

Plants recovered quickly from the terminal Cretaceous cataclysm but when we see the British flora again everything is changed. Conifers were still important in some areas, but one previously important conifer family, the Cheirolepidaceae, had disappeared. Seed-ferns and bennettites had also disappeared, and cycads were much less important. Instead the landscape was full of flowers.

Now there were many kinds of fossils similar to living plants. Their presence has encouraged the naming of fossils with the names of living genera or species as if they were the same. A careful study of a winged fossil fruit from the Palaeocene of Mull called *Calycites ardtunensis* has shown it to be very similar to *Abelia* (Crane, 1988) a genus of cultivated ornamental shrub from Mexico and the Himalayas to eastern Asia (Figure 1.40). Other Tertiary plants to which a modern name can be applied with some confidence are the mangrove palm, *Nipa*, and the Japanese red cedar, *Cryptomeria* (see below). The use of the names of modern genera and species is useful, but only because it allows us to picture the vegetation; however, plants, especially angiosperms, were rapidly evolving, and family and generic limits were very fluid. For example, a detailed examination of fossil leaves and

bracts, from the Reading Beds near Newbury in southern England, dating from the Palaeocene, has allowed the reconstruction of a fossil species *Palaeocarpinus laciniata* which combined the characters of several living genera, including *Corylus* (hazel) and *Carpinus* (hornbeam) (Figure 1.40) (Crane, 1981). Separate parts of the plant might have been assigned to separate living genera.

Many of the names of living taxa used in accounts of the Tertiary vegetation, especially for the earlier stages, might be better given the suffix '-like', or placed in inverted commas. For example one plant, which was *Sequoia*-like in its foliage, was important in some kinds of vegetation for a long time in the Early Tertiary of Britain, but seems not to have produced pollen like that of the living *Sequoia*, (Californian redwood) (Boulter, 1980). Nevertheless, in this account, familiar names have been used wherever possible.

Fossil pollen has been related to existing taxa. There are problems with this because of the lack of differentiation of the pollen of some large groups. Pollen called *Tricolpopollenites* and *Tricolporopollenites* have a

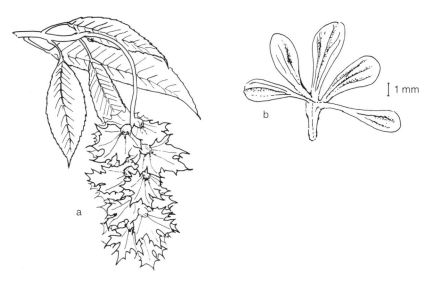

Figure 1.40 Tertiary plants with modern affinities: (a) reconstruction of part of *Palaeocarpinus*; (b) *Calycites ardtunensis* fruit body with wings, like living *Abelia*. (From Crane, 1981, 1988.)

broad affinity to trees and herbs. In addition, some important Early Tertiary taxa, such as the laurels, are under-represented because their pollen grain had a thin sporopollenin coat which did not survive as well as other kinds.

1.10 TERTIARY FLORAS

The most important Tertiary fossil flora comes from three different kinds of deposit in different geographical regions: from inter-basaltic sediments in Mull and Antrim, from marine clays in south-east England and from lake and river deposits in a series of basins in western Britain. They are all from the Early or Middle Tertiary. There are very few diverse or reliably dated plant fossil assemblages from the British Isles for the later Tertiary.

A distinction between different vegetation elements of Tertiary fossil assemblages has often been made. Statistical analysis of pollen and spore spectra from boreholes of the Palaeogene (Early Tertiary) provides evidence for a number of different kinds of vegetation which were significant at different times and in different places. Normally there is a marsh or coastal fringe element of plants which are associated because of their closeness to the catchment area of the pollen-bearing sediments. In addition, three other major assemblages have been recognized: a paratropical pollen group, a conifer-fern group and a deciduous forest group. However, there is considerable difficulty in separating distinct vegetation elements from individual assemblages and correlating these elements between sites.

1.10.1 THE PALAEOCENE FLORAS OF MULL AND ANTRIM

The Early Tertiary period was a time of extensive volcanic activity over a large area between Britain and Greenland. Volcanic activity continued on and off for many millions of years, and could be said to be continuing today in the volcanic activity of

Iceland. The North Atlantic was spreading and the Rockall shelf was sinking, opening the sea between Britain and southern Greenland. Land connection with North America was probably maintained until the Oligocene by a land bridge between northern Scandinavia and northern Greenland (Taylor, 1990). The fossiliferous plant beds of Mull and Antrim are sedimentary beds interleaved with basalt beds from different episodes of lava flows, dating from 58 million years ago (Simpson, 1961; Boulter, 1980; Boulter and Manum, 1989; Boulter and Kvaček, 1990). The dating of these inter-basaltic fossiliferous beds has, in the past, proved somewhat problematic but it is likely that those in Antrim date from the Palaeocene/Eocene boundary and the ones from Mull from a little later, but still in the Eocene.

The assemblage shares many similarities with an earlier one from Greenland (Seward, 1926) and may even have been connected with others from Alaska, Siberia and western USA, dating back to what has been called the polar broad-leaved deciduous forest of the Cretaceous (Mai, 1991). The flora was relatively uniform. It was dominated by evergreens and broad-leaved deciduous trees (Seward and Holttum, 1924). Conifers, especially the *Sequoia*-like one (*Sequoites*), dominated the lowlands with ferns (Figure 1.41). Fossil conifers included those like *Pinus* (pine), *Cedrus* (cedar), *Picea* (spruce) and *Cunninghamia* (Chinese fir), *Metasequoia* (dawn redwood) as well as *Cryptomeria*. There was also something like *Podocarpus* which is now a diverse genus of nearly 100 species distributed from New Zealand to Japan and the West Indies, including tropical mountains.

There was also a fern prairie dominated by a fern like *Onoclea* (sensitive fern) which is sometimes grown in gardens and has become naturalized in western Britain on wet soils. In the Tertiary it was sometimes associated with groves of *Platanus* (plane) trees (Figure 1.42).

Mixed in with the conifers were diverse

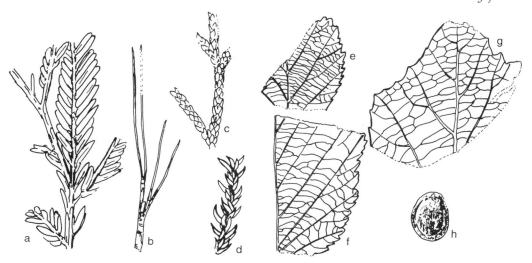

Figure 1.41 Fossil fragments from the Tertiary of Scotland: (a) *Sequoites*; (b) *Pinites*; (c) *Cupressites*; (d) *Pagiophyllum* (*Cryptomeria?*, *Araucaria?*); (e)–(f) *Corylites*; (g) *Platanus*; (h) nut of *Corylites*. (From Seward and Holttum, 1924.)

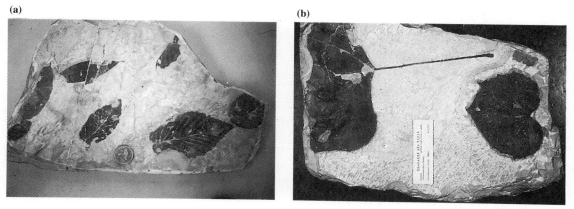

Figure 1.42 Fossils from the Eocene of Mull: (a) *Corylus* (hazel), (b) *Platanus* (plane) and *Vitis* (vine). (Courtesy of W.G. Chaloner.)

flowering trees, including many recognizable families from today. They include many possibly deciduous species such as *Quercus* (oak), *Corylus* (hazel) (Figure 1.42), *Acer* (maple), *Alnus* (alder), *Celtis* (Nettle-tree) and *Cercidiphyllum* (katsura). Others were in the magnolia family (Magnoliaceae) and the dog-wood family (Cornaceae). Several primitive woody plants in the witch-hazel family (Hamamelidaceae), or closely related to it,

were represented. At this time, the broad-leaved deciduous trees, which had first evolved in the polar broad-leaved forest in the Cretaceous, spread into middle palaeo-latitudes. Many of these Early Tertiary genera have relatives that will grow perfectly well in Britain today, but many of them, including *Quercus*, also have tropical or subtropical species. Others, including *Vitis* (vine) (Figure 1.42) are mainly low latitude in distribution today, although, in the tropics, they usually grow at moderate altitudes. One species was like *Engelhardtia* in the walnut family, which is now found from the Himalayas to Malaya. It grows at an altitude of 610–1500 m in Malaya. Another fossil was like *Casuarina* (she-oak) which grows above 1500 m in New Guinea. Other tropical plants included one like the tropical water-weed *Aponogeton*.

Some of these trees may have been growing in the uplands with conifers; the uplands here, perhaps only a few metres above the flood-plain swamps. Traces of lignite are derived from a lowland lake-margin swamp flora. *Equisetum* (horsetails) and water-lilies have been found, including *Nelumbo* (lotus) and *Euryale*, which has a single living species found from China to northern India. On the flood plain were trees like *Taxodium* (swamp-cypress), the related *Glyptostrobus* (Chinese swamp-cypress) and *Planera* (water-elm).

The fossil flora of Antrim is very similar to that of Mull. It lacks the fern *Onoclea* but has many of the trees such as *Sequoia* (Californian redwood), *Quercus* (oak), *Platanus* (plane) and *Acer* (maple). There are some other distinctive kinds. One of these is called *Macclintockia*. This species, and the general character of both the Antrim and Mull floras, link them with the Gelinden flora of Belgium, which is of Upper Palaeocene age. At least at times the broad-leaved deciduous flora reached to the south of Britain and beyond. Volcanic activity was throwing dust and smoke into the atmosphere which was perhaps causing climatic cooling.

1.10.2 THE TERTIARY GLOBAL FLORA

Many genera, such as *Ginkgo*, *Metasequoia* (dawn redwood) and *Cunninghamia* (Chinese fir) which had relatives globally distributed in the Tertiary, have a very restricted distribution today. *Metasequoia* is interesting as a living fossil, because it had been identified as a fossil *Sequoia* for over a century before it was

named by a Japanese palaeobotanist called Shikeru Miku in 1941. By chance, and separately, it was discovered in the winter of the same year by a Chinese forester called T. Kan growing in a remote valley of Szechwan Province in China. It is also interesting as one of the few deciduous conifers. Another deciduous conifer present at this time was like *Pseudolarix* (golden larch) from eastern China.

The Tertiary flora has strong affinities with the living flora of East Asia and North America. *Onoclea* is native in eastern North America and northern Asia. *Taxodium* (swamp-cypress) and *Planera* (water-elm) are now found only in North America. The counterpart of *Taxodium* in Asia is *Glyptostrobus*. Species with an East Asian affinity include *Exbucklandtia* from the eastern Himalayas, southern China, the Malayan peninsula and Sumatra; *Euptelea*, a genus of small trees or shrubs from Assam to southwestern and central China; and *Corylopsis*, which grows naturally from Bhutan to Japan. Many of the living relatives of these Early Tertiary plants from more temperate regions have been introduced to parks and gardens in the British Isles, and like *Cercidiphyllum* (katsura), the largest deciduous species native to Japan. Most grow very well here, especially in the less frosty and moister west.

The presence of a *Dicoryphe*-like species, also in the Hamamelidaceae, a genus today endemic to Madagascar and the Comoro Islands, illustrates a link with the flora of Gondwana. Another example of this Gondwanan link is the presence of members of the Proteaceae family; *Dryandra*, *Petrophila*, *Lambertia* and *Faurea*. Southern-hemisphere conifers, *Podocarpus*, were present. Two other curious plants that illustrate this Gondwanan link are the taxonomically isolated *Balanops* and *Casuarina* (she-oak). *Balanops* is a small genus of trees or shrubs, related to the oak family. It is found now only in north-eastern Queensland, Fiji and New Caledonia. *Casuarina*, with its branches looking like those of a horsetail and fruits looking like conifer cones, grows from Australia up into Asia. Both grew on the Isle of Mull in the Early Tertiary.

1.10.3 EVOLUTION IN THE EARLY TERTIARY

Despite its seeming uniformity, it is unwise to regard the fossil flora of the Early Tertiary as being of one kind. It existed over many millions of years and underwent many changes. An examination of two assemblages on Mull show remarkable differences. The older one is of pre-basaltic lignite from Arslignish Ardna. It has a high proportion of dry-habitat or poor-soil adapted species such as *Casuarina* (she-oak). It has 43% fossils with an East Asian or North American affinity. At Shiaba and Bremanour, in a later inter-basaltic plant bed, there are more temperate deciduous plants, 80% of which have an East Asian or North American affinity. Together these assemblages have 71 species, and although they share several genera they have no angiosperm species in common. Perhaps this represents evolutionary change.

With the spread of deciduous broad-leaved forest a wide range of new habitats became available, encouraging the evolution of newly adapted populations. Species consist of sets of genetically varying individuals in semi-isolated groups called populations. As populations of species were established in new areas they were often different genetically from their ancestral populations. In part this was due to the founders being genetically different by chance sampling from the average of their ancestral population. In the new geographically isolated population that difference was maintained and magnified as the new population increased in size. This evolutionary process is called the 'founder effect'.

Genetic evolution may also have been rapid in the Early Tertiary because, although the vegetation was rich in woody species, no one species ever rose to dominance. It has

been suggested that the diversity of the Co. Antrim flora may have reflected the pioneer conditions that existed (Boulter and Manum, 1989). Vegetation succession was halted frequently because of regular ash fall and lava flows. In the volcanic landscape of the Palaeocene the development of a forest climax vegetation was continually being prevented. These are the conditions, especially the many changes in population size, in which the evolutionary process was speeded. Evolution through a random sampling process by what has been called 'genetic drift' occurs most rapidly in small and fluctuating populations.

In addition, natural selection for new adaptations in the many new habitats that became available led to distinct populations. In time these populations became distinct species, reproductively isolated from ancestral-type populations. The spread of the polar broad-leaved forest into lower latitudes was accompanied by a great diversification of broad-leaved species. The morphological limits of modern genera were becoming established.

It may be because of this great evolutionary flux that the Early Tertiary vegetation cannot be compared easily to any living vegetation. There was a mixture of what are today mainly temperate or tropical and subtropical types. Although many plants from living kinds were present, they are from different species and probably had different habitat requirements. For example, one peculiar characteristic of these floras is the large size of some of the leaves, like a *Platanus* leaf that measured 40 × 27 cm. *Corylus macquarii* was a large-leaved hazel. The presence of large leaves is evidence of calm, humid conditions. Large leaves were an adaptation to the low light conditions of these high latitudes.

In some assemblages there are also plants like *Casuarina*, with its tiny leaves and photosynthetic stems, and also species such as members of the Proteaceae, with their small, hard leaves, which are adapted for dry or low-nutrient conditions. The presence of species from the Ericaceae, the heather family, with their small and needle-like leaves is also evidence of the presence of low-nutrient environments.

1.10.4 THE LATE PALAEOCENE IN SOUTHERN BRITAIN

In southern Britain the uplifted chalk was being eroded away. In the Palaeocene epoch southern Britain lay at about 40°N. On a coastal plain subject to flooding by the sea, the Thanet, Woolwich and Reading Beds were laid down (Chandler, 1961, 1964). Fossils have been recovered at Herne Bay and Cobham in Kent; at Felpham, West Sussex; Saint Pancras in London; and Newbury in Berkshire. Dating from the Late Palaeocene, these have a flora comparable in some respects to Mull and Antrim, with pine-like and *Sequoia*-like trees forming patches of woodland and thin forest on drier land. There were clearings in which shrubs and climbers flourished. The transient vegetation of flood plains included species such as *Cercidiphyllum* (katsura) and *Platanus* (plane). Here also there was *Palaeocarpinus* in the birch family (Crane, 1981), *Casholdia* in the walnut family, *Sassafras* in the laurel family and a plant with rhododendron-like seeds. A reed-swamp community was dominated by palms and the fern *Straelenipteris*. Terrestrial grasses were not important but there were other herbaceous species, such as the fern *Anemia*.

There was a mix of temperate and sub-tropical elements. These included *Cleyera/Eurya* in the camellia family, which was probably a shrub or small tree (Daley, 1972). *Iodes*, which is today a twining plant from South-East Asia, was represented often by its inner fruit-wall *Icacinicarya*. *Meliosma*, in the family Sabiaceae, and the woody climber *Tinospora* in the Menispermaceae, which are today tropical and subtropical genera from Asia and America, were also present. Other fossils were members of the Laurel family, *Mastixia* in the dogwood family, *Vitis* (vine),

and *Rubus* (blackberry), which are normally regarded as being mainly temperate in origin.

The supposed mixture of temperate and tropical elements may be due to a greater restriction of each kind of plant to a particular habitat since the Early Tertiary, rather than the presence of any peculiar habitat that combined the features of both habitats. Early Tertiary habitats were very different in one respect from those we might see today, in the range of large animals present. There was a high proportion of insectivores, some fruit-eaters but few small and unspecialized leaf-eaters (Hooker *et al.*, 1980). There were also few arboreal animals and so it seems possible that high-canopy forest was not well developed.

1.11 THE LONDON CLAY FLORA

A rich flora of about 50 million years ago is preserved in the London Clay. It is a marine clay that was laid down in calm coastal waters off-shore from a mangrove-fringed coastline (Figure 1.43). Large rafts of vegetation floated out to sea before sinking. This process has inevitably resulted in some selection of the types preserved. The mangrove palm, *Nipa*, which today grows in South-East Asia, is a characteristic fossil of the assemblages (Figure 1.44). Another species, *Ceriops*, from the mangrove family itself, was also present. Further up the creeks the mangrove gave way to marshes dominated by a sedge, '*Scirpus lacustris*'. Away from the water there was a multistoried forest.

The London Clay assemblages consist mainly of pyritized fruits and seeds, with some carbonaceous material (Figure 1.45). The most important fossil sites are on the coast where the sea has eroded the clay cliff and sorted fossils which are collected by 'surface picking, crawling along the foreshore on hands and knees, sorting through pyrite concentrates' (Collinson, 1983). The earliest fossils are from the clay cliffs at Herne Bay in Kent (>130 species). Another important site

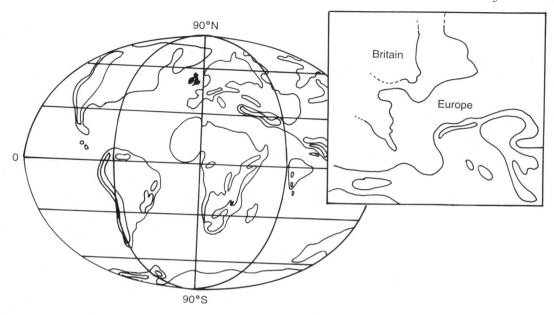

Figure 1.43 Map of the Eocene world (50 Ma), showing the positon of the British Isles. (Courtesy of C.R. Scotese, Palaeomap Project, University of Texas, Arlington.)

from slightly later is Bognor in West Sussex (>130 species). Later still, reaching into the Oligocene, are fossils from Sheppey in Kent (>300 species). There are many other sites of lesser importance. At some of these, fossils can be obtained by collecting sediment, disaggregating it and sieving the debris.

The identification of species is not easy, as Margaret Collinson describes:

> Taking an hypothetical fossilised orange preserved in pyrite, in simple cases the fossil may be the orange; a cast of all parts internal to the skin (i.e. with the outside looking like the inside of the skin); the orange minus the skin, revealing the segments; individual segments; casts of internal cavities (locules) which contain the seeds; the seeds themselves; and internal casts of the seeds

or a combination of these. Most species are known from fewer than 10 specimens and so the flora probably represents just a very

partial sample of the true total flora. Nevertheless, over 500 plant types have been collected, with 350 named species (Reid and Chandler, 1926; Collinson, 1983).

1.11.1 THE EOCENE 'TROPICAL' FLORA

In the Eocene there was a rise in temperature, perhaps because of a decline of volcanicity in the north. The precise timing of the warming varies between different authors, the methods used to detect it, and the stratigraphical correlations that have been proposed. However, the Early–Middle Eocene fossil floras show a rise in the percentage (up to 92%) of kinds of plants with tropical affinities.

With the break-up of Gondwana and the relative drift north of Africa and India, the Tethys sea-way was closing. The London Clay flora was part of a boreotropical flora, a northern hemisphere 'tropical' flora, a green fringe growing across southern Europe and

Figure 1.44 View of *Nipa* swamp in southern Thailand.

Figure 1.45 Fruits from the London clay of Sheppey. (Courtesy of W.G. Chaloner.)

North Africa along the Tethys sea-way to south-eastern Asia with an extension, either westward or eastward into North America (Tiffney, 1985). All traces of this flora have disappeared from Europe and North Africa but elements survive in Asia and America. A large proportion of the London Clay flora has close relatives that survive today in South-East Asia. Seventy per cent of Eocene genera have an affinity to today's flora of Malaysia, 20% with Central America and the West Indies and only 4% with southern Europe. There was also a selection of genera, such as *Beilschmiedia* (bolly gum) in the laurel family (Lauraceae), with living relatives from Australia and New Zealand. The flora has been called laurophyllic because of the large proportion of evergreen broad-leaved species. A relic laurophyllic flora survives precariously in the laurel forest of the Canary Islands.

The closest analogous present-day vegetation may have been a kind of paratropical (= subtropical) rainforest that grows where there is a mean annual temperature of 20–25°C and where the annual range in temperature is only 8.0°C. This occurs today in the lowlands of East Asia between 10°N and 27°N. At those latitudes, however, there is little seasonal variation in sunlight, unlike the conditions the London Clay flora experienced, and so the present vegetation of South-East Asia is probably not necessarily a good analogue. However, there was not a strong dividing line between tropical–subtropical elements and temperate elements as there is seen today (Taylor, 1990). The lack of frost and high rainfall in high latitudes allowed tropical elements to survive much further north than they can today. 'Tropical' species were present in Alaska in the Middle Eocene but were subject to extreme seasonal changes in the availability of light. Many London Clay fossils are from living tropical families, Anarcardiaceae (cashew-nut family), Annonaceae (custard-apple family), Burseraceae (frankincense family), Dilleniaceae (*Hibbertia*) the Lauraceae (*Cinnamomum*), Magnoliaceae (magnolia) and the Sapindaceae (litchi family).

We can contrast the faunal and floral evidence. The fauna showed an increase of arboreal mammals at the beginning of the Eocene (Pearson, 1964; Hooker *et al.*, 1980). There was an increase in the proportion of soft-fruit-eating rodents compared to insectivorous forms. These changes followed the closure of the forest canopy. Taller angiosperm trees replaced shrubs and there was a trend for an increase in size of ground-living browsers. In the shade of the higher tree canopy there was more open ground where they could wander. The shrub layer was grazed back by new, more specialist leaf-eaters, the Perissodactyls, ancestors of

rhinoceroses, horses and tapirs; one, *Lophiodon*, was rhinoceros-sized. The activity of the leaf-eaters, nipping the growing shoots off saplings, may have prevented the regeneration of trees in some areas, opening up light-dappled glades in the forest.

However, there was an absence of the families of very tall trees found in some tropical forests today. Several of the families represented are noted today for woody climbers. There were also specialist families of woody climbers and vines, including the vine family (Vitaceae), the Menispermaceae and the Icacinaceae. Palms were particularly abundant and included talipot-palm-like and palmetto-palm-like species. At first there was dense, low-canopy vegetation. Palms poked through the impenetrable 'bush', their trunks draped in climbers and twiners. Straggling plants draped across the canopy, hanging down where the canopy was broken by the presence of a stream or river. Woody climbers grew profusely at the river margins, taking advantage of the increased light levels to establish.

1.11.2 THE EOCENE TWIG FLORA

Interpretation of these fossil assemblages is bound to be difficult because of the haphazard and different ways in which land-derived fossils are deposited. The twig flora of the London Clay is different from the fruit and seed flora (Poole, 1992). It represents a different sample of the vegetation; a sample of those species that did not provide fruits or large, decay-resistant seeds. The twig flora of the London Clay is closer to the Mull and Antrim flora. It has up to 30% conifers, including *Pinus* (pine) and *Sequoia* (Californian redwood) which are usually regarded as being mainly temperate in distribution. Eleven per cent of the flora has temperate most-close relatives. Robust twigs may have floated down rivers and streams from distant temperate upland sites. However, the temperate elements may not have been growing in a separate vegetation, because no significant highlands were present around the London Clay Basin. Sediments came mainly from the area of Devon and Cornwall, and Brittany, which rose only to an altitude of about 300 m, compared to the 600 m of today.

However, the climate was seasonal at times. Growth rings are found in 31% of twigs, especially those of conifers. These may have come from an environment where seasonal fluctuations were experienced more extremely. Growth rings may have been more likely to form on nutrient-poor soils or soils seasonally flooded. The whole flora suffered at least a seasonal fluctuation in the amount of sunlight and also probably rainfall. Southern Britain lay at about 41°N compared to the 51°N of today. In the Eocene the winters were uniformly mild, without frost, but mean annual temperature fluctuated. At times the summers were very hot. In existing paratropical forests 'temperate' elements also recorded from the London Clay such as *Pinus*, *Parthenocissus* (virginia creeper), *Rubus* (blackberry) and *Vitis* (vine) can be found growing in more open habitats and along stream sides.

1.11.3 INTERPRETING EOCENE DIVERSITIES

An alternative explanation for Eocene vegetation diversities has come from an examination of changes in Eocene spore–pollen associations. This proposes a vegetation distribution not unlike that of the Palaeocene of Mull and Antrim, with four main communities: a distant 'montane' conifer-dominated community; a mixed lower 'mesophytic' community with broad-leaved species such as *Quercus*, *Ulmus* and *Liquidambar* (sweet gum) and conifers such as *Sequoia* (Californian redwood); a tree-swamp and freshwater fen community; and a coastal swamp community dominated by *Nyssa* (tupelo) and *Taxodium* (swamp-cypress). The

relative dominance of these communities changed with rising and falling sea levels (Figure 1.46) (Jolley and Spinner, 1991). A pioneering fern-dominated vegetation colonized the margin of localized or ephemeral coastal lagoons.

An examination of aquatic or marginal-aquatic species, which have a more constant presence near depositional basins over time, gives a better idea of a changing climate. Changes in these aquatic and marginal-aquatic floras show a marked warming at the beginning of the Eocene, although possibly this may be more the result of a change of depositional environment from the Woolwich and Reading beds to the London Clay than a result of an increase in mean annual temperature (Jolley, 1992). *Nipa* became abundant. More interesting is the rather gradual change in species composition in the London Clay after the Early–Middle Eocene, probably a result of slow climatic cooling. First there is a loss of tropical and subtropical elements, so that *Nipa*, for example, shows a marked decline. There is a gradual change-over of species of sedges and other marsh and aquatic plants. This is accompanied by an increased abundance of the spores and pollen of ferns and conifers and the development of a reed swamp dominated by *Typha* (bulrush) fringing open water which had *Potamogeton* (pondweed), *Nymphaea* (water-lily) and *Stratiotes* (water-soldier). These are plants indicative today of more temperate conditions.

Climatic cooling occurred over a period of 15 million years. Its effects lasted into the Oligocene. An opening up of the vegetation may have been general, so that woodland became confined to patches on islands within the marsh. The fruit-eating rodents declined in importance, giving way to more generalist fruit-eating/leaf-eating browsers. There was a marked increase in the proportion of ground-living herbivores of larger and larger body size, but notably these were browsers and not specialist grazers. Grasses had evolved,

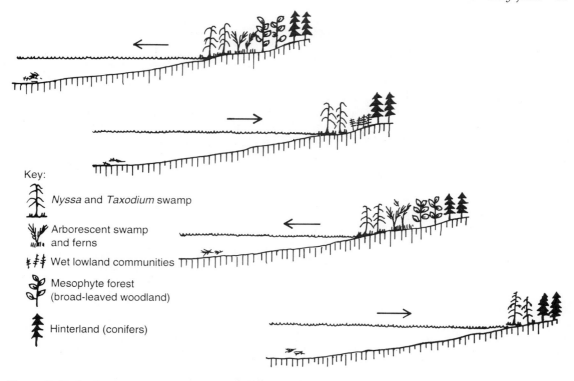

Key:

Nyssa and Taxodium swamp

Arborescent swamp and ferns

Wet lowland communities

Mesophyte forest (broad-leaved woodland)

Hinterland (conifers)

Figure 1.46 An interpretation of changing Early Tertiary vegetations at times of marine transgression and regression. (After Jolley and Spinner, 1991.)

but grasslands were not yet an important feature of the world's vegetation.

1.11.4 A COOLING AND DRYING CLIMATE

Palaeotemperature curves from oxygen isotope ratios in the North Sea sediments and global studies correspond to the changes in fauna and flora described above, showing a marked cooling at the end of the Eocene. From the Middle Oligocene there was also a decrease in humidity. The boreotropical flora was invaded by northern elements. The British Isles, connected to Europe on and off throughout this period, were moving with Europe into more northern latitudes. By the Late Eocene Europe was connected to North America only by a northern land bridge between Greenland and northern Scandinavia (Tiffney, 1985). Henceforward North America was closed to the direct spread of thermophilic elements to and from Europe.

Many elements of a world-wide northern temperate flora remained, but there was much more regionalism. Of the British fossil genera of this time, 71% have living relatives centred on China and Japan and 49% living relatives in the USA. There was an increasingly recognizable European element. After the Middle Oligocene there was the rise of a broad-leaved deciduous forest. Temperatures declined eventually to an annual mean of 18°C (Hubbard and Boulter, 1983) (Figure 1.47). The Bembridge fossil flora of the Isle of Wight includes trees, such as *Quercus* (oak), *Fagus* (beech) and *Carpinus* (hornbeam) trees

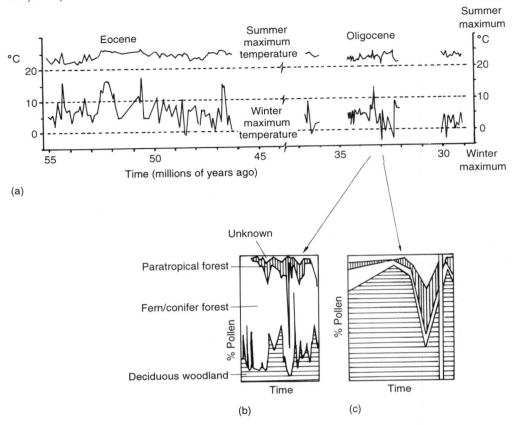

Figure 1.47 Temperature and vegetation in the European Tertiary: (a) winter and summer maximum temperatures predicted on the basis of average pollen spectra at 12 different sites, including Bovey Tracey and Mochras; (b) pollen assemblages at Bovey Tracey; (c) pollen assemblages at Mochras. (From Hubbard and Boulter, 1983.)

and several herbaceous genera such as *Ranunculus* (buttercup). *Papaver* (poppy) and *Carex* (sedge) (Reid and Chandler, 1926).

Temperatures declined to a temporary minimum in the Late Oligocene, but rose again at the beginning of the Miocene, which started 25 million years ago, only to decline again and further by the Middle Miocene. Most importantly, the world now experienced a marked latitudinal difference, so that we can now truly talk about a temperate as well as a tropical region. Glacial ice was now present in Antarctica. A period of mountain

uplifting was well under way. Sea levels fell to perhaps 50 m below that of today. The Tethys Sea was shrinking as Africa moved north coming into contact with Europe about 17 million years ago. In many continental areas there was a seasonally dry climate. In some parts of the world the grass, which may have first arisen in the Palaeocene on forest margins, now became abundant. The grasslands encouraged the evolution of large mammalian herbivores. A drier and colder climate was gradually extirpating the European connection with the flora of South-East Asia.

1.12 OLIGOCENE FLORAS BESIDE MOUNTAIN LAKES AND STREAMS

Perhaps the beginnings of a northern temperate flora can be seen in an Oligocene assemblage preserved in clays at Bovey Tracey in Devon (Chandler, 1957) and also in a middle Oligocene assemblage from the Bristol Channel (Boulter and Craig, 1979). The clays were laid down in a lake surrounded by mountains. Mountain torrents from Dartmoor carried a mass of wood and peat, which was fossilized as lignite.

The most important fossil is a conifer called '*Sequoia couttsiae*', although either its pollen did not survive or it produced a different kind of pollen to living *Sequoia*. *Nyssa* (tupelo) and a kind of *Cinnamomum* (cinnamon) were also abundant (Figure 1.48). *Nyssa* grows in southern North America and in East Asia, and *Cinnamomum* also grows in South-East and East Asia today. *Quercus* (oaks), *Carpinus* (hornbeam) and *Corylus* (hazel) were also present, and *Osmunda* (royal fern) grew in more open areas. A single cone-scale of *Taxodium* (swamp-cypress) shows that the lake had some swamp vegetation on its margins. The lake was also fringed by forest, containing species like *Ficus* (fig), *Symplocos*, *Laurus* (laurel), and *Calamus* (rattan palm) (Figure 1.49), the tropical climbing palm. One

of the few other surviving elements from the boreotropical flora of the Eocene was *Mastixia*, in the dogwood family (Cornaceae), native today in South-East Asia.

The Bovey Tracey flora has similarities to a Late Oligocene flora from clays around Lough Neagh (Chandler, 1957; Wilkinson, Bailey and Boulter, 1980), another from the Bristol Channel (Boulter and Craig, 1979) and also a more diverse brown-coal flora of Germany of the same age. This indicates a 'temperate' woodland dominated by the conifers '*Sequoia couttsiae*', '*Pinus*', and a kind of hemlock spruce like *Cathaya*, which was discovered living in China in 1955. Ferns were abundant. This graded into a more mixed woodland with *Magnolia*, palms and broad-leaved species like *Castanea* (chestnut), *Quercus* (oak) and *Juglans* (walnut). This vegetation gave way to a zone of plants such as *Myrica* (bog myrtle), *Salix* (willow) and *Ficus* (fig) in wetter areas, and finally a zone of *Taxodium*. In places there was a profusion of woody climbers. The fossil assemblage is a mixture of plant types we would not expect to find together today in Europe. Perhaps living analogues are the broad-leaved evergreen forests found in south-eastern North America, southern Japan and central China, parts of New Zealand and Tasmania and the southern tip of South America, sometimes

(a)

(b)

Figure 1.48 Living analogues of the Tertiary flora: (a) *Magnolia*; (b) *Nyssa* (copyright Jon B. Wilson).

Figure 1.49 'Tropical' elements of the Tertiary flora: (a) *Calamus* (rattan palm); (b) *Ficus* (fig).

called temperate rainforests. Summers are warm and wet and winters are mild with only rare frosts. In the North American version *Taxodium* grows in the swamps and trees are festooned with *Tillandsia* (Spanish moss). The southern hemisphere versions point back to an earlier Tertiary flora, the Palaeocene vegetation of Mull and Antrim. They are particularly diverse but include familiar Early Tertiary plants such as *Araucaria* and *Podocarpus*.

Similar conditions to contemporary broad-leaved evergreen forests are suggested at Bovey Tracey and Lough Neagh by the presence of palm pollen, *Arecipites* and *Monocolpopollenites*, throughout the pollen spectra. Although winters could be cold, frosts must have been rare. At this time the climate was uniformly cooler than the Eocene, but the mean annual temperature was probably 10–11°C warmer than today, with an annual average of about 21°C (Collinson, Fowler and Boulter, 1981). From pollen evid-

ence, the fern and conifer forest had a mean annual temperature of 14°C and range of 14°C (Figure 1.47). Summer temperatures were stable but mean winter temperatures fluctuated. Temperate rainforests experience very high rainfall, in excess of 1500 mm annually. At Bovey Tracey the presence of the podzolized soils which resulted from the high rainfall, is indicated by the importance of *Myrica*.

The pollen flora of the Mochras borehole in the Tremadoc Basin of North Wales, overlaps in age with the Bovey Tracey and Lough Neagh deposits, but contrasts with them (Wilkinson and Boulter, 1980). Its pollen spectrum indicates a lower mean annual temperature. There is a greater proportion of the pollen of broad-leaved flowering species. This may indicate the vicinity of a highland, supporting perhaps even a deciduous flora, where mean annual temperature was 10°C with a range of 30°C.

1.12.1 THE LATE TERTIARY, A GAP IN THE FOSSIL RECORD

The British Isles are almost entirely silent about the Late Tertiary, with no readily datable fossil assemblages for the entire period of between 30 and 3 million years ago. Britain was being uplifted, so there were no onshore sediments deposited for the Miocene and for most of the Pliocene. Fossil assemblages come from redeposited clays filling caves and sinkholes in limestone. The fossil assemblage from Bee's Nest pit, near Brassington, and Kenslow Top, near Friden, in the limestone of Derbyshire probably comes from the Miocene/Pliocene boundary

(Boulter and Chaloner, 1970; Boulter, 1971). It includes wood, seeds, leaves, spores and pollen of 63 species.

The flora is probably of a mixed woodland and heathland. The woodland vegetation was dominated by a variety of 10 different conifers, including a kind of *Podocarpus*, *Cryptomeria* (Japanese red cedar), *Sciadopitys* (umbrella pine), two kinds of *Pinus* (pine), *Abies* (fir) and *Tsuga* (hemlock spruce) (Figure 1.50). There was a range of ferns, but one called *Gleicheniidites* was most abundant. Flowering plants in the mixed woodland included plants like *Pterocarya* (wingnut), *Corylus* (hazel), *Carpinus* (hornbeam), *Juglans* (walnut), *Salix* (willow), *Liquidambar* (sweet

Figure 1.50 Conifers of the Tertiary flora: (a) *Cryptomeria*; (b) *Sciadopitys*; (c) *Cunninghamia*; (d) *Abies*.

gum), *Ulmus* (elm) and *Quercus* (oak, probably the pollen species *Tricolpopollenites microhenrici*). Growing as an understorey was a kind of *Ilex* (holly). *Nyssa* (tupelo), *Alnus* (alder) and a species of the Taxodiaceae were probably growing in swampy conditions on the margins of a lake.

Heathers of various sorts, *Empetrum* (crowberry), *Myrica* (bog myrtle) and even *Rhododendron ponticum* may have formed a heathland community, or alternatively may have grown as an understorey. The presence of some open habitats is indicated by the presence of herbs such as *Bellis* (daisy), *Epilobium* (willow herb) and members of the lily and grass families (Liliaceae and Poaceae). There were also *Sphagnum* (bog moss) and clubmosses. The presence of heathers (Ericaceae) and *Empetrum* may indicate that the Derbyshire flora was subject to a strongly oceanic climate, although the assemblage is largely similar to a central European flora of a similar age from Germany. The climate was warm enough to allow the growth of warm elements like *Aralia* (*Tricolpopollenites edmundii*), Symplocaceae and Sapotaceae.

A pollen assemblage from a limestone depression at Hollymount in Ireland has been dated variously to the Miocene/Early Pliocene or Upper Pliocene. It had pollen types such as *Pinus*, *Quercus*, *Corylus*, *Myrica*, Ericaceae, *Taxodium*, *Sciadopitys*, *Liquidambar* and palms.

Profound changes in the European flora occurred in the Pliocene, starting 7 million years ago. Climatic cooling and greater aridity seem to have opened the forests. This was a continuation of a trend first seen in the Middle Tertiary. Steppes and deserts spread at the expense of the forests, finally cutting off Europe from south-eastern Asia. From the Middle to Late Tertiary there was a greatly increased diversity of herbivores of all sizes, including larger, long-limbed, specialist grazers similar to those of today. Grasslands became very important. A savannah reached into south-eastern Europe and there was an

extremely rich 'Hipparion fauna', so called after the large herds of three-toed horses which ran across the plains. There were mastodons, tapirs, okapi-like giraffes, antelopes and rhinoceroses. Big cats, such as sabre-toothed tigers, and wolf-hyenas preyed on the herbivores. This kind of faunal diversity is characteristic today of open savannah-like habitats in Africa.

At the same time there was a 'mixed forest' of conifers and deciduous trees over much of northern Europe. On the western fringe, the oceanic climate allowed a vegetation dominated by conifers to develop. A fossil flora from The Netherlands had *Taxodium* (swamp-cypress) (Zagwijn, 1960). *Taxodium* requires long, hot summers to do very well. It flourishes in many parks or gardens in southern England today, and although it can grow perfectly well away from the water-logged soils of lakesides and creek bottoms, it prefers these soils. Its disappearance from the north-western European flora in the Pliocene, where it was replaced by broad-leaved trees such as *Carpinus* (hornbeam), *Quercus* (oak) and *Fagus* (beech), and later *Castanea* (chestnut), marks an important change in conditions. At the same time some other tree species such as *Aesculus* (horse-chestnut), *Juglans* (walnut) and some species of *Acer* (maple) and *Ulmus* (elm) seem to have retreated to the Caucasus.

At Cromer in Norfolk, preserved in the Coralline Crag rock formation, there is a similar vegetation which may date from the Late Pliocene. It had *Pinus* (pine) and *Picea* (spruce) with *Quercus*, *Betula* (birch), *Salix* (willow), *Acer* (maple) and *Taxus* (yew). Fossils, spiny fruits, of the *Trapa natans* (water-chestnut) are abundant. Later a deciduous woodland became dominant.

In many ways the overall look of the present European vegetation, Europe's native flora, was established at this time. The eastern Asiatic/American element was replaced by a European/western Asiatic element. Today the flora of eastern North

America has closer affinities to the flora of eastern Asia than to, the geographically closer, Europe. An important fossil assemblage from Kroscienko in Poland in the mid-Pliocene indicates a January mean temperature 11°C higher and a July mean 9°C higher than today, with twice the present rainfall. A flora at Frankfurt provides evidence of lower temperatures, with an overall annual mean of 14°C.

From now on the plant kinds seen in fossil assemblages are very similar to living plants familiar to us, so that it is possible, with some caution, to use the names of living equivalent genera, and even species. In the mid-Pliocene, 7 million years ago, the first living wholly 'European' plant taxa are recorded. They account for 12% of species, 59% of the genera of the fossil flora. By the Late Pliocene, 3 million years ago, 18% are living species. However, the vegetation still had a strong exotic look to it.

1.13 THE QUATERNARY PERIOD

The Quaternary period succeeded the Tertiary period. It is divided into the Pleistocene epoch which began 2.5 million years ago and the Holocene, Flandrian or Postglacial epoch, which started 10 000 years ago. The Pleistocene is an age of dramatic change, the Ice Age (Kurten, 1972). In fact, as early as 1909 the geologists Penck and Bruckner showed that there had been at least four glaciations. The Pleistocene is called the Ice Age but the glaciations in the British Isles are confined to the later part. The study of the succession of warm and cold periods in the Pleistocene is like a detective story where many different kinds of evidence have been used, not least the fossil remains of plants. The most important task has been to correlate the results, the stratigraphy between different areas, so that an overall chronology can be achieved.

There is still some controversy between the interpretation of different kinds of results, but there is good overall agreement.

1.13.1 THE OSCILLATING CLIMATE

The story of the Pleistocene is one of many cold stages, latterly glacial stages, with intervening warm periods, latterly interglacial stages. Evidence for rapid climatic change comes from isotopic changes in ocean carbonate sediments of benthic and planktonic foraminiferans. There are several difficulties in interpreting the changing isotope ratios. For example, the sediments are subject to mixing because of the burrowing activities of the bottom-living fauna. They are a better record of the amount of glacial ice than of actual air temperatures.

A change in the planktonic fauna from polar to subpolar marks the position of the changing North Atlantic Polar Front, of cold surface waters. In addition, cold periods are marked in ocean sediments by the presence of material of terrestrial origin rafted out to sea on icebergs (Bond *et al.*, 1992). Shifts in the position of the front in the Late Pleistocene were accompanied by rapid climatic change, as if a switch from cold to warm and back again happened in a period of a few hundred years (Ruddiman, Sancetta and McIntyre, 1977) (Figure 2.2). Changes in seawater temperature are not necessarily very closely correlated to the terrestrial environment, but they do illustrate the rapidity of climatic change, perhaps so fast that, for most of the time, the changing vegetation was out of synchrony with the changing climate. Soil development also lagged behind, and soil development was periodically disrupted by cold periods and erosion. Some elements of the fauna, such as beetles, responded much more rapidly to climate change than the flora.

Oxygen isotope ratios show that sea temperatures fell from 20°C in the Cretaceous, 130 million years ago, to 10°C in the Oligocene, 36 million years ago, and then further to 2°C in the Pleistocene. The reasons for the decline are complex. There is evidence of a previous period of ice ages 300 million years ago, and another one 300 million years before that. The regular periodicity may be accidental. There are many theories. The most recent ice age was associated with a period of mountain building and uplift of the Tibetan Plateau. In addition, land masses had moved by continental drift, bringing changes such as the closure of the Panamanian land bridge that affected ocean currents. Land masses moved into high latitudes, providing areas for permanent ice to form. The consequence was that climate systems that had operated for many millions of years were disrupted, and strong latitudinal variation in climate, like we have today, was established.

Superimposed on the gradual decline in temperature was a relatively regular oscillation from warm to cold and back. The oscillation is first noticeable in the Oligocene as a change in temperature of a degree or two. By the time of the Miocene the fluctuation was of 4–5°C. Such short-term periodicity of a few hundreds of thousand years cannot be due to geographical change. Rather, a combination of various perturbations of the Earth's orbit has imposed cyclic changes, called 'Milankovitch Cycles' (Imbrie and Imbrie, 1979):

1. Variation in the eccentricity (0–5.3%) of the orbit around the sun has a periodicity of *c.* 100 000 years.
2. Variation in the date of perihelion, the date when the sun is closest to the Earth each year, has a periodicity of *c.* 21 000 years. This is called the procession of equinoxes and has an opposite effect in the northern and southern hemispheres.
3. Variation in the obliqueness of the angle of rotation of the Earth to the sun's rays (24°36′–21°39′) has a periodicity of *c.* 41 000 years. The greater the obliqueness the greater the difference between seasons.

Each of these factors alters the distribution of the sun's rays over the Earth. Together they can act to increase the seasonality of the climate. In the last 700 000 years of the Pleistocene the longest Milankovitch cycle determined the most extreme changes, so that there was approximately a 100 000 year cycle of glacial and interglacial periods. For much of the cycle, during the glacial stage, there was potential for snow to accumulate so that, with positive feedback, the sun's radiant energy being reflected, eventually glaciers formed. The warm interglacial stage after the glaciers melted was relatively short, perhaps as short as 10 000 years. Superimposed on the longer cycle were fluctuations with a periodicity of 30 000 and 50 000 years, giving relatively cold stadial and warm interstadial phases within the glacial period. Within the Quaternary there may have been at least 21 full warm/cold cycles.

For the first half of the Pleistocene there is no evidence of glaciations in Britain, though probably there was at least one glaciation in the Alps, the Donau glaciation. There were cold periods of increasing severity, culminating in the first glaciation to leave its mark on our landscape, the Anglian glaciation, which started 430 000 years ago. Each glacial and interglacial period is given the name of the type site where it was first described. Hence, the Ipswichian interglacial from a type site near Ipswich in Suffolk. However, the terminology is complex because the same period may have been studied in northern Europe, the Alps and America as well as Britain but may have been named differently. The Ipswichian interglacial probably covers the same interglacial as the Eemian interglacial in northern Europe and the Riss-Wurm interglacial in the Alps. The different names are maintained because of the different character of the glaciation in each place.

It is difficult to correlate the glaciations from different areas. They have not all been dated. The whole sequence of glacial and interglacials is preserved nowhere (Bowen *et al.*, 1986). In East Anglia large parts of the record are missing between the Pastonian and Baventian. In The Netherlands three or four different periods have been named from this hiatus. Fossil faunas collected from the Middle Terrace of the Thames may come from one or two intervening interglacials between the Hoxnian and Ipswichian which are missing in the palaeobotanical sequence. There is also the problem of determining whether phases represent true glacials and interglacials or stadials and interstadials. Was the temperate period discovered from oceanic cores 100 000–70 000 years ago a true interglacial (the Ipswichian) or an interstadial in the Wurm/Weichselian/Devensian glacial stage? However, there is general agreement that the last glacial period, the Devensian glacial, in Britain is equivalent to the Weichselian glacial in northern Europe, the Wurm glacial in the Alps and the Wisconsinan glacial in North America.

1.13.2 FRAGMENTARY EVIDENCE OF THE EARLY PLEISTOCENE

The Early Pleistocene is poorly represented in Britain. One fossil assemblage from the Pliocene or Early Pleistocene has been described from Poulnahallia, near Headford in County Galway (Coxon and Flegg, 1987). A borehole into deposits which collected within a network of limestone gorges and caves demonstrates a diverse vegetation. In some areas there was a kind of heath, with areas of *Sphagnum* (bog moss), with trees like *Pinus* (pine), *Picea* (spruce) and *Abies* (fir) growing above the heathers and *Empetrum* (crowberry). Elsewhere there were woodlands of *Betula* (birch), *Ulmus* (elm), *Quercus* (oak), *Tsuga* (hemlock spruce), *Alnus* (alder), *Carpinus* (hornbeam), *Pterocarya* (wingnut), *Carya* (hickory), *Sciadopitys* (umbrella pine) and *Sequoia* (Californian redwood). Shrubs or small trees such as *Juniperus* (juniper), *Corylus*

(hazel) and *Ilex* (holly) were present. Karst limestone was widespread, and on the thin soils developed over the limestone grew a grassland with sedges (Cyperaceae), sorrels (*Rumex*) and *Urtica* (nettle) and flowers like *Helianthemum* (rock-rose), *Filipendula vulgaris* (dropwort) with umbellifers and roses.

A fossil deposit from near the Blackhall Colliery, Castle Eden in Co. Durham, has been variously dated as Late Pliocene to Middle Pleistocene (Reid, 1920). It consists of fillings in a fissure in the Magnesian limestone which contain plants, mammalian bones, insects and freshwater shells. There are 58 species of plants, including many water plants and mosses. The vegetation included trees, *Carpinus*, *Betula*, two species of *Alnus*, *Crataegus* (hawthorn), *Ilex*, *Liquidambar* (sweet gum), *Aralia* and *Rhus* (sumach). Shrubs were *Hypericum* (St. John's wort), *Rubus* (blackberry) and a species of the heather family (Ericaceae). There were many herbs.

The Castle Eden flora consists of nearly 31% of species having an affinity with present-day East Asian or North American species, though less than 10% having an affinity with North American species. A mid-European and Eurasiatic affinity was shown by 40% of species. In comparison to the flora of today, 64% was exotic or is now extinct. This botanical evidence was used by Szafer (1946–47) to place the assemblage in a series of deposits from across Europe in a stratigraphical sequence that showed the gradual decline of first North American and then the East Asiatic elements of the European flora in the Pliocene and Early Pleistocene. This certainly happened, but Szafer's argument is confounded by the lack of absolute dates for many samples, by the fluctuations in the climate, and the wide geographical origin of the sample sites.

The Castle Eden assemblage requires more study before reliable conclusions can be made.

1.13.3 A PATTERN OF EXTINCTIONS IN THE EARLY PLEISTOCENE

Most of the evidence of the Early Pleistocene of the British Isles comes from the East Anglian Crags, though not including the Coralline Crag which is probably Pliocene in age (Beck, Funnell and Lord, 1972). The East Anglian Crags are shelly marine deposits which have accumulated in shallow seas. They are poor botanically and are difficult to interpret because, although they also contain animal fossils, the material has been mixed through erosion and in-wash from older fossil deposits to include fossils of widely different ages. Fossils contemporaneous with the age of the deposit can only be picked out by their good condition. They have not been smoothed by rolling around in streams and rivers.

The oldest Lower Pleistocene Crag formation is the Red Crag. Deposits at Walton-on-the-Naze, in Essex indicate a cool temperate period which has been named the Waltonian or Pre-Ludhamian. A borehole at Stradbroke in Suffolk produced a pollen flora of this stage which is entirely dominated by pine. During this cold period much of the warm temperate Tertiary flora may have been eliminated from the British Isles, including the trees with affinities to those of East Asia and North America today, which had been common throughout the Tertiary. About 70% of the fossils were probably closely related to species living today in Europe.

After this fossil assemblage there is a hiatus in the fossil record, of perhaps half a million years. In the film of our changing flora it is as if the screen goes blank. Pollen from a borehole at Ludham in Norfolk begins the fossil record

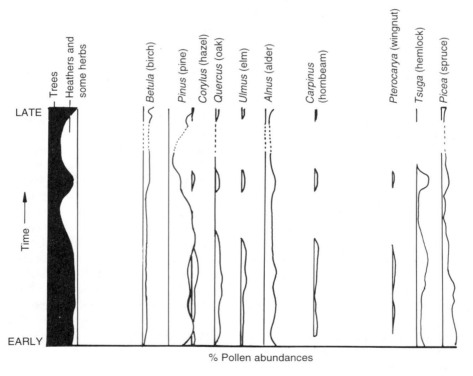

% Pollen abundances

Figure 1.51 Schematic pollen diagram through the crag at Ludham with selected taxa. (After West, 1980b.)

again about 2 million years ago (Phillips, 1976). Pollen analysis of the borehole deposits show an alternation of a wooded landscape with one dominated by grasses and heathers, perhaps relating to warm temperate and cool temperate stages (Figure 1.51) (West, 1970, 1980a, b). There was an increasing amplitude of change.

The stages are named Ludhamian (warm)/ Thurnian (cold)/Antian (warm)/Baventian (cold) up to the Pastonian (warm). The Thurnian and Antian stages were named after rivers to the east and west of Ludham. The Baventian stage was named after Easton Bavents on the Suffolk coast. Other stages, the Bramertonian (warm) and the Pre-Pastonian (cold) were inserted in this sequence later between the Baventian and the Pastonian (Table 1.2).

The Ludham borehole record pierces the Icenian Crag formation which provides a fossil fauna where it outcrops. The Icenian Crag is of two sorts, the Norwich Crag and the Weybourne Crag, laid down in different coastal conditions, with the latter typified by the presence of a mollusc *Macoma balthica*.

By this time, in the first half of the Pleistocene, the flora was very like that of today, with few exotic species. Some cold stages were not very marked but the succession of cool and cold stages continued to make warmth-loving species extinct in Europe, establishing the fully European flora of today. The cold Thurnian showed decreased frequencies of warmth-loving species such as *Tsuga* (hemlock spruce), *Quercus* (oak), *Carpinus* (hornbeam) and *Pterocarya* (wingnut). *Ulmus* (elm) seems to

Table 1.2 Pleistocene cold and warm stages, dates according to Wymer (1985) and various sources in Jones and Keen (1993)

		Started, thousand years BP[a]
Flandrian/Littletonian/Holocene	Warm	10
Devensian/Weichselian	Glacial	75
Ipswichian/Eemian	Warm	128
Wolstonian 3	Glacial	195
Ilfordian	Warm	240
Wolstonian 2/Saalian	Glacial	297
Wolstonian 1/2	Warm	330
Wolstonian 1	Permafrost	367
Hoxnian/Gortian/Holsteinian	Warm	400
Anglian/Pre-Gortian/Elsterian	Glacial	450
Cromerian	Warm	700
Beestonian	Periglacial in south	
Pastonian	Warm	
Pre-Pastonian	Cold	
Bramertonian	Warm	
Baventian	Cold	
Antian	Warm	
Thurnian	Cold	1900
Ludhamian	Warm	
Pre-Ludhamian	Cold	2400
Reuverian (= Pliocene)	Sub-tropical	

[a]Older dates are very approximate.

have been absent. *Betula* (birch), *Picea* (spruce) and *Alnus* (alder) remained well represented, and *Pinus* was the most important tree. Most importantly, the decline in warmth-loving species led to a higher proportion of pollen of herbaceous species, 40–65% compared to 25–35% in the previous Ludhamian stage.

1.13.4 HEATHLANDS AND OCEANICITY

A large part of the non-arboreal pollen in the cooler periods came from shrubby species in the heather family (Ericaceae) and other related families, especially *Empetrum* (crowberry), which rose from less than 20% pollen abundance in the Ludhamian to more than 50% in the Thurnian.

After the warm Antian stage, the next cold stage was the Baventian stage, about 1.6 million years ago. It had a more marked decline of tree pollen, down to about 21%. Warmth-loving species, including *Quercus* (oak), all but disappeared. *Pinus* (pine) was still the most important tree, but was less abundant than before. In the middle of the stage, shrubby heath species, especially *Empetrum* became even more important than previously, indicating the spread of wet, oceanic conditions. The extent of wet soils is shown by the importance of *Alnus* (alder). Permafrost formed in parts of northern Europe, and possibly the first glaciation took place. There is evidence for one, called the Donau glaciation, in the Alps.

In the intervening warm periods, called the Ludhamian and Antian stages, there was a mixed coniferous/deciduous woodland with *Quercus* and *Pinus* as the dominant species (West, 1961). The dominance of these species, which are relatively early successional trees, and good colonizers, probably indicates that for much of the time the vegetation was out of equilibrium. However, a later successional tree, *Picea* (spruce), was also common, with its pollen percentage rising on one occasion to a frequency of 27% (Figure 1.51). Another

important species was *Tsuga* (hemlock spruce), which rose to contribute 15% of the pollen at times. Rarer trees were *Ulmus* (elm), *Pterocarya* (wingnut) and *Carpinus* (hornbeam), each contributing 1–3% of the pollen. *Abies alba* (European silver fir), *Acer* (maple), *Fagus* (beech) and *Tilia cordata* (small-leaved lime) were also present. The warm Ludhamian and the next warm stage, the Antian stage, are directly comparable, except the Antian had more *Betula* (birch), and less *Pinus* and *Quercus*. The presence of abundant *Sphagnum* (bog moss) and clubmoss spores indicates the widespread presence of acid heath and bog communities in the British Isles. The climate throughout Europe was generally wetter than it had been in the Pliocene. High oceanicity of climate is indicated by the abundance of *Empetrum* and *Tsuga*.

One interesting Tertiary relic surviving in Britain in the Ludhamian was *Sciadopitys* (umbrella pine). There is only a single living species, *S. verticillata*, which is grown around temples in the Far East, and is restricted in nature to small areas of forest between 1000 and 2000 m on the Japanese islands of Honshu, Shikoku and Kyushu. It is a peculiar conifer, not closely related to others, with its whorls of grooved leaves, each formed from two needles fused together. It was reintroduced to the British Isles in the nineteenth century but only grows well in the wetter and warmer south-west.

The Ludhamian landscape may have looked very familiar to us, rather like a parkland with open, wet grassy glades, heaths and bogs, with groves of trees. However, wandering through this very English countryside there were some strange and very unexpected animals. Imagine a safari park established in the grounds of one of our country houses with lions, elephants and gazelle wandering beneath the oaks and pines. There was a woodland fauna with species such as the fallow deer, four-tined deer, elk and mastodon. The mastodon had tubercular teeth for chopping leaves and juicy foliage. There were also species of more open habitats such as the horse, giant deer, gazelle, and the southern elephant. Preying on these herbivores were the sabre-toothed cat and wolf.

1.13.5 SUCCESSION AND THE VEGETATION CYCLE

The end of the Lower Pleistocene is sometimes given as the end of the Pastonian stage, perhaps 1 million years ago. The transition from Lower to Middle Pleistocene is recorded in the Cromer Forest-Bed Series, in deposits laid down in estuarine and freshwater conditions and now exposed on the cliffs of Norfolk, between Runton and Corton (Duigan, 1963; West, 1980a, b). The deposits gained their name because they sometimes include the stumps of trees, though these were not preserved *in situ*, but after having rafted down estuaries before sinking. The Cromer Forest-Bed Series provides one of the best records of the vegetation in north-western Europe prior to the glaciation. However, even though glaciers had yet to form in the British Isles, already in the oldest part of Forest-Bed Series many characteristics of succeeding interglacial/glacial cycles can be observed in the alternation of the warm and cold stages, from the warm Pastonian stage, the following cold Beestonian stage, and so on.

A four-phase system has been adopted to describe a glacial/interglacial cycle (Iversen, 1958; Watts, 1985; Birks, 1986). The first phase was the cryocratic phase which existed in the ice-free parts of a glacial period. It was a period of immature soils, churned by freezing and thawing and eroded by ice, water and wind. The vegetation was a sparse, low, shrub or herb community of arctic–alpine species. The protocratic was a transitional phase when the vegetation and soils responded to a climatic warming. The soils were immature but fixed and began to

stratify. Warmth-loving species established, and there was a succession as light-loving species on raw soils gave way to shade-tolerant species or highly competitive shade-casters on more mature soils. Woodland developed and became more continuous. The mesocratic phase represented the development of a closed deciduous forest, often a 'mixed oak woodland'. It was the time of maximum temperatures. Vegetation and soil were relatively stable and mature. However, as leaching of the soils continued and climates deteriorated the soils became podzolized and base poor. In this oligocratic or telocratic phase there was a shift to heathland- and moorland-type communities, or in past interglacials to spruce-dominated woodland. Finally, the cryocratic phase was entered as climatic cooling led to the elimination of trees and other warmth-loving species. Watts (1988) postulates 'Milankovitch cycles' of enhanced seasonal contrast as driving the individualistic species response of populations in their post- (inter-) glacial migrations.

Iversen's model suggests a gradual regular change in climate, vegetation and soil, but it seems certain now that changes were both much faster and more haphazard (Delcourt and Delcourt, 1991). Nevertheless, the recognition of this cycle has some value because it emphasizes the biologically important characteristics of the climatic oscillations that occurred in the Quaternary (Table 1.3). However, it overemphasizes the similarities between the interglacial stages. Different interglacials were of very different lengths and conditions. For example the last, the Ipswichian, may have been 2–3°C warmer than the present one. A measure of the differences between interglacials is the way some species were important in one interglacial and not another. For example, *Tsuga* (hemlock spruce), *Abies alba* (European silver fir) and *Picea abies* (Norway spruce) became, in turn, less and less important in the British flora with each succeeding interglacial.

Table 1.3 Biological characteristics of phases of an interglacial stage (after Watts, 1985)

	Protocratic	Mesocratic	Telocratic
Important species	*Betula* (birch)	*Quercus* (oak)	*Picea* (spruce)
	Populus (aspen)	*Ulmus* (elm)	*Carpinus* (hornbeam)
	Salix (willow)	*Tilia* (lime)	*Fagus* (beech)
Reproductive rate	High	Low	Medium
Age at first seeding	Young	Mature	Mature
Frequency of seed setting	High	Low	Low
Propagule-dispersal	Good	Poor	?
Migration rate (m/year)	>1000	500–1000	<500
Competitive tolerance	Low	High	High
Longevity	Short	High	High
Seedling tolerances	Light-demanding	Shade-tolerant	Shade-tolerant
Seed production	High	High or low	High or low
Seedling mortality	High	?	Low
Growth rate	Fast	Slow	?
Regeneration under own canopy	Rare	Rare	Common
Shade production	Light	Dense	Dense
Invasion	Large gaps essential	Small gaps possible	?
Crown geometry	Multilayered	Monolayered	Multilayered
Rate of increase	High	Medium or low	Medium or low
Soil preferences	Fertile, unleached	Brown earths, mull humus	Podsols, mor humus
Life history	*r*-selected	*K*-selected	*K*-selected
Ecological traits	Ruderal	Competitive	Stress-tolerant

1.13.6 THE PASTONIAN, AN EARLY TEMPERATE STAGE

The Pastonian temperate stage opened with a woodland of a *Pinus* (pine) and *Betula* (birch) which developed into a coniferous woodland of *Pinus* and *Picea* (spruce). Warmth-loving species such as *Cladium mariscus* (saw sedge) were present. In time the coniferous woodland gave way to a mixed deciduous woodland with *Quercus*, *Carpinus* (hornbeam) and *Ulmus* (elm), as well as *Pinus*, as the most abundant species. In the understorey there was *Corylus* (hazel), but not in large quantities, and also *Hedera* (ivy). This woodland is recognizably north-west European but was more diverse than at present. There were scattered specimens or rare groves of a number of other species; *Tilia cordata* (small-leaved lime), *Acer* (maple), *Juglans* (walnut) and *Abies alba* (European silver fir).

Nevertheless, two pollen species of *Tsuga* (spruce hemlock) were present in the Pastonian. *Tsuga* has been reintroduced for forestry not *T. carolinianum/canadensis* which may have grown in Britain in the Pastonian but *T. heterophylla* from the west coast of North America. As a forestry tree it thrives, especially in the cool, damp areas of Scotland and least well in eastern England. *Tsuga* has extant living species in North America and East Asia, where it can still be found forming dominant stands in a few places. There were only scattered occurrences of *Tsuga* in the Pastonian, with a maximum of 4% pollen only at one site. Even here this represented a decline from the levels of 7% reached in the Ludhamian stage. Possibly as each cold, dry period intervened in Europe it lost some genetic diversity which would have allowed it to recolonize northern Europe successfully in competition with other species. *Tsuga* was one of the last survivors of the Tertiary forests.

Another Pastonian tree with North American or East Asian affinities was probably a species of cypress, like *Chamaecyparis thyoides* which today grows in coastal swamps in North America. In the Pastonian it was mainly present only at one site, Sidestrand near Cromer, where it was growing with two different species of *Osmunda* (royal fern) in a coastal swamp. A link with the East Asian flora was provided by the presence in these swamps of *Eucommia*, a small tree which grows today only in China. A slightly different community is indicated at Happisburgh, a fen woodland, with high levels of *Alnus* (alder) and *Osmunda*, fringing marshland. Present also was an unidentified species with a rather undistinguished pollen, but which may have been a species of palm, like *Sabal adansonii* which is found today beside streams in south-eastern USA.

As the Pastonian entered its oligocratic phase there was a rise in heathland communities, with some tree species declining. *Carpinus* (hornbeam), which had been present in the former deciduous woodland, remained important for a while, foreshadowing its more obvious behaviour as an oligocratic species in later interglacials. An interesting species of the late Pastonian stage was the *Pterocarya* (wingnut). *Pterocarya fraxinifolia*, perhaps the Pastonian species, native to the Caucasus, is today a vigorous suckering tree which grows well in southern Britain.

The deciduous woodland was eventually replaced by one dominated by *Pinus* (pine) and *Picea* (spruce), but with birch woodlands and heathlands becoming more and more important as time passed. It is not possible to say whether the pine present was our native *Pinus sylvestris* or any of the other many species, several native to Europe. Good candidates are *P. mugo*, *P. cembra* and *P. nigra*. However, there were two species of spruce present; *Picea abies* (Norway spruce), the familiar Christmas tree, and *P. omorika*

(Serbian spruce) (Figure 1.51). The latter is interesting as a late link with the East Asian/American Tertiary flora. It has flat needles, unlike the square-sectioned ones familiar to us in all other European and eastern American spruce species. Other flat-needled *Picea* species are found from western North America and China to the Himalayas. *Picea omorika* was discovered high in the Drina valley in the Balkans in 1875. Another Pastonian link with the Balkans was the presence in late interglacial heaths of *Bruckenthalia spiculifolia* (spike heath) which grows today in non-calcareous situations in woods and pastures in the Balkan mountains. The heaths also included two species of heather not native today: *Erica umbellata* and *E. lusitanica* from present-day south-west Europe.

The temperate Pastonian was followed by the cold Beestonian stage, which was very cold at times. Permafrost formed in Britain and it was even cold enough for ice-wedge casts to be formed. However, at other times, there was a herb-rich grassland with *Artemisia* (mugworts) and sedges (Cyperaceae). Low shrub communities of *Betula* and *Salix* were also present, and at times a birch and pine woodland developed. These communities, which reappeared in each following cold/glacial stage, are described in detail in for the last cold stage, the Devensian glaciation.

1.13.7 PLANTS AND ANIMALS IN THE CROMERIAN STAGE

The next warm stage was the Cromerian stage. The type site for the Cromerian is West Runton in Norfolk. This is one of the few sites of the Pleistocene where there are both fossil plants and animals. The development of the vegetation showed some similarities to that in the Pastonian stage, but the fauna was remarkably changed. The four-tined deer, which was such an important component of Pastonian faunas, was not present. The species of elk had changed from *Libralces gallicus* to *L. latifrons*, having antlers with a narrower span. The narrow-nosed ox was replaced by the bison and the early hyena by the spotted hyena. Species recorded for the first time in Britain were the hippopotamus, the Etruscan rhinoceros and new species of beaver, bear, hamster and vole. One survivor from the Pastonian was the southern elephant, but the mastodon was absent.

The Cromerian vegetation had lost some of the 'exotic' species of the Pastonian such as *Eucommia*, *Tsuga* (spruce hemlock) and *Picea omorika* (Serbian spruce). The fully developed mesocratic woodland had a somewhat different character (Figure 1.52). *Ulmus* (elm) was a significant part of the vegetation and *Tilia* (lime) was much more important than before. *Carpinus* (hornbeam) behaved much more strongly as an oligocratic species than before; its pollen frequency not rising until quite late in the stage. Then it reached a peak of 12–13%. *Abies alba* (European silver fir) was

more important than before, following the rise of *Carpinus* (hornbeam), its pollen frequencies rising to above 10% in some deposits late in the oligocratic phase. *Fagus* (beech) and *Acer* (maple) were present but in low quantity and late in the succession. Another rare species was *Juglans* (walnut).

The behaviour of *Picea abies* is interesting. It grew in importance throughout the meso- and oligocratic phases to reach pollen frequencies of about 15%. With the decline of other trees it became, with *Pinus sylvestris*, the most important tree species. The decline of other tree species was probably not related to a decline in temperature because warmth-loving species, such as *Cornus sanguinea* (dogwood), *Trapa natans* (water-chestnut) and a kind of water fern called *Salvinia*, continued to grow. The increasing dominance of *Picea abies* (Norway spruce) and then *Pinus* may indicate the leaching of soils from neutral

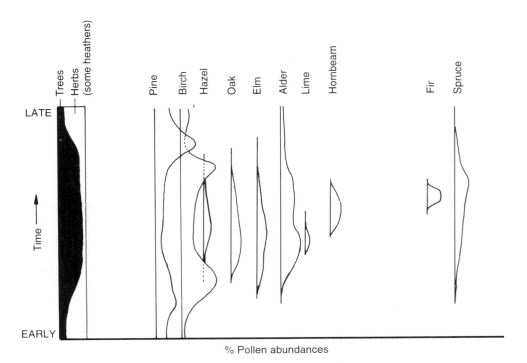

Figure 1.52 Schematic pollen diagram of the Cromerian interglacial. (After West, 1980b.)

brown earths with a mull humus to acid brown earths and podzols with a mor humus. This is indicated, as in the Pastonian stage, by the rise of *Calluna* (heather) and various heath species, *Bruckenthalia* (spike heath) and *Empetrum* (crowberry).

1.14 THE ANGLIAN GLACIATION

The first glacial stage in the British Isles for which there is direct evidence was the Anglian stage, which has a type site at Corton Cliffs in Suffolk. As the stage progressed there was a spread of low-growing arctic–alpine species. There was a fauna of reindeer and ground squirrels. Remains of steppe lemmings, hamsters and wolverine have been discovered in Tornewton Cave in Devon, which may date from this time. Later the Forest Bed became overlain by glacial deposits as glacial ice reached down and into East Anglia.

In this and the succeeding glacial stages the flesh was shaped on the skeleton of the British Isles. U-shaped valleys were cut into the uplands. Cirque glaciers cut amphitheatres in the sides of highlands. While the uplands were cut and shaped, the lowlands were filled with debris, glacial till, deposited by ice or water. Terminal and lateral moraines were left at the margin of the ice. Glaciers melting *in situ* dumped material in a more haphazard manner. Rivers and melt streams carried a rich load of sand and gravel. The glaciations changed patterns of drainage, leading to the cutting of new river valleys. Lobes of ice in the Anglian glaciation blocked the outflow of the Thames, which had run through the vale of St. Albans somewhere to the north over East Anglia. Its flow was diverted into its present channel. In the southern North Sea water from the Thames and Rhine filled a large freshwater lake, dammed to the north by ice. On its southern margin water from the lake spilled over in a wide series of waterfalls and rapids, eroding its way through the chalk ridge between England and France, making the Strait of Dover.

We have inherited a glacial landscape which still helps to determine the variation in our vegetation today. The glaciations changed the land for plants, by destroying old soils and exposing rock strata. Alternatively, older strata were hidden beneath thick beds of deposits of very diverse origins and qualities. The new landscape had a different potential. Soils of different types bearing a different vegetation could develop. It is also from the Anglian stage that the fossil record of the Pleistocene changes. When the ice retreated it left a landscape rich in deposits, glacial till, gravel and alluvium, pockmarked with hollows and dips. In these hollows and dips organic sediments were laid down, preserving in their stratigraphy a rich pollen record of the developing vegetation.

At first the soggy late Anglian landscape was colonized by a herb-rich grassland, shrubby willows, some *Betula* (birch) and *Pinus sylvestris* (Scots pine). *Hippophae rhamnoides* (sea buckthorn) was also present. This is highly frost tolerant as a mature plant and its seeds can survive temperatures of −200°C for 12 weeks. It rapidly colonized the glacial till, flourishing on the newly ice-free terrain south of the ice front (Pearson and Rogers, 1962). In Britain and north-west Europe *Hippophae* only grows today on coastal sand dunes but it is widespread in Eurasia as a species of calcareous gravel deposits, sand banks beside alpine streams and on valley-side moraines. It spreads vegetatively very effectively by rhizomes and lateral roots. In calcareous situations, which have low nutrient availability, its nitrogen-fixing nodules give it a measure of advantage. However, it requires open situations and dies in only moderate shade.

1.15 LATE PLEISTOCENE INTERGLACIALS

Between the Anglian stage and the final glacial stage, the Devensian stage, two interglacials have left an extensive fossil flora. They are the Hoxnian or Gortian interglacial stage and then the Ipswichian interglacial stage, with an intervening Wolstonian glacial stage. A fossil assemblage of Ipswichian age has not yet been discovered in Ireland. There may be at least two more glacial/interglacial cycles after the Anglian glacial stage between the Hoxnian and Ipswichian interglacials. They have left a scant floristic record but are recorded by features of the landscape and geological deposits and sometimes by a fossil fauna.

Several interglacials are recorded by a series of Thames alluvial river terraces in the London area. These record the approximate highest levels reached by tides during different interglacials. The famous site at Greenhithe in Kent, which preserved the earliest fossils of man in Britain, fragments of skull of Swanscombe Man, lies on the Upper Terrace of the Thames, 25 m above present sea level. It probably comes from a later part of the Hoxnian interglacial (Jones and Keen, 1993) but has also been placed after the Anglian glaciation and before the Hoxnian (Bowen *et al.*, 1986).

Rich fossil mammal localities on the Middle Terrace, about 10 m above present sea level probably come from a temperate period between the Hoxnian and Ipswichian interglacials, called the Ilfordian stage from one site with a rich mammalian fossil fauna. The fauna contains many cold or steppe elements, such as mammoth, bison and woolly rhinoceros, as well as others which might indicate warmer conditions, such as lion, straight-tusked elephant and horse. Warm climates are suggested by the widespread presence of the mollusc Corbicula fluminalis (Sutcliffe, 1985).

The Floodplain Terrace, associated with a sea level only marginally higher than at present, is probably from the Ipswichian interglacial.

1.15.1 THE HOXNIAN INTERGLACIAL

The Hoxnian has been dated to the period 400 000–367 000 years BP. The beginning of the Hoxnian interglacial stage is marked by a decline in *Hippophae rhamnoides* (sea buckthorn) as a woodland of *Betula* (birch) and *Pinus sylvestris* (Scots pine) became established. The type site is Hoxne (West, 1956) in Suffolk, but the most complete record of about 20 000 years comes from Marks Tey in Essex (Turner, 1970). The Hoxnian deposits at Woodston near Peterborough, deposited in an estuarine environment, have pollen, plant macrofossils, molluscs, ostracods, insects and mammals (Horton *et al.*, 1992). There are minor differences between sites of this age, but also remarkable agreements (Coxon, 1985) (Figure 1.53). A curious comparison between some Hoxnian sites, including Beetley (Roosting Hill) in Norfolk, Fisher's Green in Hertfordshire and Mark's Tey in Essex, is a synchronous change in water levels accompanied by the slumping of lake margins (Gibbard and Aalto, 1977).

At first a wide range of herbs survived the spread of *Betula* and *Pinus*. There was a brief expansion of *Ulmus* (elm) ahead of *Quercus* (oak), but the latter soon became dominant, accompanied by *Corylus* (hazel), *Fraxinus* (ash) and *Alnus* (alder). *Ulmus* remained high. The mixed oak wood had as understorey shrubs, *Euonymus europaeus* (spindle), *Viburnum lantana* (wayfaring tree), *V. opulus* (guelder rose), *Cornus sanguinea* (dogwood), *Sambucus nigra/racemosa* (elder) and *Hedera helix* (ivy). As conditions became wetter *Alnus* spread at the expense of *Quercus* on the damper soils, and on drier soils a thick understorey, and perhaps even a canopy of *Corylus*, developed so that a ground flora was all but eliminated. *Ilex aquifolium* (holly) spread and there is a pollen-type, Type X, which may have come from something like *Phillyrea* (Figure 1.54), a Mediterranean genus related to *Olea* (olive). The tree flora remained diverse with *Picea abies* (Norway spruce),

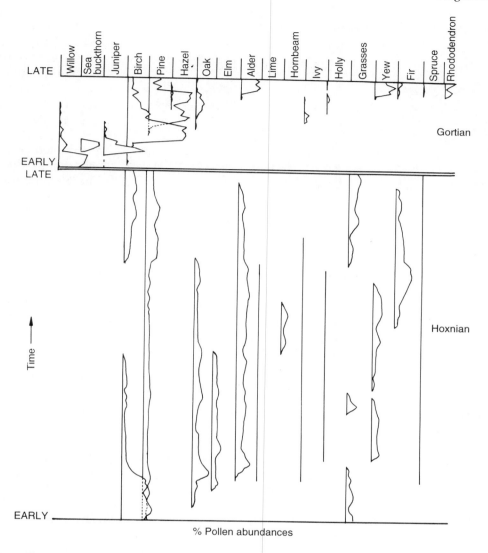

Figure 1.53 The Hoxnian and Gortian interglacials compared; schematic pollen diagrams. (After Watts, 1964; Turner, 1970.)

Fraxinus (ash), *Pinus*, *Abies* (fir), *Carpinus* (hornbeam) and *Acer* (maple). Several species of *Acer* may have been present, probably including the *A. platanoides* (Norway-maple), which is not native in our present flora, as well as native *A. campestre* (field maple). Both *Tilia cordata* and *T. platyphyllos* (small-leaved and large-leaved limes) were present, though more abundantly at Hoxne than Marks Tey. Later, as conditions became drier again, *Alnus* declined and *Corylus* exceeded all other trees. It was, almost certainly, now growing as a canopy tree. *Ulmus* and *Taxus baccata* (yew) became more abundant.

The flora of the Hoxnian interglacial was unremarkable, except it may indicate that the

Figure 1.54 Plants of the Late Tertiary and Early Pleistocene:(a) *Acer platanoides* (Norway maple); (b) *Phillyrea*, possibly the source of Type 'X' pollen. ((a) Copyright Jon B. Wilson.)

climate was a little warmer than at present. Plant remains and land molluscs from the Woodston deposits provide evidence of a closed tree canopy (Horton *et al.*, 1992), as does the presence of roe and fallow deer and beaver in several sites. However, some elements of the fauna were remarkable. Macaque monkeys chattered in the trees. Large herbivores, like the straight-tusked elephant, would have been immensely destructive of young saplings (Stuart, 1982). Trees with the ability to sucker from the base, such as *Corylus*, might have been favoured. In the 'elephant bed' at Clacton, with bones of elephants, there are fruits of *Sonchus asper* (prickly sow-thistle) and *Urtica dioica* (nettle). These species were taking advantage of the churned and manured ground. The plant macrofossils of Woodston include herbs of probably several different unshaded assemblages. There were woodland-margin species, such as *Lapsana communis* (nipplewort) and *Silene dioica* (red campion). The largest assemblage was of species which probably grew in open areas of the flood plain and were subject to disturbance. They included *Aethusa cynapium* (fool's parsley), *Lamium* sp. (dead-nettle), *Chenopodium* sp. (goosefoot), *Atriplex* sp. (orache), *Stellaria media* (common chickweed) and *Sonchus asper* (prickly sow-thistle). Grassland herbs

included *Ranunculus* sp. (buttercups) and *Plantago media* (hoary plantain). Fen and marsh species included *Lycopus europaeus* (gypsywort), *Bidens tripartita* (trifid burmarigold) and *Eupatorium cannabinum* (hemp agrimony). Horse, giant ox and the narrow-nosed rhinoceros, preyed on by lion, probably grazed the flood plain.

A particular feature of several Hoxnian sites, but not Woodston, is a sudden and extended period of deforestation. Grass pollen rose to 25% of the total and *Taxus*, *Ulmus* and *Corylus* pollen declined. At Marks Tey the deposits are laid down in annual lamina and so it has been possible to show that this phase lasted 300 years. Perhaps there was a huge forest fire or several forest fires covering much of southern Britain. Indeed, traces of charcoal are found at this time. Man may have started the fires to help concentrate his chosen prey species.

Once cleared the grassland may have been maintained by grazing, especially by the abundant horses and giant deer. In time the clearing closed as *Betula* and *Pinus sylvestris* spread and eventually the mixed oak woodland was re-established much as before, though *Taxus* never achieved its former importance. The later part of the Hoxnian is marked by the spread of *Carpinus* (hornbeam) at the expense of *Quercus* and *Ulmus*. It was accompanied by *Cornus sanguinea* (dogwood), *Prunus spinosa* (blackthorn) and *Lonicera periclymenum* (honeysuckle). The climate was probably more strongly continental. Further west, in the region of Birmingham, where *Carpinus* was uncommon, *Picea abies* became more dominant. The dominance of *Carpinus* and *Picea* did not last very long because quickly *Abies alba* (European silver fir) spread to dominate the forest.

The importance of *Abies* is characteristic of the Hoxnian. It is recorded as native for the last time in this stage. *Abies alba* is now a southern continental species. It has similar requirements to *Fagus* (beech), with which it is often found on the continent (Polunin and Walters, 1985). *Abies alba* grows on the Pyrenees, Vosges, Jura, Alps and elsewhere but it is more sensitive to frost than *Fagus sylvatica* and prefers richer soils. It is also easily damaged by grazing. In the nineteenth century it was widely planted in Britain, but in about 1900 an aphid, *Adelges nordmannianae*, was accidentally introduced from eastern Europe. In the mild, humid climate of Britain the aphid seriously damages young trees. *Abies grandis* (giant fir) from North America, is now planted by foresters in preference. *Fagus* is also recorded at this part of the Hoxnian but only as a rare species. Two species growing with *Abies* were *Vitis* (vine) and *Pterocarya fraxinifolia* (Caucasian wingnut).

Changes in pollen in the Upper Sequence at Hoxne and elsewhere provide evidence for climatic cooling and the development of open conditions (Gladfelter, 1975; Wymer, 1985). This represents the start of the Wolstonian glacial stage, called the Munsterian in Ireland. At first, between 367 000 and 330 000 years BP, (Wolstonian 1) cold restricted tree growth, but it was not cold enough for periglaciation (Jones and Keen, 1993). Then between 330 000 and 297 000 years BP, tree cover increased again (Wolstonian 1/2) before, 297 000–240 000 years BP, cold periglacial conditions arose, and now glaciers formed in

the north and west (Wolstonian 2). The Wolstonian has been named after a site in the West Midlands.

Intervening between Wolstonian 2 and Wolstonian 3 was the Ilfordian temperate stage, which has been dated to 245 000–195 000 years BP. Wolstonian 3 was a period of intense periglaciation which came to an end 128 000 years BP.

1.15.2 AN IRISH INTERGLACIAL, THE GORTIAN

A very distinctive interglacial stage has been described from several sites in Ireland, including Gort, Spa-Fenit, Kildromin, Baggotstown and Kilbeg (Jessen, Andersen and Farrington, 1959; Watts, 1959, 1964, 1967; Coxon and Flegg, 1985). Recently it has been dated, from a site in Cork, to the same time as the Hoxnian stage in England (Scourse *et al.*, 1992). Differences between the Gortian and Hoxnian illustrate the kind of regional variation that must have existed in the British Isles at all times.

The protocratic phase of the Gortian is marked by *Salix repens* (creeping willow) and herb vegetation, giving way to a shrub vegetation dominated by *Juniperus* (juniper) and *Hippophae rhamnoides* (sea buckthorn) (Figure 1.53). Then, as tree species colonized, a patchy woodland of *Betula* (birch) and *Pinus* (pine) developed with a little *Quercus* (oak). There were many surviving open and marshy areas in which there was a grassland or fen, rich in sedges and tall herbs, including a species of *Astrantia*. Conditions were warm enough for *Ilex* (holly), *Hedera* (ivy) and *Lonicera* (honeysuckle) to grow. As pine and mixed oak woodland spread, in the mesocratic phase, *Betula* and the herb flora declined. The continuing importance of *Pinus* is unusual for an interglacial and contrasts with the Hoxnian which had a mixed deciduous woodland. In eastern Europe *Picea* (spruce) had risen to dominance. In Ireland there may have been large areas of poor, thin or peaty soils unsuitable for *Quercus*, which was therefore restricted to deeper soils in the valleys. *Corylus* (hazel) and *Fraxinus* (ash) were a component of the vegetation, perhaps especially on the base-rich limestone areas, and *Ulmus* was a significant minor component at some sites.

The most distinctive Gortian phase was the telocratic phase, and differences between Gortian sites are also most obvious then. With the rise of *Alnus* (alder), *Quercus*, *Pinus* and *Hedera* declined. What has been called an Atlantic-climate forest arose with *Taxus* (yew), *Abies* (fir), *Picea* (spruce) and *Rhododendron ponticum*. *Taxus* may have formed a dense woodland on the calcareous soils in which little else could grow. *Buxus sempervirens* (box) was a common shrub. At Cork the pollen assemblage may come from the transition between mesocratic and telocratic forests when *Rhododendron ponticum* (Figure 1.55) and *Abies* were not yet significant elements of the vegetation. At Ballyline there may have been greater areas of leached acid soils. *Abies* became especially important here and there were significant quantities of *Pterocarya* (wingnut) and *Carpinus* (hornbeam), while *Taxus* was less important. *Fagus* (beech) was present at Ballyline and Gort. The presence of *Pterocarya* (wingnut), *Buxus* (box), *Rhododendron* and some other species provides evidence of some similarity between Gortian conditions and those of the Black Sea coast today. If so, the climate was wet with more than 2000 mm of rainfall distributed rather evenly through the year, with much warmer summers. *Rhododendron ponticum* is not recorded in England in the Hoxnian.

In the Gortian, *Rhododendron* may have grown as an understorey in the woodlands. Later, towards the end of the Gortian, there was a decline in trees, perhaps because of a period of greater cold. *Rhododendron* increased in importance and may have formed extensive thickets on a heathland. *Rhododendron ponticum* is not native in the British Isles today but was introduced to England in 1763 and to Ireland by 1843 (Cross, 1975). There are two extant subspecies. *R. ponticum* ssp. *baeticum*, from a few places in Portugal and Spain, has a hairy inflorescence and short, broad leaves. Plants of *R. ponticum* ssp. *ponticum*, from eastern Europe, which have narrow leaves and a more or less hairless inflorescence, have been widely planted and are well naturalized, spreading by seeding and suckering. These are so vigorous that they are regarded as a serious problem, because they form dense thickets in western oak woodlands, like the Killarney oakwoods. This variant is native to parts of Turkey, Lebanon and the Caucasus. Winter conditions can be very cold there but the south-facing slopes are much milder.

There is an extensive list of heath species growing in the Gortian, including *Erica mackaiana* (Mackay's heath), *Bruckenthalia spiculifolia* (spike heath) (also present in the Hoxnian), *Daboecia cantabrica* (St. Dabeoc's heath), something like *Erica ciliaris* (Dorset heath), *Erica scoparia*, as well as the familiar *Erica tetralix* (cross-leaved heath), *Erica cinerea* (bell heather), *Calluna vulgaris* (heather), and *Empetrum nigrum* (crowberry) (Figure 1.55). They may have formed part of an extensive Late Gortian heathland but several could also have grown as an understorey shrub layer. *Daboecia cantabrica* grows in pine and oak-woods in Spain (Woodell, 1958). It prefers only weakly acid soils but can also grow in the thin, leached superficial soil over limestone. *Erica mackaiana* is the least tolerant of shading (Webb, 1955).

Some workers have identified a special category of today's flora called the 'Lusitanian element', which includes some species present in the Gortian, such as *Erica mackaiana*, *Erica ciliaris*, *Erica scoparia* and *Daboecia cantabrica*. These species have distributions centred on north-western Spain, Portugal, south-western France and reach up to western Britain and especially western Ireland (Figure 1.56). These species share,

(a)

(c)

(d)

Figure 1.55 Ericaceous plants of the Gortian interglacial and Lusitanian plants of the Holocene: (a)
Rhododendron ponticum; (b) *Erica ciliaris*; (c) *Daboecia
cantabrica*; (d) *Arbutus unedo*. ((a) and (b) Copyright
Jon B. Wilson.)

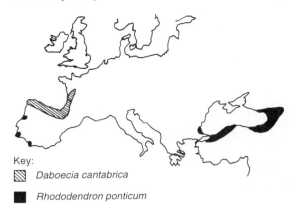

Key:

▨ *Daboecia cantabrica*

■ *Rhododendron ponticum*

Figure 1.56 Distribution of *Daboecia cantabrica* and *Rhododendron ponticum*.

with perhaps the exception of *Erica scoparia*, a common liking for high rainfall or high humidity, but they do not necessarily share an identical climatic requirement. *Daboecia cantabrica* (St. Dabeoc's heath) has been recorded at an altitude of 1300 m in Spain but grows outside the area of severe frost. The other Lusitanian heathers are found in coastal heathlands where there is little likelihood of frost. *Erica scoparia* is no longer native. It grows in dry acidic sites on sandy soils from south-western France to Morocco. Other ericaceous species, part of a Lusitanian/southern European element, but not recorded from the Gortian, are *Erica erigena* (Irish heath), *Erica vagans* (Cornish heath) and *Arbutus unedo* (strawberry tree) (Figure 1.55). The last grows in Spain and south-western France and is also widespread in the Mediterranean region (Sealy and Webb, 1950).

Other Lusitanian plants include *S. spathularis* (St. Patrick's cabbage), which has been tentatively recorded from the Gortian (Watts, 1964). Lusitanian species not recorded in the Gortian but now native are, *Minuartia recurva* (recurved sandwort), *Pinguicula grandiflora* (dense-flowered butterwort) and *Saxifraga hirsuta* (kidney saxifrage). There is also a Lusitanian fauna.

Another interesting group of plants found in the Gortian are those which are mainly North American; *Azolla filiculoides* (waterfern) (also found in the Hoxnian), *Nymphiodes cordata* (fringed water-lily) and *Eriocaulon aquaticum* (pipewort). Only the last is native in the British Isles today, but it is very locally distributed in Scotland and Ireland. *Azolla* subsequently became extinct in western Europe until its reintroduction in recent times, probably as an aquarium or pond plant. The presence of *Eriocaulon* illustrates the potential of long-distance dispersal.

1.15.3 THE LAST TRUE INTERGLACIAL STAGE, THE IPSWICHIAN

The next interglacial, the Ipswichian stage, probably ran from about 128 000 to 110 000 BP (Phillips, 1974; Jones and Keen, 1993). The type site is at Bobbitshole near Ipswich. Another important site is at Beetley in Norfolk. As usual, the beginning of the period was marked by the abundance of *Pinus sylvestris*, but this time it remained abundant for longer. The conditions were mild and oceanic. It was warmest about 125 000 years ago. *Acer* (maple) was important in the mixed oak woodland. This was probably mainly *Acer monspessulanum* (Montpellier maple) which is today a plant of the Mediterranean region and central Europe. *Picea abies* (Norway spruce) was also present.

There was considerable variation in vegetation development even within East Anglia (Figure 1.57). For example, at Beetley and at Barrington an open vegetation of grasses and herbs is recorded. In many other deposits trees and shrubs dominate the pollen spectra, as at Histon Road, Cambridge, where there was a dense forest. Different tree species were important at different sites. At Trafalgar Square the vegetation was more open with many shrubs.

A clue to this variability may be that a different fauna is recorded in the Ipswichian from the Hoxnian stage (Stuart, 1976). Horses,

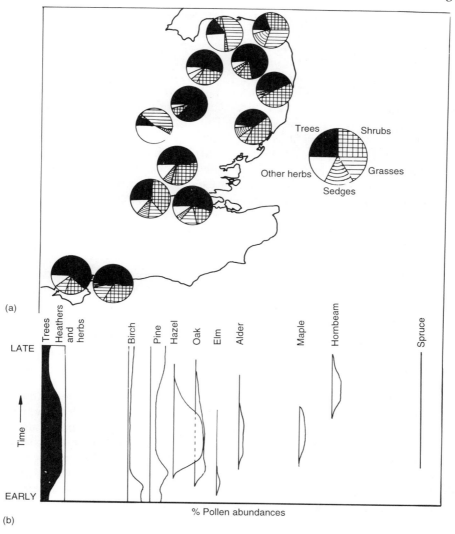

Figure 1.57 The Ipswichian interglacial in East Anglia: (a) local variation between sites (after Stuart, 1976); (b) schematic pollen diagram. (After West, 1980b.)

for example, were absent and there was no evidence of humans. There were two Ipswichian faunas. At first, hippopotamuses swam in the Thames and there were straight-tusked elephants and some large cattle, like the European bison, a woodland species. Around water-holes and trampled river margins, *Plantago* (plantain) grew particularly abundantly. It accounted for 41% of pollen around a shallow hippopotamus pool at Wick in Worcestershire. Rhinoceroses were common. There were spotted hyenas, a different species from that present in the Hoxnian. Later in the stage, bison became very important. They grazed open herb-rich grasslands on the river terraces, preventing the growth of trees. Herbs associated with these open sandy or wet grasslands, such as *Plantago lanceolata* (ribwort plantain), *Leontodon* spp. (hawkbits) and *Taraxacum* spp. (dandelions) were common. Animals we now regard as indicating a hot climate were roaming in a temperate vegetation not very unlike today's Britain. Towards the end of the Ipswichian, in its oligocratic phase, *Carpinus* (hornbeam) became dominant. Eventually it was displaced as usual by *Pinus sylvestris* and *Betula* as the climate cooled towards the last glacial period, the Devensian stage.

1.16 THE LAST COLD STAGE, THE DEVENSIAN STAGE

The last glacial stage, the Devensian, is the best understood of all glacial stages. However, there is no single, complete stratigraphic sequence, and so there are the same problems of reconstructing the jigsaw puzzle as are seen in other stages. In addition, the beginning of the stage lies beyond the reach of radiocarbon dating methods, and other dating methods do not have the necessary precision. Nevertheless, much is agreed. In fact, recent evidence from Greenland ice cores (Taylor *et al.*, 1993) provides a more complex picture. Electrical conductivity measurements of ice cores show fluctuations because of the differential input of calcium carbonate dust, which is up to 40 times greater in glacial periods, due to changes in atmospheric circulation patterns. The record is complicated by the input of volcanic dust, which causes isolated peaks, as if the climate had warmed. However, there is a good correspondence between different ice cores. The thickness of the ice deposited each year, 3 cm during cold, high-dust-input periods and 6 cm in warmer periods, allows fluctuations to be recorded down to individual years. The ice record can be dated by counting annual layers. Within the stage there were several phases of warming and cooling, and shifts between an oceanic and continental climate (Table 1.4) (Ruddiman, Sancetta and McIntyre, 1977; Shotton, 1977). What is revealing about ice cores is that they show a particularly unstable climate, flickering repeatedly between cold and warm during the full glacial period. There were several relatively warm interstadial events in the middle of the coldest part of the Devensian (Figure 1.58).

Probably many of these changes were instigated by shifts in the pattern of circulation of North Atlantic Ocean currents. These changes in circulation may have happened very rapidly, even over the course of a few hundred years. In the coldest periods polar waters reached far south of Britain and mild south-westerly winds were absent from the British Isles. The first southward excursion of

Table 1.4 Episodes of the Devensian stage (information from Jones and Keen, 1993)

			Vegetation	Date
Late	Loch Lomond	Stadial	Glaciation in Scotland tundra	11 000–10 000
	Windermere	Interstadial	Meadow, birch woodland	13 000–11 000
	Dimlington	Stadial	Glaciation, tundra	c.28 000–13 000
Middle	Upton Warren	Interstadial	Herbs, dwarf shrubs	c.41 000
	Brimpton	Interstadial	Boreal woodland	c.41 000
Early	Chelford	Interstadial	Boreal woodland	c.60 000
	Wretton	Interstadial		

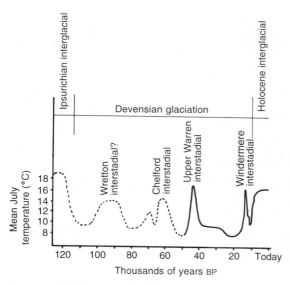

Figure 1.58 Temperature variation in the Late Devensian. (From Coope, 1977.)

polar waters may mark the beginning of the Devensian at around 110 000 years ago, dated from oxygen isotope ratios. In the period 70 000–12 000 years ago, six episodes of cold (Heinrich events) have been recognized in North Atlantic sediments, as layers poor in foraminiferans and rich in rock fragments rafted out to sea on icebergs (Heinrich, 1988; Andrews and Tedesco, 1992; Bond *et al.* 1992). In Britain three separate Devensian glaciations have been recognized from surviving glacial deposits and evidence of sea-level changes (Bowen and Sykes, 1988).

1.16.1 THE TUNDRA

Pollen evidence shows that there was at first an open herbaceous vegetation, a damp grassland with many sedges (Cyperaceae) (Watson, 1977) in which *Artemisia* (mugworts) and *Selaginella selaginoides* (lesser clubmoss) were important species. The latter is the only member of its genus to have an arctic–alpine distribution. Cold climatic conditions are also indicated by the range of beetle species. The climate was cool and moderately continental. In sheltered sun-warmed places, relic pockets of the old vegetation of *Pinus sylvestris* (pine) and *Betula* (birch) remained. The initial cooling did not last long but gave way to a relatively long warmer period, the Wretton interstadial, which lasted until 90 000 years ago. Wretton is in East Anglia near Kings Lynn. A birch/pine woodland developed, giving way later to a more complex vegetation with *Alnus* (alder) and abundant grasses.

This was followed by a marked cooling between 80 000 and 65 000 years ago, with a decline in tree species, giving a treeless tundra. Winter temperatures were well below 0°C. There were phases of cool climatic conditions interspersed with more severe coldness. There was a glaciation in Scotland and Ireland, but it was not as extensive as later (Bowen and Sykes, 1988). The presence of ice wedges in southern Britain shows that at times mean annual temperatures were below −8°C with continuous areas of permafrost. Erosion created thin, fresh soils with little organic content. Frost-heave, churning of the soil by alternating cycles of freezing and thawing, destroyed soil stratification, bringing nutrients to the surface. Mammoths, horses and bison wandered over steppe and tundra.

A peculiar feature of tundra environments today is the low level of summer precipitation and high level of evaporation. This can make the soils markedly saline. A conspicuous element of the flora of the cold periods in the Devensian were halophytic species, adapted to coping with these peculiar nutrient conditions (West, 1977). Today we regard many of these species as coastal plants (Figure 1.59). They include *Silene vulgaris* and the related *S. uniflora* (bladder and sea campion), *Suaeda maritima* (annual sea-blite), *Armeria maritima* (thrift), *Plantago maritima* (sea plantain) and *Artemisia* spp. (mugworts). *Artemisia campestris* (field wormwood) is important in

Figure 1.59 Plants of the Early Devensian: (a) *Armeria maritima* (thrift); (b) *Seriphidium maritimum* (sea wormwood); (c) *Glaux maritima*; (d) *Silene uniflora*.

dry central European grasslands, and *A. vulgaris* plays an important part in the arctic tundra vegetation, but perhaps the mugworts here also included the entirely coastal species *Seriphidium maritimum* (sea wormwood).

Other common plants of the Devensian tundra are now part of our arctic–alpine flora, including *Dryas octopetala* (mountain avens), and *Thalictrum alpinum* (alpine meadow-rue). They formed a continuous tundra vegetation in the area south of the Ice Sheet. Some might even have survived on rocky outcrops, nunataks poking through the ice sheet. Today the arctic–alpine element is restricted to relict, geographically isolated populations.

A third element comprised the 'weedy' species, ruderals of open habitats like *Polygonum aviculare* (knotgrass), *Persicaria maculosa* (redshank) and *Stellaria media* (common chickweed) (Sobey, 1981). This third element is familiar today from agricultural habitats and waste ground, so much so that they were once regarded by Buchli (Simmonds, 1945) as 'archaeophytes', confined to man-made habitats. However, they are also found even today in the natural communities of shingle banks, river banks and strand lines where bare ground is exposed and soils are often nitrogen-enriched by debris. Ruderals are normally self-pollinating, ensuring reproduction in habitats where pollinators may not have been abundant. In *Persicaria maculosa* and *Polygonum aviculare* self-pollination is achieved by incurving of the stamens on to the stigma.

Mosses, liverworts and lichens were also important members of the tundra. They included many calcicoles and species of unshaded habitats.

1.16.2 TAIGA IN THE CHELFORD INTERSTADIAL

About 65 000 to 60 000 years ago there was a warm interstadial, the Chelford interstadial, known from sections exposed in sand pits between Chelford and Congleton in Cheshire (Simpson and West, 1958). The organic mud includes macrofossils of trees. Mean July temperatures were in the range 12–16°C, January temperatures between −15°C and −10°C. There was a more continental-type climate. Preserved in cave earths there are large quantities of vertebrate remains of this age, including the brown bear, the fox, the red deer, spotted hyena, reindeer, woolly rhinoceros, horse and elk (Sutcliffe, 1985). Some of these are definitely forest creatures.

Somehow in this short period a boreal forest developed of the kind now found in southern Finland. It included *Pinus sylvestris* (Scots pine), *Betula* (birch) and, remarkably, *Picea abies* (Norway spruce), the latter recorded here for the last time as native in the British Isles. After the last glaciation it was apparently even a late arrival in Norway where the naturally dominant conifer is the *Pinus sylvestris*. Nevertheless it evidently

managed to colonize Britain within the middle of the Devensian stage or had survived the cold in some sheltered site. It grows as native over large parts of northern and central Europe, but natural woodlands are restricted to montane and subalpine habitats where it is often mixed with *Pinus sylvestris* in the north, and with *Fagus sylvatica* (beech) in the south. A clue to its behaviour is that it is particularly frost tolerant. It casts a deep shade and so spruce woodland has a poorly developed shrub and field layer, with *Vaccinium myrtilus* (bilberry) and *Oxalis acetosella* (wood sorrel) among other species.

1.16.3 A TREELESS LANDSCAPE AND THE UPTON WARREN INTERSTADIAL

In the cold period following the Chelford interstadial there were arctic conditions and a treeless landscape with occasional dwarf shrubs. As before, drought-adapted and salt-tolerant species such as *Glaux maritima* (sea milkwort) were present. Soils were frozen for 7 months of the year.

Then between 43 000 and 42 000 years ago, there was a short warm interstadial, the Upton Warren interstadial, perhaps equivalent to the long interstadial noticed in Greenland ice cores from 40 000 years ago. Upton Warren is south of Birmingham. The fossils from this and other sites of a similar age record a peculiar flora and fauna. The community existed around a series of slightly brackish pools, indicated by the presence of species of ostracod. The landscape of Upton Warren remained treeless. The flora included a mixture of halophytes, calcareous grassland herbs and southern continental species, such as *Lycopus europaeus* (gipsywort). There were high summer temperatures, up to a mean July temperature of 16°C, allowing some 'steppe' species to grow, but there were also dwarf shrubs such as *Salix herbacea*, *S. reticulata* (dwarf- and net-leaved willows) and *Betula nana* (dwarf birch), which are now common only in northern latitudes. Beetle evidence supports the idea that this was a very mixed community with many different elements (Coope, Shotton and Strachan, 1961; Coope 1977). Beetles with non-overlapping northern and southern distributions today clambered over the same plants at Upton Warren.

The peculiarities of the community may have been the result of the extreme exposed continental-type climate. Winters were very cold, and it is possible that there were discontinuous areas of permafrost. Alternatively, because the stage was so short it is likely that, because of differential migration rates, trees did not have a chance to become established. Only species that could migrate rapidly, with a short life cycle, took advantage of the rapid climatic amelioration. Horses, bison, mammoths and reindeer have been found. They must have been plagued by clouds of flies and midges. They may have migrated back and forward with the season. Possibly, their grazing prevented the establishment of woodland.

Following the Upton Warren interstadial there was an overall decline in temperatures. The Greenland ice record shows several shorter warm periods, giving way finally to a long period of intense cold (Taylor *et al.*, 1993).

1.16.4 AN ARCTIC DESERT AND GLACIATION

From about 40 000 years ago arctic conditions prevailed over Britain again. Climatic conditions became markedly continental with mean July temperatures of about 10°C and winter means of −30°C to −20°C with only 250–350 mm of precipitation. In the whole period from 40 000 years ago up to 11 000 years ago annual average temperatures in Scotland were on average at least 9°C less than at present (Miller *et al.*, 1987). At first the typical species of previous cold periods – grasses, sedges, arctic–alpine, halophytes and ruderals – were found. The vegetation was sufficient to support a fauna of lemmings

and reindeer, preyed upon by wolf, arctic fox and palaeolithic man. The very large herbivores such as the giant deer, woolly rhinoceros and woolly mammoth had been forced south where they were eventually to become extinct, probably eliminated by hunting. In parts of eastern Britain tundra gradually gave way to a drier polar desert with an even sparser vegetation. In the snowy north and west, winter snow continued to lie through the summer and glaciers started to form and spread.

The formation and spread of glaciers marks a change to a more oceanic, but still cold, climate with higher precipitation (Lockwood, 1979). There was strong cyclonic activity centred off north-western Britain, but with levels of precipitation falling off very rapidly to the south and east. The expansion of the ice may not have started until after 27 000 years ago, but then very rapidly grew to reach down as far as Norfolk and the Gower (Figure 1.60). In Scotland it may have been very thick, up to 450 m thick over Ben Nevis. There was a great lobe of glacial ice, 1200 m thick, over the Irish Sea. By 20 000 years ago sea ice was forming as far south as Portugal in winter.

In the parts of Britain not covered by ice, conditions were very cold. The mean annual temperature may have been −6°C or lower. There were marked geographical differences. Periglacial patterning of the soil is concentrated in south-eastern England. Deep freezing of the soils sorted soil constituents into polygons, garlands and stripes. It was warmer and wetter in the south-west than in East Anglia. Sea levels were perhaps 130 to 160 m lower and a considerable part of the western approaches off southern Ireland would have been dry land. East Anglia was a long way from the sea and influenced by easterlies, giving it a markedly continental climate.

Dryness allowed loess and sand to be shifted and deposited by the wind. These wind-blown glacial deposits are widespread

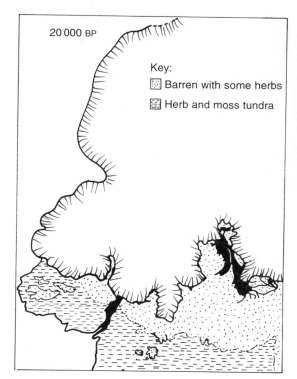

Figure 1.60 Britain and Ireland at the time of maximum Devensian glaciation *c.* 20 000 years ago. (Adapted after Campbell, 1977.)

Figure 1.61 *Saxifraga oppositifolia* (purple saxifrage). (Copyright Jon B. Wilson.)

in south-eastern England. In this desert landscape few plants could survive. One might have been *Saxifraga oppositifolia* (purple saxifrage) one of our most beautiful arctic–alpines (Figure 1.61). In the Arctic *Saxifraga oppositifolia* is sometimes the only vascular plant species growing over large areas. It survives because it is one of the chief stabilizers of mobile screes, alluvial sands and mobile dunes, and can bind sand into hummocks up to 15 cm high. This ability probably explains the presence of *Saxifraga oppositifolia* in Huntingdonshire and Cambridgeshire in the full glacial. Today in Britain it is confined to quite different habitats, damp, base-rich soils on ledges and steep stony slopes on our mountains. It is climatically limited in Britain. In its southern localities, such as the Brecon Beacons, it is confined to north-facing cliffs and slopes. However, in the Alps and Arctic, where the climate is less marginal, it has a broader ecological range, growing on both calcareous and non-calcareous substrates (Jones and Richards, 1957).

1.17 A RELIC GLACIAL VEGETATION?

An impression of the bleak tundra landscape south of the ice sheet may perhaps be seen today in the coldest and most exposed parts of our hills and mountains. The largest area of high plateau-lands in Scotland is the Caringorms (Figure 1.62, see plate section). Here there is the most extensive late snow-bed and fell-field vegetation in the British Isles, reminiscent of the arctic islands such as Spitsbergen. The soil is immature and the action of frost has patterned the terrain. There is soil creep, solifluction, on slopes. The flora has low diversity but several different assemblages have been described (Rodwell, 1992b) whose distribution is related to altitude and degree of exposure.

The most exposed areas are where snow does not accumulate and the gravelly soil experiences extensive churning because of frost–thaw cycles. The extreme cold, with short summers and relatively low precipitation, are the kind of conditions experienced in south-eastern England at the time of the maximum Devensian glaciation. Here today there is a *Juncus trifidus–Racomitrium lanuginosum* rush–heath community (U9; Rodwell, 1992b) found in the harshest environments above 900 m. The vegetation is a patchy mosaic of the tussock plants, which can resist the cold-scorching winds, alternating with areas of bare sand and gravel. Species are *Juncus trifidus* (three-leaved rush), *Festuca vivipara* (viviparous sheep's fescue), *Luzula spicata* (spiked wood-rush) and *Deschampsia flexuosa* (wavy hair grass). *Racomitrium lanuginosum* (woolly fringe moss) may form extensive beds but at the higher altitudes the lichens *Cladonia* spp. and *Cetraria* spp. (reindeer moss and Iceland moss) become more abundant (Figure 1.63). Here *Juncus trifidus* becomes more common in a patchy mosaic, with *Racomitrium lanuginosum*, *Cladonia* spp. and *Carex bigelowii* (stiff sedge) alternating with areas of bare stones and gravel. *Carex bigelowii* is one of a range of arctic–alpines which may be found in slightly more sheltered places. They include *Salix herbacea* (dwarf willow), *Alchemilla alpina* (alpine Lady's mantle), *Diphasiastrum alpinum* (alpine clubmoss) and the moss, *Polytrichum alpinum*.

With higher humidities and also on reaching lower altitudes, *Racomitrium* becomes

more abundant as part of a *Carex bigelowii–Racomitrium* moss–heath (U10l; Rodwell, 1992b). Here there is an increased range of arctic–alpine plants, some of which have been recorded from the full glacial: as well as those already mentioned there is *Armeria maritima* (thrift), *Silene acaulis* (moss campion), *Sibbaldia procumbens* (sibbaldia), *Minuartia sedioides* (cyphel), *Polygonum viviparum* (alpine bistort) and *Thymus polytrichus* (wild thyme). Shrubby species such as *Empetrum nigrum* subsp. *hermaphroditum* (crowberry) and *Vaccinium myrtilus* (bilberry) are found. A transition can be seen between this community and heather bog communities, but in the sheltered areas, between the high spurs of the Cairngorms, snow lies longer and *Calluna* (heather) performs less well. Here *Vaccinium myrtilus* with *Carex bigelowii* become relatively more important.

Snow lies longest in hollows and even on undulating ground in the deep upper corries. More diverse assemblages are associated with snow beds, varying with the length of snow lie. Snow cover provides protection from the worst winter conditions and snow melt irrigates the vegetation. With increasing snow cover *Nardus stricta* (mat grass) becomes important, sometimes accompanied by *Trichophorum cespitosum* (deergrass) in a *Nardus stricta–Carex bigelowii* grass–heath (U7; Rodwell, 1992b). The vegetation includes *Salix herbacea* (dwarf willow), *Alchemilla alpina* (alpine Lady's mantle), *Diphasiastrum alpinum* (alpine clubmoss) and the moss, *Polytrichum alpinum*. Where snow lasts for a long time, keeping the soil waterlogged, the mosses *Polytrichum alpinum*, *Dicranum fuscescens* and the liverwort *Ptilidium ciliare* become more abundant (*Carex bigelowii–Polytrichum alpinum* sedge–heath). In the longest-lasting snow beds other bryophytes, such as the mosses *Polytrichum sexangulare*, *Kiaeria starkei*, *Racomitrium heterostichum*, *Conostomum tetragonum* and *Oligotrichum hercynicum* occur. *Gymnomitrium concinnatum*, a liverwort, and some other bryophytes can also be found, or

Figure 1.63 *Racomitrium lanuginosum* (woolly fringe moss) and reindeer moss.

are common at low altitudes in the milder western coastal fringe, but at high altitude they survive where they are protected from cold and desiccation by the long-lasting snow cover. Mosses and liverworts lack the ability to control effectively water loss from the plant, and although many can survive some drying out and revive when water becomes available, repeated dehydration and rehydration uses up their energy. Two communities have been described: *Salix herbacea–Racomitrium heterostichum* snow bed (U12; Rodwell, 1992b) on slopes subject to solifluction, and *Polytrichum sexangulare–Kiaeria starkei* snow bed (U11; Rodwell, 1992b). Herbs associated with snow beds are *Saxifraga stellaris* (starry saxifrage), *Alchemilla alpina* and *Gnaphalium supinum* (dwarf cudweed).

Melting of the snow provides a continuous supply of water in summer so that on screes and elsewhere there may be a rich growth of *Cryptogramma crispa* and *Athyrium distentifolium* (parsley fern and alpine lady-fern) (U18). Few herbs can survive the very unstable conditions here.

Similar high-altitude communities can be found outside the Cairngorms but are restricted in extent. For example in North Wales in Cwm Idwal, the east-facing slopes of Y Garn show a good altitudinal zonation: dominant *Calluna vulgaris* (heather) gives way to a steep, boulder-strewn slope at about 600 m, then to a zone in which *Vaccinium myrtilus* (bilberry) and *Empetrum nigrum* (crowberry) are co-dominant, and that, in turn, gives way on the summit ridge, at about 850 m, to a wind-eroded open community of *Racomitrium lanuginosum* (woolly fringe moss) and *Salix herbacea* (dwarf willow).

On calcareous soils, which were constantly being renewed because of erosion, but perhaps only in somewhat more sheltered conditions to those of the fell-field, a richer arctic–alpine flora may have survived the full glacial conditions. Similar arctic–alpine assemblages adapted to cold and exposed conditions survive today on a few calcareous crags and ledges where grazing is restricted. They have a range of species adapted to the cold, short season, including *Dryas octopetala* (mountain avens), *Selaginella selaginoides* (lesser clubmoss), *Saxifraga oppositifolia* (purple saxifrage), *Myosotis alpestris* (alpine forget-me-not), *Sagina saginoides* (alpine pearlwort) and *Cerastium alpinum* (alpine mouse-ear). One particularly interesting community is a *Dryas–Empetrum* heath which has been described on rotten eroding sugar limestone at the Cairnwell, Perthshire.

1.18 THE ICE MELTS

At some time about 13 000 ^{14}C years BP the climate improved dramatically (Dansgaard, White and Johnsen, 1989). Greenland ice cores provide an absolute date of 14 600 years ago for climatic amelioration. Polar waters retreated. In southern England mean July temperatures reached 18°C and in southern Scotland 14–15°C. Very rapidly the huge volume of ice, 346 000 km^3, disappeared from everywhere, including most of the Scottish highlands.

THE 'NATURAL' VEGETATION OF THE BRITISH ISLES: 14 000 TO 5000 YEARS AGO AND ITS SURVIVAL TODAY

2.1 THE LATE DEVENSIAN

In studying more recent times the landscape itself bears the smudged print of the past. Land forms record periods of erosion and deposition. The formation of ice has left its fingerprint on the land. There are moraines, dumped material, marking the limits of glaciers. There are eroded valleys, U-shaped and scoured by ice. Detailed features of soils and rocks, their mineralogy, particle size distribution and chemical composition, record the conditions in which they were produced. Sea-level changes have left raised beaches or rock platforms, or eroded and deposited new coastlines. River terraces, each cutting through the previous strata, can be markers to particular ages. The landscape is a three-dimensional record, layer upon layer, print on print. It is an engraving which has been worked and reworked many times. The picture may be faded, blurred or smudged, but it is extraordinary how much of the past history can be revealed by the skilled expert.

This Late Devensian interval, from about 14 000 to about 10 000 years ago, is particularly confusing to understand even though there is a lot of evidence from many pollen sites. The timing of major changes in the vegetation is different in different parts of the country. In the south and west climatic conditions seem to have improved earlier than in the north.

Alternatively, plants' refuges were nearer so that colonization was quicker. There is enough evidence to provide many contrasts and contradictions but not enough to be conclusive about the development of our vegetation. Today's vegetation is very varied and we should not expect the vegetation of the past to be more homogeneous. True, there were fewer species at first, but there was the added uncertainty because of recent colonization. Studies of colonization of volcanic islands or glacial moraines in our lifetime have emphasized the importance of chance in determining the pattern of vegetation development (Burrows, 1990).

Until 8500 years ago, Britain, but not Ireland, was part of mainland Europe, and the vegetation of Europe reached into it. A great plain stretched between eastern England and The Netherlands (Figure 2.1).

The plain was dissected by great rivers, the Rhine and Thames. Some British and continental rivers shared part of their lower reaches or estuaries so that they now share a native fauna, including the fish species barbel, eelpout and white bream. To some extent southern and eastern England shared a vegetation and climate with the rest of Europe. In contrast, Ireland was connected to Britain by a land bridge for only the first 2000–3000 years of this period, and in its extreme westerly situation might have had a very different climate.

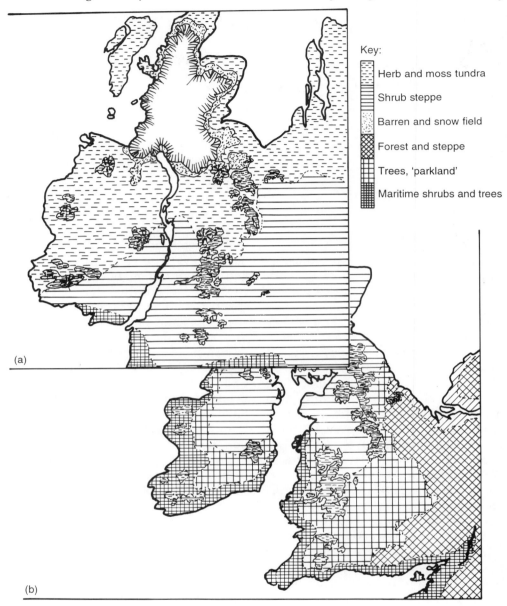

Figure 2.1 Britain and Ireland in the Windermere/Woodgrange interstadial: (a) 14 000 years BP; (b) 11 500 years BP. (Adapted from Campbell, 1977.)

2.1.1 THE ICE RETREATS

About 14 000 years ago the development of the present British vegetation began in earnest. The last pulse of the Ice Age was nearly over. The bleak tundra landscape filled with flowers. At first the species were limited to the same range of steppe-tundra and alpine species which had survived just to the south of the ice sheet. It was still relatively colder than at present, but it was drier so that the glaciers started to dwindle. Then, more and more, warmth-loving species colonized, especially after 13 000 years ago, when it became warmer and wetter. As the ice melted, the sea level rose, breaching land bridges, but the land also rebounded isostatically from the weight of the ice removed. Isostatic rebound was slower than the rate at which the sea was filling with water, and in Scotland and northern Ireland coastal areas were inundated. Shorelines were created 40 m above the present mean sea level around the Firth of Forth. Whales were swimming in the sea around the rock on which Stirling Castle was later built. Later, as isostatic rebound continued, the sea drained away.

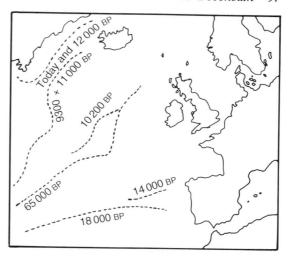

Figure 2.2 North Atlantic position of Maritime Polar Front at various ages in the Late Devensian. (From Ruddiman, Sancetta and McIntyre, 1977.)

2.1.2 THE FLUCTUATING CLIMATE

Part of the difficulty of understanding what happened is that the climate fluctuated several times between 14 000 and 10 000 years ago, as the polar waters of the Arctic withdrew back from the Atlantic, and then reached southward again, only to retreat north suddenly, for the final time, about 10 000 ^{14}C years ago (Figure 2.2).

In Europe and in south-eastern England a floristic change, recorded in the pollen record, marks a double oscillation in the climate, with two false starts before it warmed up permanently. The first change was a minor one at 14 000 ^{14}C years ago, called the Bölling interstadial on the continent. Then 12 600 ^{14}C years ago a more important warm phase started, called the Allerød interstadial on the continent. This warm phase has been called the Windermere interstadial in Britain and the Woodgrange interstadial in Ireland. In northern and western Britain the Bölling interstadial did not register clearly. It was too far from the source of colonizing plants for them to reach north-western Britain quickly enough to mark a change. In northern and western Britain the changes in both the flora

and fauna were more gradual rather than the sharp fluctuations seen in south-eastern Britain.

The Greenland ice core documents with annual resolution an independent chronology for the Late Devensian Interstadial warm period, starting about 14 600 calendar years ago and lasting until about 12 900 years ago (Taylor *et al.*, 1993). Most significantly, the ice core shows that throughout this time the climate remained unstable. Within the Late Devensian stage the climate flickered between warm and cold. One slightly longer period of cold lasted between 14 060 and about 14 000 years ago. It may represent the cool period between the Bölling and the Allerød interstadials. More importantly, within the warm Allerød interstadial *sensu stricto* there were several other cold periods which lasted for several years or more at a time. One cold snap in the late Allerød lasted for nearly 200 years.

Later, about 11 000 ^{14}C years ago (12 900 calender years ago in the Greenland ice core), it became very cold again. There was greater precipitation of snow in the north and west, so that the glaciers reformed. This period has been called the Loch Lomond stadial in Britain and the Nahanagan stadial in Ireland. It lasted until 10 000 ^{14}C years ago (11 600 calender years ago in the Greenland ice cores) when it suddenly became warm again (Dansgaard, White and Johnsen, 1989). This was the start of the present Holocene interglacial, also called the Flandrian or Post-glacial.

2.2 PLANTS, CLIMATE AND SOILS

The vegetation was developing not in a vacuum but in a very particular setting. The position of the British Isles on the western margin of Europe has given it a unique setting because of its climate. The whole of the British Isles lies in the climatic region influenced by the North Atlantic Drift in the Atlantic Ocean. The sea moderates seasonal changes in temperature. In summer, sea breezes cool the land. In winter, especially near the sea, the nights of frost are few because of warm sea breezes. In our maritime climate there is little change in mean temperature from summer to winter, only about 8°C in western Ireland compared with 20°C in parts of continental Europe (Chandler and Gregory, 1976). The climate of Britain is wet, very wet in the north and west.

Within the British Isles there are important regional and local differences in climate because of latitude, altitude and topography. In the area of Britain with the most continental-type climate, East Anglia, the annual temperature range between January and July mean daily maximum temperatures is about 16°C. Northern Britain has colder winters and cooler summers than the south. There are about 4 annual days of frost over Scilly Isles and over 119 days at Balmoral in the Scottish Highlands. However, within the British Isles altitude has a more profound effect than latitude. Close to sea level in Fort William the July mean is 14°C while in southern England it is 17°C in places. At the

same time, on top of Ben Nevis, at 1344 m, the July mean temperature is 5°C and the yearly mean temperature is below freezing. The temperature drops about 1°C for every 150 m.

Altitude has an effect on rainfall as well as geographical location. In the north and west of Britain, where the Highlands are concentrated, there is much higher rainfall, with up to 5000 mm/year, than in the south-east. Rain-laden clouds are forced up and rain precipitates out. Much of eastern Britain and Central England has 750 mm or less rain each year. In western Ireland and upland areas elsewhere annual rainfall is 1250 mm or more. In contrast, the low flatlands of East Anglia are the driest part of the British Isles, with less than 550 mm rainfall per year. However, not just the total amount of rainfall but also the timing of it is important. In most of Ireland and western Britain, including southern England, most rainfall is in the winter. Elsewhere it falls mainly throughout the second half of the year, except in East Anglia where there is a tendency for a summer rainfall maximum.

There is great variation from year to year and from place to place, even within a region. Cambridge Botanic Garden had 65 days of frost on average between 1956 and 1965, while at Santon Downham, only 45 km away, there were 102 days of frost. There is also a great deal of variation from year to year. There are only 15 days of frost in Falmouth on average but in one year (1963) there were 74 days of ground frost. There is an interaction with the topography and the soil. However, there is a time lag for soils to warm or cool at different depths. Roots are insulated to some extent. In February, soil temperatures at a depth of 0.31 m are 3°C in the north-east and 6°C in the south-west. By August they are 14°C in north-east and 18°C in the south-west.

Dry soils are more sensitive to air temperature changes. At the surface of the soil, especially a dry soil, there may be an extreme range of temperature. A diurnal range of 35°C has been measured in a dry soil compared to 20°C in a wet soil. Diurnal fluctuations like this are used by some species to time germination.

Differences in topography creates local differences in climate. The slope and aspect of the terrain is critical, as well as the nature of the vegetation itself. At night the air over slopes cools quicker than that over a valley bottom, but then the cold air may slide down into the valley bottom, making it a frost hollow. Cold air can also collect behind obstructions such as hedges and walls and embankments. Local climates develop most strongly on clear, calm nights and days. Local variation in the strength and duration of sunlight is complex. A south-facing slope of 50° achieves nearly maximum sunlight, but the period of insolation is actually shorter, with a later sunrise and earlier sunset than on slopes facing east or west, respectively. There is seasonal and diurnal variation. Flatter slopes receive more sunlight in mid-summer and steeper slopes more in mid-winter. The microtopography can have a profound effect. There can be as much as a 3°C difference in temperature between a south-facing potato ridge and its furrow in June. The vegetation itself has a profound effect. Only 20% of the sunlight reaches the ground below meadow grass.

The other component of the unique setting of the British Isles is the young soils. Only 15 000 years ago glacial action destroyed most pre-existing soils over much of the British Isles. The ice sheets scoured the land and glacial deposits were dumped on the surface. Although the southern part of Britain was unglaciated, it was very cold. Old soils did survive in some places but they were disturbed or overlain by later deposits. The history of soil development here can be said to have started 15 000 years ago. Soils are the result of a complex interaction between climatological, geological and biological influences and can take many thousands of years

to mature. In cold, upland situations soils mature very slowly (see section 2.3). Our soils are young, immature and still changing. Plants grow in a soil which they have helped to create. In turn the type of vegetation has been determined by the soil. As the vegetation developed, the soil developed with it.

2.2.1 CLIMATIC LIMITS

Climate influences the vegetation in several ways. The most important characteristics for plant growth are rainfall and temperature. Many species have a distribution very closely correlated to a climatic factor such as temperature. This is perhaps most closely seen in some introduced species which can be grown only near the southern and western coasts but not elsewhere. The Canary and Chusan palms (*Phoenix canariensis, Trachycarpus fortunei*) are grown in several coastal resorts and even in places up the west coast of the Highlands. The family Aizoaceae has a range of succulents, dew plants (*Lampranthus, Disphyma, Drosanthemum*), sea figs (*Erepsia*) and Hottentot figs (*Carpobrotus*), introduced mainly from South Africa but now naturalized beside the coast in southern and western Britain. They flourish in hot summers and are cut back by frosty winters. Perennial succulents are particularly susceptible to frost damage.

There is a group of species for which the British Isles is at the northern limit of their natural distribution. In the British Isles their distribution is south-western or coastal. *Limonium binervosum* (rock sea-lavender) for example, is limited to southern Britain where the mean daily maximum temperature for November is about 10°C or more (Figure 2.3).

It grows further north on western coasts than on the colder eastern coast of Britain. There are several possible reasons for this. In areas further north either its seed cannot mature before winter, or it cannot establish as a seedling quickly enough in autumn, or there is too much competition from grasses in

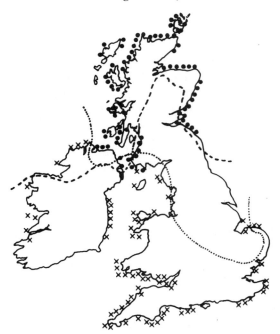

Figure 2.3 Contrasting geographical distribution of two coastal plants related to temperature limits: x, *Limonium binervosum* agg. (rock sea-lavender); ●, *Ligusticum scoticum* (Scots lovage).

May–16 October. Overall productivity of upland grasslands may be less than a third that of lowland grasslands. There is nearly three times the productivity of *Lolium perenne* (perennial rye-grass) at an altitude of 50 m as at 300 m in the Bowland Fells. Upland grasslands start to turn brown in October, while those at sea level remain green and become ragged with winter growth by spring.

Plants adapted to cold climates, such as *Ligusticum scoticum* (Scots lovage) (Figure 2.3) have a remarkable ability to photosynthesize at low temperatures, but they must make best use of the short growing season (Crawford, 1989). Most are perennials, and they accumulate carbohydrate reserves which allow them to grow very quickly when temperatures warm up in spring. However, to do this they have to respire very quickly. The ability to respire quickly is disadvantageous in warm winters because it wastes the energy they have gained in photosynthesis. *Trientalis europaea* (chickweed wintergreen) (Figure 2.48) has forms of enzymes designed to work at low temperatures. Some widespread species show evolutionary differentiation in response to different climates. At 5°C *Dactylis glomerata* (cocksfoot) from Norway becomes dormant and is protected from frost, while that from south-western Britain, where frosts are rarer, remains active. However, the latter is frost sensitive.

Differences in the length of winter conditions, especially the number of nights of frost, have an influence on the timing of leaf-flushing in the spring and even on the time of flowering. There is about a 4 week delay between the onset of spring in south-western British Isles and upland areas in Scotland. Floral development in grasses is delayed by 4.3 days/100 m in altitude. A classic study was carried out by Pearsall (1971) (Figure 2.5).

He showed that length of flower stalk, the number of flowers and the degree of maturity of seeds of *Juncus squarrosus* (heath rush)

the wetter north. In Britain *Tilia cordata* (small-leaved lime) rarely produces fertile seed in the northern parts of its range, though it evidently could in the past.

A useful way of looking at regional variation in climate is by the direct effect it has on plant growth (Figure 2.4).

This can be characterized in several ways. One way is by the length of the growing season. An arbitrary definition of this is the number of months where the accumulated temperature is over 6°C. Using this measure the south-western coastal areas of the British Isles have a growing season of 9 or more months while in most of upland Britain the season is only 5 or 6 months. In the highest parts of the Scottish Highlands the growing season is 4 months or less. In Upper Teesdale the growing season at 450 m is from 18 April–23 October, while at 670 m it is from 4

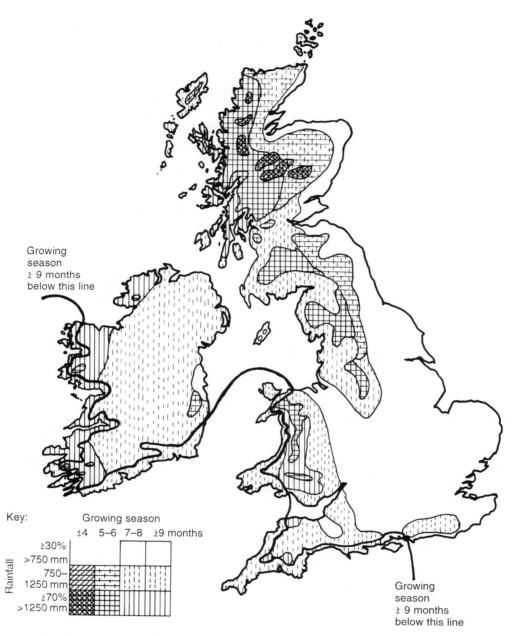

Figure 2.4 Climatic zones in the British Isles, based on length of growing season and rainfall. (Adapted from Chandler and Gregory, 1976, Fig. 15.4.)

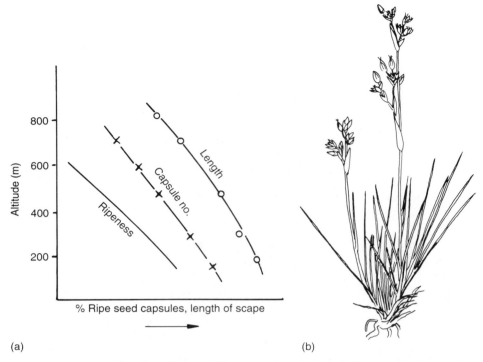

(a)   (b)

Figure 2.5 *Juncus squarrosus* (heath rush): (a) different performance at different altitudes. (Adapted from Pearsall, 1971); (b) plant in fruit.

decline regularly with altitude. Alternatively, a large proportion of high-altitude and high-latitude plants are polyploid, including rare arctic–alpines such as *Saxifraga cespitosa* (tufted saxifrage) and *S. nivalis* (alpine saxifrage) (Figure 2.7). They have more than one set of chromosomes. This allows rapid vegetative expansion in spring.

The relationship of climate and plant distribution is complex. Rainfall is as important as temperature. The availability of water to plants for growth is complicated by the evaporation of water from the soil and through the drying of the soil by plants. Plants take up water in their roots and transpire it through their leaves. Over much of England more than 50% of rainfall is lost through evaporation and transpiration. Different kinds of vegetation will dry out soils at different rates. Evapotranspiration is

greater from woodland than grassland, which, in turn, is much greater than that from bare soil. Evapotranspiration is influenced by temperature and the amount of sunlight. In the summer evapotranspiration may be greater than rainfall. In North Wales rainfall usually exceeds evapotranspiration in every month. In East Anglia rainfall exceeds evapotranspiration only from October to March. There is a soil moisture deficit. The soil dries out and this may lead to a reduction in plant growth. South-east of a line from the Humber to Lyme Bay irrigation is required during 7 years out of 10 to achieve the maximum potential growth of crops. The number of 'grass growing days' is defined as the number of days from the beginning of April to the end of September when the soil moisture deficit does not exceed 50.8 mm. There are less than 130 'grass growing days' in eastern England and more than 170 days in upland Britain.

2.2.2 THE ORIGINS OF THE FLORA

There is an obvious north-west to south-east gradient in climate and flora. At one time it was thought that the elements were associated with the separate origins of floral elements colonizing the British Isles by separate routes or at different times. This may be partly true, but the whole western European vegetation was very disturbed in the glacial period and the majority of our species have colonized from rather a few refuges over a very short period. An important glacial refuge for the least thermophilic elements of the British flora, including some woody species such as *Corylus avellana* (hazel) and *Quercus* (oak), was in western France and northern Spain, especially in the area of the Bay of Biscay (Huntley and Birks, 1983). More thermophilic elements, such as *Tilia cordata* (small-leaved lime) and *Fagus sylvatica* (beech), had survived in refuges in southern and south-eastern Europe. Watts (1988) has emphasized the importance of climate change, the Milankovitch cycles, in driving vegetation changes by determining the abilities of different plant species to migrate and successfully establish. The onset of an interglacial is heralded by a phase of cold winters and warm summers, with an extreme difference between winter and summer mean temperatures. In this phase herbaceous species and those tree species which were limited by low spring and summer temperatures but not by cold winters, can establish. Only later, when seasonal differences were less extreme and winter temperatures ameliorated as well, are other species, intolerant of winter cold, allowed to spread (Delcourt and Delcourt, 1991).

It is possible to compare present patterns of distribution of species to identify similarly climatically adapted floristic elements. Matthews (1955) divided the flora into seven main elements, subdivided into a total of 16 elements (Table 2.1) with different geographical distributions which are largely climatically defined.

The allocation of some species to different elements is somewhat arbitrary, but the percentages provide an estimate of the relative importance of differently adapted plants in our vegetation. The elements may be grouped into three major categories; the widespread species (1–3), those from southern or western Europe (4–8) and those from northern Europe (9–10).

Species that have a distribution centred on southern and western Europe prefer mild winters. In particular, the Oceanic species are tolerant of cool, wet summers but rather intolerant of cold winters. Their distribution is centred on western Europe. *Erica cinerea* (bell heather) is common in western Britain and north-western France, but gets no further east than parts of Belgium. Other Atlantic species are *Hyacinthoides non-scripta* (bluebell) and *Digitalis purpurea* (foxglove). Bluebell woods are a particular feature of the British Isles and elsewhere are only found in northern France and Belgium.

Coastal plants have a similar, but more extreme, dislike of cold. They are very intolerant of frost. In Britain they may be found only near the coast. Further south on the continent they may be found inland. Some are found only in the extreme south and west of the British Isles or in the Channel Isles. Good examples are *Hypericum androsaemum* (tutsan), *Scilla autumnalis* (autumn squill) and *Romulea columnae* (sand crocus). More extreme still are the few Mediterranean species which have very restricted distributions in Britain. They include *Gladiolus illyricus* (wild gladiolus), *Frankenia laevis* (sea heath) and *Arbutus unedo* (strawberry tree) (Figure 2.6).

Thirty-one of these Mediterranean species are found in the Channel Isles, 22 in Cornwall, 19 in Dorset and all of the five that reach Scotland are littoral species.

Table 2.1 Geographical elements of the flora according to Matthews (1955)

	Geographical element	Contribution to flora	Example species
1.	Widespread	14%	*Poa annua* (annual meadow grass)
2.	Eurasian	32%	*Daucus carota* (wild carrot)
3.	European	9%	*Ranunculus bulbosus* (bulbous buttercup)
		Subtotal 55%	
4.	Mediterranean	2%	*Arbutus unedo* (strawberry tree)
5.	Oceanic southern	5%	*Rubia peregrina* (wild madder)
6.	Oceanic west	5%	*Erica tetralix* (cross-leaved heath)
7.	Continental southern	9%	*Acer campestre* (field maple)
8.	Continental	6%	*Quercus robur* (pedunculate oak)
		Subtotal 27%	
9.	Continental northern	6%	*Betula pubescens* (downy birch)
10.	Northern–montane	2%	*Potentilla fruticosa* (shrubby cinquefoil)
11.	Oceanic northern	1%	*Myrica gale* (bog myrtle)
12.	North American	<1%	*Piranthes romanzoffiana* (Irish lady's-tresses)
13.	Arctic–subarctic	2%	*Rubus chamaemorus* (cloudberry)
14.	Arctic–alpine	5%	*Dryas octopetala* (mountain avens)
15.	Alpine	<1%	*Gentiana verna* (spring gentian)
		Subtotal 17%	
16.	Endemic species	<1%	*Coincya wrightii* (Lundy cabbage)

The Continental-element species are tolerant of frost. Perhaps all of our native deciduous trees, except for some variants of elm, belong to this element. Within this category there are those species common in southern Europe, such as *Arum maculatum* (lords and ladies) and *Ruscus aculeatus* (butcher's broom). Species more common in northern Europe include *Drosera longifolia* (great sundew), *Galium boreale* (northern bedstraw) and *Teucrium scorodonia* (wood sage).

The arctic–alpine element in our flora is a relict of a past glacial and early postglacial vegetation. Some species, including *Salix lanata* (woolly willow), are mainly distributed in the subarctic; others, including *Minuartia sediodes* (cyphel), mainly on the mountains in southern Europe; a third category, including *Oxyria digyna* (mountain sorrel), are found in both areas. The saxifrages (*Saxifrage* spp.) (Figure 2.7) are remarkable for including many arctic–alpine species.

Arctic–alpine species tend to be very rare and are confined to a few sites. They lack competitive ability in our normal lowland climate. Now they are largely confined to the uplands and some areas of limestone at lower altitude. They tolerate the stress of cold climates and short growing seasons.

The American element includes a small number of species which are distinctive because they are common in America and today rare in the British Isles. Some, like *Sisyrinchium bermudiana* (blue-eyed grass), may have been spread by migrating birds, or in historical times by humans. Others are perhaps more like the widespread northern species, some of which even have a circumpolar distribution. Their present distribution

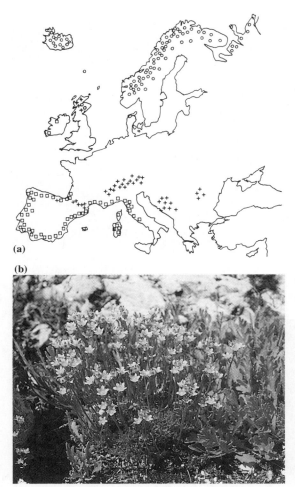

(a)

(b)

Figure 2.6 (a) Contrasting European distributions of three native species: 0, *Salix lanata* (woolly willow); +, *Minuartia sediodes* (cyphel); and □, *Arbutus unedo* (strawberry tree) (see Fig. 1.55); (b) *Minuartia sediodes* (copyright Jon B. Wilson).

is merely the contracted range of an ancient widespread distribution.

The largest elements in the flora under Matthew's climatological/geographical classification scheme are the widespread species. These are the species that have wide environmental tolerances. They are also species that have a strong ability to migrate and colonize new areas, an important ability because our

flora is almost entirely composed of relatively recent immigrants.

2.2.3 THE DEVELOPMENT OF SOILS

The most fundamental factor determining the nature of the soil is the nature of the underlying bedrock, whether soft (like clay) or hard (like granite), and calcareous (like chalk and limestone) or siliceous (like granite, sandstone and quartzite) (Avery, 1990). In some areas the bedrock was exposed by glacial action. In many other areas the bedrock was overlain by glacial deposits, dumped by retreating glaciers, loess (fine wind-blown dust sometimes called brick-earth) and alluvial sand or silt produced by glacial action or by glacial meltwater streams. Glacial soils can be very heterogeneous.

Water-borne sediments have given rise to alluvial soils, wherever conditions were still enough to allow the sediments to settle. Later they have been drained, as in coastal areas where the sea has retreated and/or in temporarily flooded areas inland, like river terraces and the margins of lakes. Colluvial soils have arisen from erosion of existing soils. Rainwash or downslope creep, often accelerated after the clearing of vegetation, has resulted in colluvial deposits in dry valley bottoms or against obstructions such as hedgebanks. In very cold periods freeze–thaw cycles have partially liquefied the soil so that at the bottom of even quite shallow slopes gelifluction deposits have accumulated.

Periglacial action has resulted in many kinds of soil patterning: components within the soil have been churned and sorted by cycles of freeze and thaw. Ice wedges have broken the soil open and have then melted to allow wind-blown material, different from the surrounding soil, to fill them as ice-wedge casts. Patterns of stones in polygons or garlands or stripes have resulted from sorting processes. Other kinds of patterning are those due to the jointing in bedrock or

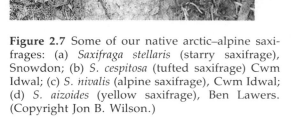

Figure 2.7 Some of our native arctic–alpine saxifrages: (a) *Saxifraga stellaris* (starry saxifrage), Snowdon; (b) *S. cespitosa* (tufted saxifrage) Cwm Idwal; (c) *S. nivalis* (alpine saxifrage), Cwm Idwal; (d) *S. aizoides* (yellow saxifrage), Ben Lawers. (Copyright Jon B. Wilson.)

because of contrasting rock types. Relic topographical features such as creeks and river banks may be preserved in a pattern of contrasting soil types. These features exert an influence on the vegetation because of the different mineral content and permeability of the different soils.

One of the most important environmental variables determining plant distribution now and in the past has been soil moisture. There is variation in moisture in relation to the topography. Even in the absence of differences in the bedrock, seepage of water from higher ground can create a change in the moisture conditions at lower levels. Alternatively, porous rock may overly impermeable rock strata so that springs arise on the lower slopes. In both cases the lower slopes may be waterlogged while upper slopes are free-draining. In a valley bottom the water table may be permanently at the surface.

There is an interaction between the climate and the rock. Different soil types can develop over the same rock type in different climatological regimes. The climate determines the movement of water and soluble materials in the soil. The soluble materials include mineral bases, dissolved rock, and also nitrogen-containing compounds from biological activity of soil organisms, and humic substance from the breakdown of organic materials. Acidic soils generally have pH 4–7, basic soils pH 7–9. The chemical nature of the bedrock may have determined the kind of soil which developed; its mineral content, its acid or base status (pH). Mineral weathering releases large amounts of bases. Rainwater is normally neutral or slightly acidic, or, if polluted, it can be highly acidic. 'Acid rain' damages plants by acidifying the soil as well as by directly damaging the aerial parts of the plant.

The nutrient availability in soils is very varied. Studies of the colonization of glacial moraines in New Zealand and North America have shown a very strong correlation between long-term vegetation changes and soil changes (Burrows, 1990). As soils mature they become stratified through leaching. In British weather conditions of relatively high rainfall, soluble substances are leached out of surface layers and deposited lower down; sometimes distinct layers differing in colour can be recognized. Typically there is a surface 'A-horizon' where humus accumulates, and out of which humic substances and bases are leached. These are deposited in the lower 'B-horizon', which

sometimes has distinct sublayers. Below this there is the 'C-horizon', the basal subsoil and the bedrock. Many different sublayers can be recognized in different soils (Figure 2.8).

Plants exert an influence on soils through the leaf litter and other organic matter they deposit on the soil surface. Three contrasting situations can be recognized. The first, called mor, is where raw humus is sharply separated from the soil. Within the humus there is a layer at the top of recently deposited, hardly decomposed material, above a fermentation layer where fungi and arthropods such as springtails and mites, are active. At the base there is a dark, well-decomposed (humified) layer. In a mull soil, organic material is rapidly incorporated into the top soil layers by the activity of earthworms, so that there is only a thin litter layer and a deep humified 'A-horizon' in the soil. A moder is intermediate between a mull and a mor.

Humification produces some bases but also large quantities of organic acids. The leaf litter of some kinds of plants, especially those with small, hard leaves, such as the heathers and conifers, produce a more acid humus than others. The breakdown of humus is a strongly seasonal process, limited by the cold. In cold, wet climates there is a more partial breakdown of dead plant material because the activity of soil organisms is restricted. In a waterlogged soil, low oxygen availability inhibits biological activity, which can also result in a more acid humus. The soil is also influenced by the roots of plants, which extract water and provide an environment for micro-organisms. Respiration of soil organisms and the roots of plants produces carbon dioxide which acidifies the soil solution. The uptake of certain nutrients, especially the ammonium ion, by plant roots will acidify the soil by leading to the release of hydrogen ions.

There are many kinds of soils in Britain but they have been categorized into six major types by Avery (1990): lithomorphic soils,

Brown soil   Humic gley   Gley podzol   Rendzina

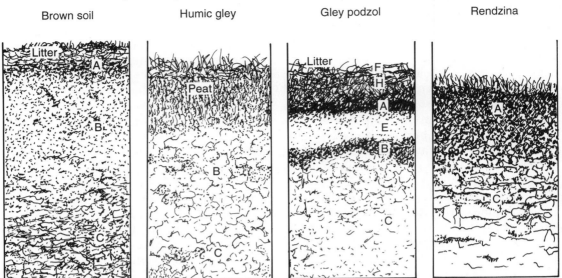

Figure 2.8 Diagrams of contrasting soil profiles; A–F indicate soil horizons. (Drawn from photographs in Avery, 1990.)

brown soils, podzols, gley soils, peat soils and man-made soils. Lithomorphic soils were particularly common after the retreat of the ice. They were unstratified soils which, through leaching and the development of a vegetation cover, have become stratified, converted to other soil types. Surviving lithomorphic soils tend to be thin soils which develop directly over, and are greatly influenced by the chemical characteristics of the bedrock. Two contrasting stratified soils are brown soils and podzols. Brown soils are relatively deep and nutrient rich with a moderate acid or base status. They developed under broad-leaved deciduous woodland. Podzols are leached acidic soils which have developed under heathland or conifer woodland. Gley soils are periodically or permanently saturated with water or were formed under wet conditions. Such soils were widespread in the Late Devensian before drainage patterns had become well established. The texture of the soil, the range of particle sizes and their proportion helps to determine the permeability. Soil particle sizes range through stones and gravel, >2 mm; coarse sand, 2–0.2 mm; fine sand, 0.2–0.02 mm; silt, 0.02–0.002 mm; and clay, <0.002 mm. Glacial drift, which has a high proportion of fine sand, is very liable to waterlogging.

The influence of the bedrock is very great in lithomorphic soils, so, for example, the soil over chalk and limestone is base-rich. Rendzinas are lithomorphic soils overlying calcareous bedrock, which are very rich in calcium (Ca^{2+}) ions. Soil creep and the action of soil fauna keeps the soil well mixed and prevents much stratification. Other lithomorphic soils are called rankers. The thinnest lithomorphic soils are called lithosols. Lithomorphic soils of all sorts may be porous and have a low water content. They may fluctuate in temperature to a greater extent than other soils. These characteristics are as important determinants of the vegetation on a rendzina as the concentration of calcium ions. On the other hand, plants growing in a rendzina over chalk may never be limited by lack of water because chalk holds water well.

2.2.4 CALCICOLE AND CALCIFUGE

One of the most important determinants of vegetation is the soil pH, the acidity or alkalinity of the soil (Jeffrey, 1987). The solubilities of metallic ions vary markedly with pH and this can profoundly alter the species composition of a vegetation. The acidity of soil is normally closely related to the amount of calcium in the soil, so that two classes of plants can be recognized; the calcicoles, or acidofuges, which grow on neutral or basic (alkaline) soils, and the calcifuges, or acidophiles, which grow on acidic soils. The effect on plants is due to the different availability of nutrients and concentration of toxic ions at different acidities.

In particular, aluminium may be solubilized to toxic concentrations of aluminium (Al^{3+}) ions at acidic pH (<pH 4). It can interfere with the uptake and utilization of nutrients and inhibit root growth in calcicole species. Manganese (Mn^{2+}) ions may also be present in toxic concentrations. The calcifuge species like *Deschampsia flexuosa* (wavy hair grass) can compete better at high soil concentrations of aluminium ions (Figure 2.9).

They have mechanisms to trap the aluminium ions and render them harmless. Calcifuge species include a range of very strong competitors with high potential growth rates, so that acid soils are often dominated by one or few species.

Alternatively, at high pH calcifuges suffer from a lack of the necessary nutrients such as ferric (Fe^{3+}) ions, magnesium (Mg^{2+}) ions, phosphate ions and trace elements. They become chlorotic. Ferric ions are less available to the plant because of the high concentration of calcium (Ca^{2+}) ions. With other nutrients they become complexed with phenolic derivatives of soil organic matter. The ability to cope with aluminium ions in low pH spoils may work against calcifuges in high pH soils because the same mechanisms that normally protect them in acidic soils here reduce the availability of mineral nutrients. Calcicoles do not necessarily require high calcium concentrations. They have a better ability to mobilize ferrous ions from iron complexes and are adapted to the very low nutrient availability of calcareous soils. Calcicoles are not normally as highly competitive as calcifuges. Many have the lower potential growth rates of stress-adapted plants. One of the consequences of this is that a calcicole vegetation is floristically more diverse than a calcifuge one.

Even in areas where there are thin soils overlying calcareous rocks, a mosaic of calcicole and calcifuge vegetation is often found. In the conditions of high rainfall generally found in the British Isles, bases can be leached out of the surface layers of a rendzina. The thin vegetation is then converted into a closed turf as nutrients like phosphate become more available and as more competitive grasses invade. Typically, a thin *Festuca ovina* (sheep's fescue) grassland is found on the unstable rendzina. This is a very interesting species because it can thrive on both acidic and calcareous soils and has different subspecies in either situation. However, it shares with the calcicoles the lower growth rate of stress-adapted species, so it is eliminated as more competitive grasses, at first *Festuca rubra* (red fescue) and then *Agrostis canina* (velvet bent) invade. As the soil becomes more acidic *Deschampsia flexuosa* (wavy hair grass) becomes dominant. It may also invade a *Festuca ovina* (sheep's fescue) grassland directly. The lower pH and the greater production of leaf litter of coarse calcifuge grasses can lead to the production of a mor humus to create a peaty ranker. Soil development seems inevitably to go from base-rich to acidic soils in our climate, but erosion of a peaty ranker, when the vegetation cover is lost after fire or drought, can restore a rendzina.

Figure 2.9 Calcicole and calcifuge grasses: (a) *Briza media* (quaking grass), mainly a calcicole species; (b) *Deschampsia flexuosa* (wavy hair grass), a calcifuge species; (c) *Festuca ovina* (sheep's fescue), which has different subspecies tolerant of acid or base soils.

2.3 THE WINDERMERE AND WOODGRANGE INTERSTADIALS

The vegetation of the Windermere/ Woodgrange interstadial was very different from anything that can be seen today. The climate was changing very rapidly, some estimates say it improved at up to 1°C per decade in one period, and more importantly the climate was still fluctuating rather widely. Soils were fresh, and in some places took a long time to mature. Even when the climate would have permitted a mature vegetation there was a long lag before it developed while the soils matured (Pennington, 1986). Everything was in flux.

2.3.1 *DRYAS*-HEATH

The fossiliferous clays of southern Scandinavia and northern Germany of the Late Devensian have been labelled 'Dryas clay' from the abundance of *Dryas octopetala* (mountain avens) fragments they contain. The leaves of *Dryas* are readily recognizable. There are also widespread scattered Late Devensian records of *Dryas* from northern and western Britain.

Dryas octopetala is a species that took advantage of the fresh soils made available by the retreating ice and ameliorating climate (Figure 2.10).

Dryas is a component of the arctic–alpine flora, adapted to a short, cold growing season. The plants of the high Arctic tundra have small leaves 0.3–1.0 cm long. This small-leaved variant has been detected in one Late Devensian flora in Co. Monaghan in Ireland (Mitchell, 1942). A larger-leaved variant survives on some of our mountains and hills today (Elkington, 1971). However, its presence in a lowland sedge-heath vegetation in north-east Scotland and the west of Ireland, on calcareous soils, is particularly interesting, because a similar kind of community may have been widespread in the Late Devensian landscape. The climate within the area of the *Dryas*-heath is cool but winters are relatively mild; perhaps similar conditions were widespread as the ice fields melted away in the Late Devensian. A *Dryas octopetala–Carex flacca* heath (Rodwell, 1992b; CG13) has been described from Scotland; on the Durness limestone of Sutherland and near the north coast of Sutherland *Dryas* is found in a mosaic

(a)

(b)

Figure 2.10 *Dryas octopetala*: (a) flower, Ichnadamph, Sutherland; (b) view of *Dryas*-heath north of Balanakeil Bay, Sutherland. (Copyright Jon B. Wilson.)

with acid-bog plants on soils enriched by wind-blown shell sand. Other arctic–alpines are rare in this assemblage but the presence of *Carex flacca* (glaucous sedge) and *Linum catharticum* (fairy flax) is evidence that, in some respects, this assemblage is the northern equivalent of southern calcareous grasslands. A similar assemblage is found on Carboniferous limestone of the Burren in Co. Clare. Other important species in this assemblage are *Salix reticulata* (net-leaved willow) and *Carex rupestris* (rock sedge).

Dryas octopetala is sensitive to grazing. Although it can grow vigorously when protected, under grazing it grows prostrate, and its flowers are grazed off before producing seed. Its abundance in the Late Devensian is evidence for the low intensity of grazing at this time. It seems that in many areas the large herds of herbivores had been eliminated by hunting. Eventually *Dryas*-heaths can be converted to herb-rich calcareous grassland by grazing. *Dryas* and other members of the *Dryas*-heathland have a low growth rate, so that, as the climate continued to warm, they could not compete against a broad range of colonizing species. Only in the cooler and moister north of Scotland did open conditions, perhaps a kind of *Dryas*-heathland, continue to cover large areas before the advent of sheep grazing and before leaching made soils more acid (Peglar, 1979). When that happened, *Dryas* was probably replaced at first by *Arctostaphylos uva-ursi* (bearberry).

2.3.2 A 'TALL-HERB' GRASSLAND

The summers 13 000 ^{14}C years ago were warm, up to a July mean of 18°C in southern England and 14–15°C in southern Scotland, but winters were probably sometimes very cold, enough to prevent the rise to dominance of tree species, even tree birch. At least in eastern Britain it was quite dry. In the west of Britain and Ireland close to the sea the climate was wetter. The retreating ice left a fresh landscape over which a carpet of flowers unrolled. At first the vegetation was rather like that of parts of the Early Devensian (see Chapter 1). There was an open vegetation dominated by sedges (Cyperaceae) with *Artemisia* (mugwort) and shrubby willows (*Salix herbacea*, *S. repens*) and *Betula nana* (dwarf birch). *Artemisia* species (mugworts) took advantage of the early dry conditions. They included *A. norvegica*, which is now very rare (Godwin, 1984). Mugworts are species of unconsolidated but well-drained mineral soils. They have a very high level of seed production, 9000 seeds/flowering stem have been recorded for *A. vulgaris* (mugwort), which must have helped them to colonize new ground rapidly.

The vegetation was in some respects like that of the usual protocratic phase of previous interglacials. The landscape must have looked something like a glorious wild-flower meadow stretching as far as the horizon (Table 2.2).

Grazing was not a very important factor, with herbivores limited by hunting (Pennington, 1977). There was a seasonal migration of herds followed by bands of palaeolithic hunters. Cave sites, like those at Cheddar Gorge and Creswell Crags, were occupied in spring and autumn. It is possible that the use of reindeer and horses was very sophisticated, verging on domestication. The widespread presence of charcoal in southern England may indicate that fires were used to open areas for grazing and direct the herds.

The flora was not limited to arctic–alpines like *Thalictrum alpinum* (alpine meadow-rue) but included species like *Linum perenne* (perennial flax) (Figure 2.11). Weedy or ruderal species were particularly important. Godwin (1984) provided extensive lists of Late Devensian herbs. Most were annuals that require open, low-competition habitats for their seedlings to establish. In today's landscape these species are familiar to us from many kinds of disturbed man-made habitats, but in the Late Devensian landscape

Table 2.2 Examples of plants from the Windermere interstadial (from Godwin, 1984)

Montane and subarctic plants	Ruderals and weeds	Marsh and fen plants
Bearberry	Yarrow	Marsh marigold
Thrift	Parsley-piert	Saw, hare's-foot, greater pond,
Stiff, black alpine, hair, hare's-foot sedges	Spear-leaved orache	cyperus, bottle, carnation,
	Shepherd's purse	distant, large yellow, pale,
Alpine mouse-ear	Musk thistle	water, and flea sedges
Cyphel	Field mouse-ear	Cowbane
Fringed, arctic, mountain, Teesdale and spring sandworts	Fat hen	Common, slender and few-flowered spikerush
	Red goosefoot	
Common and Danish scurvy grass	Smooth hawk's-beard	Meadowsweet
	Perennial wall rocket	Gypsywort
Hoary whitlow grass	Treacle mustard	Bogbean
Mountain avens	Ground ivy	Water chickweed
Hoary rock-rose	Nipplewort	Grass-of-Parnassus
Spiked woodrush	Autumn hawkbit	Water-pepper
Iceland purslane	Sainfoin	Marsh cinquefoil
Alpine catchfly	Parsnip	Greater and lesser spearwort
Stag's-horn clubmoss	Annual meadow grass	Celery-leaved buttercup
Mountain sorrel	Knotgrass	Northern yellow-cress
Alpine bistort	Pale persicaria	Marsh and bog stitchwort
Alpine cinquefoil	Silverweed	Devil's bit scabious
Bird's eye and Scottish primroses	Creeping cinquefoil	Common meadow-rue
	Meadow, bulbous and creeping buttercup	Marsh valerian
Aconite-leaved buttercup		Marsh violet
Net-leaved, dwarf, woolly, downy, tea-leaved and polaris willows	Curled dock	
	Common and sheep's sorrel	
	Annual knawel	
Alpine saw-wort	Bladder campion	
Purple, kidney, mossy and rue-leaved saxifrages	Common chickweed	
	Dandelion	
Lesser clubmoss	Sea mayweed	
Moss and bladder campions	Stinging nettle	
Alpine meadow rue	Narrow-fruited cornsalad	
Mountain pansy	Wild pansy	
	Bush vetch	

they formed the natural vegetation of the British Isles. Meadow-like communities, rich in grasses (Poaceae) and sedges (Cyperaceae) and species of *Rumex* (sorrel and dock), also became important. *Oxyria digyna* (mountain sorrel), an arctic–alpine species, survived but more widespread species, such as *Rumex acetosa* (common sorrel), were also present, exploiting the open, disturbed habitats that became available.

Other important species were in the Caryophyllaceae, the 'pinks' family, probably species such as the sandworts (*Arenaria, Minuartia*), mouse-ears (*Cerastium*) and pearlworts (*Sagina*), including many species that grow in open ground, on rocky ledges and cliffs on mountains. Some species in the family had been present throughout the colder stages and had a resurgence, *Minuartia verna* (spring sandwort) (Figure 2.11) was one of these.

This plant is interesting because it is closely

Figure 2.11 Herbs of the Windermere/Woodgrange interstadial: (a) *Minuartia verna*; (b) *Polemonium caeruleum*; (c) *Linum perenne* subsp. *anglicum*; (d) *Filipendula ulmaria*. ((a) and (d) Copyright Jon B. Wilson.)

associated with basic soils, including those contaminated with heavy metals. These kinds of very fresh soils were abundant in a landscape that had been subject relatively recently to glacial erosion and was still subject intermittently to deep frost. The mineral soils had not had time to become leached or covered by peat. There were also species such as *Stellaria media* (common chickweed), one of the world's most widespread weeds (Sobey, 1981). This is a ruderal which took advantage of the opening up of the vegetation. It has ecotypes adapted to different climatic regimes.

To liken the 'tall-herb community' to a meadow underestimates its peculiarity. *Polemonium caeruleum* (Jacob's ladder), one of its constituents, was widespread in lowland England, in places distant from the upland limestone sites it is restricted to today (Pigott, 1958). Grazing and microclimate appear to be very important in limiting its present distribution. It is often present on north-facing slopes where there is a reduction of temperature compared to surrounding areas of 5–8°C.

One member of the tall-herb community which is very familiar to us was *Chamerion angustifolium*, (rosebay willow herb). We know it today as a colonist of disturbed wastelands of all sorts, but it also grows within the Arctic Circle, and can be abundant there, especially on south-facing cliffs. It has a wide tolerance of different soils, both acid and basic, but it does not like waterlogged conditions (Myerscough, 1980; Grime, Hodgson and Hunt, 1988). Its success is due to its great potential for seed production: a single upright shoot can produce tens of thousands of plumed seeds. However, the seedlings are small and are very weak competitors. Once established it can spread laterally, dying back to rhizomes in winter, and it can survive for several decades. Rather surprisingly in historical times it was rather rare, so that botanists listed individual localities for it, and it was even grown as a garden flower. However, in the twentieth century it

has become an abundant weed, taking advantage of the many derelict open habitats made available by human activities.

Surviving natural tall-herb communities today are rare, but widespread in calcareous upland areas, wherever the terrain is craggy enough to provide protection from grazing and where peat has not developed. McVean and Ratcliffe (1962) recorded 225 species from 22 surveyed sites in Scotland, with an average of 47 species in each site. These communities are species rich and have a wide range of possible dominant species. Especially important in the survival of floristically rich herb assemblages has been the refuge they found from the later spread of woodland. Most members of the 'tall-herb' community are intolerant of shade.

Another related floristic element, very significant in the Late Devensian vegetation, was that of aquatic, marsh and fen plants. *Filipendula ulmaria* (meadowsweet) was particularly abundant (Figure 2.11). It is common today on river banks and floodplain terraces. Glacial erosion and the dumping of glacial and fluvial sediments had disrupted established patterns of drainage, so there were extensive areas which were liable to flooding. *Filipendula ulmaria* is sensitive to grazing. Like many herbs common in the Late Devensian stage, it has the strong capacity to spread by creeping rhizomes, a capacity that allowed it to survive in the fluctuating cold years.

2.3.3 A PARK LANDSCAPE

After a time, scrub and then woodland vegetation replaced the tall-herb communities of the immediate postglacial. *Juniperus communis* (juniper) and *Empetrum nigrum* (crowberry) were the most important species. *Juniperus* was more important in the drier areas, especially in the south and east but also in Wales. *Empetrum* was more abundant in some parts of the north and west, where the oceanic influence was the greatest. It created a kind of heathland. It is confined today to damp habitats and its abundance indicates the presence of a climate of high rainfall and high humidity.

Empetrum was not geographically limited by the widespread fresh calcareous soils. Although it has been described as a calcifuge, it is tolerant of a wide range of soil pH, from 2.5 to 7.7. It even occurs on highly calcareous soils in the Burren, Co. Clare and in Sutherland. Other shrubby species, of heathers (*Calluna* and *Erica*) and bilberries (*Vaccinium*), were present at this time, but were less important, restricted by the lack of the acidic soils that they require. *Empetrum nigrum* has a circumpolar distribution reaching to northernmost parts of Scandinavia. In Britain it grows from sea level to altitudes above 1000 m but is unable to survive severe winters without an insulating cover of snow (Bell and Tallis, 1973). In Scotland today it can be an important component of communities with *Vaccinium myrtilus* (bilberry), the moss *Racomitrium lanuginosum* and other mosses, liverworts and lichens. A rare species which is also sometimes found here is *Cornus suecica* (dwarf cornel), probably a more important species in the Windermere interstadial.

It is interesting that there are two subspecies of *Empetrum nigrum*: the typical *E. nigrum* subsp. *nigrum* is replaced by *E. nigrum* subsp. *hermaphroditum* at higher altitudes (Bell and Tallis, 1973). The high-altitude subspecies is a smaller plant with bisexual flowers. This allows it to self-pollinate and thereby ensures the more regular production of seed in a habitat where pollinators are rare. The other subspecies, which grows at lower altitudes, is dioecious, having separate male and female plants; it has to cross-pollinate in order to produce seed. It compensates for the greater risks of this strategy by growing larger, and reproducing vegetatively with slender, procumbent stems that root where they touch the ground.

Empetrum nigrum is intolerant of shade. At first in the Windermere interstadial there were only scattered groves of birch trees, but in time a more continuous woodland established. There was also *Populus tremula* (aspen) and *Sorbus aucuparia* (rowan). Where *Betula* became dominant *E. nigrum*, *J. communis* and many of the herbs, such as *Rumex* (dock) and *Thalictrum* (meadow rue), declined in relative importance. Increased wetness is indicated by the rise of *Filipendula ulmaria* (meadowsweet). Later a birch–pine woodland became established in the south as an extension of a birch–conifer woodland that stretched across much of central Europe. In Europe the conifer growing with birch was *Picea abies* (Norway spruce) but in the British Isles *Pinus sylvestris* (Scots pine) was the dominant conifer. In parts of northern and western British Isles, *Pinus* may have been completely absent; pollen levels for *Pinus* were low and, apart from a few sites, there is a lack of macrofossils.

It is more difficult to be sure about the presence of three other important tree species in the British Isles. The pollen record indicates the presence of *Corylus avellana* (hazel) in some places, but no macrofossils have yet been discovered. *Corylus* could have been an early colonist. In successive interglacials it became abundant earlier each time. The climate would not have restricted it: it extends far to the north, as a common tree up to 62°N, and then rarely up to 67°N, where it is found particularly in calcareous conditions.

Similarly, there are low levels of *Quercus* (oak) and *Ulmus* (elm) pollen in a few places in the Windermere interstadial. These records have been explained as being due to long-distance dispersal of pollen, but there are some fairly high records and it seems strange that high records are so patchy. They may record contamination by later deposits. Alternatively, perhaps there were widely scattered small groves of oak and elm in favourable situations, prevented from spreading further by the fluctuating climate. By some arguments much of the Irish flora and fauna must have colonized by this time, including *Quercus* and *Ulmus*, because, later, land bridges were not available for colonization.

Birch remained the dominant species in the north and uplands, but the development of a continuous birch woodland on the uplands may have been limited in the Windermere interstadial in some areas by grazing herds of reindeer and horses. Here many elements of the earlier tall-herb vegetation survived. There were many meadowland species, such as *Saxifraga granulata* (meadow saxifrage) and *Ranunculus* (buttercups). *Helianthemum* (rockroses) were important species. The vegetation of this time has been called a parktundra, emphasizing its open aspect. The persistence of open conditions are shown by the presence of the arctic fox. In the valleys there was more dense woodland. In the wetter areas, at river margins, and in patches of woodland elk could be found.

2.3.4 COLONIZING AN IRISH LANDSCAPE

An interesting comparison can be made between the vegetation of Ireland and Britain in the Late Devensian interstadial (Watts, 1977). The Irish counterpart of the Windermere interstadial is the Woodgrange interstadial. One important difference in Ireland was that there was no resident human population at this time. In addition, only isolated stands of birch woodland developed (Barnowsky, 1988). Nevertheless, about 11 900 ^{14}C years ago *Juniperus communis* (juniper) declined and a new kind of grassland was established. This marked the beginning of a cooler phase, as temperatures declined by 3–4°C. On the limestone, the grassland included *Helianthemum* (rock–rose) and other calcicoles. *Empetrum nigrum* (crowberry) remained an important species.

This later phase of the Woodgrange interstadial is marked by the abundance of the giant deer (*Megaloceros giganteus*) (Barnowsky, 1986). This large deer stood 1.8 m high at the shoulder and had 4 m wide antlers. It grazed the rich calcareous grasslands and used the scattered stands of *Betula* for winter browse. It had taken advantage of a land bridge to colonize Ireland and here it flourished in the absence of any hunting. From the prevalence of remains, the giant deer seems to have flourished especially as conditions were starting to get cold again in the late Woodgrange, but this may be an artefact of the conditions of its fossilization. A clue comes from the greater preponderance of males preserved than females. The males were heavier and more likely to fall through thin ice or get mired in bogs as the winters became colder again.

There are many remarkable differences between the floras and faunas of Ireland and Britain (Stuart and Wijngaarden-Bakker, 1985). The timing of the presence or absence of land bridges has been critical in determining the contrasting pattern of diversity of flora and fauna in Ireland and Britain. The evidence for the presence of land bridges is contradictory (Devoy, 1985). The situation is complicated by the counteracting processes of isostatic rise of the land rebounding from the weight of ice which had lain on it, and the eustatic rise of sea level as more and more ice melted to fill the seas (Tooley, 1982). Synge (1985) suggested that lowest sea levels in the south were about −130 m below mean sea levels at about 15 000 years ago due to the late expansion of the North American Ice and the persistence of a large mass of ice in the Baltic. A broad land bridge would then have connected Brittany, south-western Britain and southern Ireland. The Irish Sea was flooded from the north and gradually the land bridge was narrowed from the south as sea levels rose eustatically. The last connection was by a morainic ridge or series of ridges between the Lleyn Peninsula and Wicklow. Unfortunately much of this ridge was eroded away at the time it was breached, about 12 000 ^{14}C years ago. A later land bridge of sorts, temporary islands and morainic ridges, may have connected Co.

Derry with Islay in western Scotland between 11 400 and 10 200 years ago, but this was at a time of greater cold, the Nahanagan stadial, when it is very unlikely that thermophilous plant species could have used it to colonize Ireland.

Colonization of Ireland was restricted to a very short time at the earliest warm period so that chance has had a great effect on determining the plant assemblages that arose. This is especially obvious when the faunas of Ireland and Britain are compared (Table 2.3).

Several 'native' animals may have been introduced later. The facility with which small animals can be introduced to islands by man is demonstrated by the distribution of several species of vole and shrew on our more isolated islands. The early flooding of

land bridges to Ireland may also explain the peculiar distribution of a number of species widespread in Britain but very rare in Ireland, including *Helianthemum nummularium* (common rock-rose), *Ceratocapnos claviculata* (climbing corydalis), *Cardamine amara* (large bitter-cress), *Geranium pratense* (meadow cranesbill), and *Cruciata laevipes* (cruciata). These must have been late arrivals in Britain, too late to reach Ireland. They have colonized Ireland in a very limited way much more recently from late introductions.

However, it is not the differences between the floras of Britain and Ireland which are difficult to understand but the much greater similarities. Although Ireland has a smaller flora and fauna than Britain, most of the flora and fauna is common to both islands. The smaller fauna and flora may be simply explained by a narrower range of habitats. The shared flora is difficult to understand if Ireland has been isolated since the middle of the Nahanagan stadial. In contrast, the colonization of Britain from Europe continued for many thousands of years, because Britain remained connected to the continent until 8300 years ago. Even if much of the Irish flora and fauna had established a foothold in the Woodgrange interstadial, it must also have established in Britain, and largely without trace on either island. Even so, the foothold was a very precarious one because almost immediately, about 11 000 years ago, the climate became very cold again. There was renewed formation of glaciers in Scotland. The Irish vegetation may have been particularly vulnerable because polar surface water reached down to the same latitude as southern Ireland. Perhaps only a fringe of warmth-loving vegetation survived in southern Ireland on land now drowned by the sea. The sea level was about 15 m lower than at present. Hard evidence for the perglacial survival of *Quercus*, *Ulmus*, *Corylus avellana* and *Pinus sylvestris*, and warmth-loving herbs in Ireland is absent, but these species were quick to (re)establish later.

Table 2.3 British and Irish mammals, reptiles and amphibians

Native to Ireland and Britain	Native to Britain but not Ireland
Fox	Red squirrel
Badger	Brown hare
Stoat	Fallow deer
Pine marten	Roe deer
Otter	Elk
Mountain hare	Auroch
Pygmy shrew	Bank vole
Root vole	Field vole
Red deer	Water vole
Reindeer	Beaver
Brown bear	Common shrew
Wild cat	Water shrew
Wild boar	Wood mouse
Wolf	Yellow-necked field
Common newt	mouse
Common lizard	Hedgehog
	Mole
	Weasel
	Polecat
	Horse
	Adder
	Common toad
	Grass snake
	Crested newt
	Common frog

The Holocene pattern of vegetation development in Ireland from sites in the south, and especially the presence of southern species of land snails which have never been discovered in Scotland, supports Synge's suggestion (Synge, 1985) of an early Holocene southern land bridge, after the end of the Nahanagan stadial. Sea-level changes in the Irish Sea region have been very complex (Tooley, 1982).

The vegetation history of the Isle of Man provides a useful comparison to that of Ireland. It had a similar flora in many respects to that of Ireland in the period of the Nahanagan/Windermere interstadial. Giant deer were present and, although some tall-herb species were present, the vegetation was probably an open grazed grassland, with a few scattered groves of tree birch. A diverse herb flora of over 80 recorded species quickly established (Dickson, Dickson and Mitchell, 1970). In this period and subsequently there is no evidence of any restriction on colonization. In the early Holocene the vegetation development followed that on the mainland very closely. Is it possible that the morainic ridges in the Irish Sea region, one from the Lleyn to Wicklow, lasted into the early Postglacial period and were then only destroyed by storm surge as the sea rose?

2.4 A COLD SNAP, THE LOCH LOMOND AND NAHANAGAN STADIALS

The average annual temperature declined 11 000 years ago by about 4°C. Winter temperatures had dropped further. This temperature decline was marked everywhere by a decrease in tree birch. The giant deer, after its brief population explosion in Ireland, became extinct. Perhaps the development of permafrost and cryoturbation destroyed the birch groves and grasslands upon which it had relied for food. In some places the start of the cold period is marked by a short phase when weedy species of the cabbage family (Brassicaceae), like *Rorippa*, *Cardamine* and

Arabis (yellow, bitter and rock cresses), took advantage of the opening habitats and decline of the grasslands (Pennington, 1977).

The cold was felt most severely in the north-west (Sissons, 1979). Glaciers reformed in the Scottish Highlands and also to a lesser extent elsewhere, with small cirque glaciers in Wicklow Mountains of Ireland and in North Wales (Lowe and Lowe, 1989) (Figure 2.12).

This cold period has been called the Loch Lomond stadial from the area of main glacier formation, and the Nahanagan stadial in Ireland. The growth of glaciers in the west indicates higher snowfall, there was snow-patch vegetation in places, but in other areas, especially in the east, there was still rather low precipitation.

Over some parts of Britain a tundra

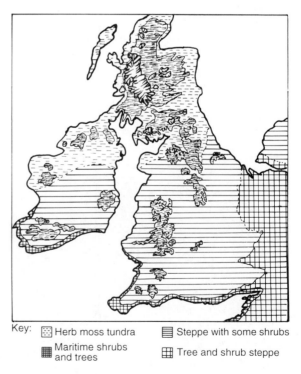

Key: ⬚ Herb moss tundra   ▤ Steppe with some shrubs
▦ Maritime shrubs and trees   ▦ Tree and shrub steppe

Figure 2.12 Britain and Ireland in the Loch Lomond/Nahanagan stadial *c.* 11 000 years ago. (Adapted from Campbell, 1977.)

developed with patches of permafrost in southern Britain and more continuous permafrost in some areas further north. Average annual temperatures were at least 0°C in southern Britain and perhaps much colder. Weedy species predominated, but included montane plants such as the clubmosses (*Huperzia*, *Diphasiastrum*) (Figure 2.13) and *Salix herbacea* (dwarf willow).

Soon *Artemisia* (mugworts), sorrels (Polygonaceae) and pinks (Caryophyllaceae), flourished in a mixture like the vegetation of 2000 years earlier. *Dryas octopetala* was widespread and so this period has been called the Younger Dryas time in Scandinavia, to distinguish it from the earlier Older Dryas time.

Although polar waters reached south again as far as southern Ireland, there may have been a surviving warmth-loving vegetation close to the sea in the west. The higher levels of precipitation which helped glaciers to grow may also have allowed patches of a sheltered scrub or wooded vegetation to survive. There is circumstantial evidence for survival of *Pinus sylvestris* in the west. It was later to colonize the west of Scotland as if from a north-western refuge, at the same time as it spread from southern Britain (Huntley and Birks, 1983). In south-eastern England there was probably more extensive survival of tree birch and perhaps other trees. In East Anglia pollen records for *Betula* are high throughout the Loch Lomond re-advance. Part of this is certainly *B. nana* (dwarf birch) but the tree species *B. pubescens* (downy birch) was also present, and at Hockham Mere there were even hybrids between the two. There is also evidence for the survival of *Juniperus communis* in many areas. A single fruit of *Potentilla fruticosa* (shrubby cinquefoil) has been found at Nazeing in Hertfordshire (Allison, Godwin and Warren, 1952).

Towards the end of the Loch Lomond stadial the glaciers, no longer fed by snowfall, began to wither, and then suddenly about 10 000 ^{14}C years ago (dated to 11 600 calender years ago in the Greenland ice cap) the climate warmed dramatically, so that the remnants of the glaciers melted almost overnight. This marked the start of the Holocene epoch. July mean temperatures rose from 10°C to 17°C. Evidence from the Greenland

(a)   (b)

Figure 2.13 Clubmosses of the Loch Lomond/Nahanagan stadial *c.* 11 000 years ago: (a) *Huperzia selago* (fir clubmoss); (b) *Diphasiastrum alpinum* (alpine clubmoss).

ice core shows that this warming occurred within 50 years (Dansgaard, White and Johnsen, 1989). The maritime polar front retreated from 35°N to 55°N in 20 years; subpolar oceanic waters retreated in turn, tracking the northward shifting atmospheric margin of the polar jet stream. The coldest part of the Loch Lomond stadial had not lasted long, little more than 1000 years, and the most extreme climate was only in the few hundred years before the beginning of the Holocene epoch.

2.5 POSTGLACIAL WARMTH

There are differences in the patterns of vegetation development between the Holocene/Postglacial and previous interglacials. They may be the result of the false starts of the Late Devensian interstadial, 'abortive interglacials'. The Holocene/Flandrian has been divided into a number of phases (Table 2.4).

The first and shortest phase, the Pre-Boreal differs from a normal protocratic phase because many species were already present in Britain or were very near. For example there was a sudden abundance of *Juniperus communis* (juniper) pollen from the rapid expansion and prolific flowering of surviving juniper. It probably survived in refuges throughout the Loch Lomond stadial as a prostrate, weakly flowering shrub. The maximum cold of the stadial had only lasted a few hundred years and it is even possible that individual plants of juniper survived the whole length of this time. Today juniper has a scattered distribution on chalk and limestone, and as an understorey plant in drier pine woods in Scotland. It is tolerant of both acid soils and limestone. Two subspecies have been recognized in Britain. One is a procumbent shrub which is found in upland areas. The other is a more upright shrub or even a small tree.

In some places the beginning of the Pre-Boreal stage is also marked by a minor resurgence of *Empetrum nigrum* (crowberry) followed by a rapid decline as it was replaced in many areas by *Calluna vulgaris* (heather) (Bell and Tallis, 1973).

The soils in the Pre-Boreal stage were fresh mineral soils. Over large areas the soil was either neutral or basic, with little stratification. Calcicole herbs were widespread. Conditions were probably relatively moist. Drainage patterns were not well established, the landscape was full of braided rivers, and many

Table 2.4 Stages of the Holocene

Date starts years BP	Period	Climate	Vegetation	Soils
2500	Sub-Atlantic	Wetter	Development of moor	Podzols, bogs
5000	Sub-Boreal	Drier and cooler	Decline of woodland rise of heathland	Mor humus, podzols
8500	Atlantic	Warm and wet	Continuous deciduous woodland	Brown earths, mull humus
9000	Boreal	Warm and dry	Pine and hazel woodland	Maturing
10 300	Pre-Boreal	Warm and moist	Open birch woodland	Unleached, high nutrient

areas became seasonally flooded. A pollen record for Tarn Moss (Malham) indicates a meadowland with *Filipendula ulmaria* (meadowsweet) fringed with *Typha latifolia* (bulrushes) (Piggott and Piggott, 1963).

Now tree birch (*Betula pendula* and *B. pubescens*) took advantage of the warmth. Tree birch was already widely present in many parts of south-eastern England and here, very soon, a continuous birch woodland developed. In northern Britain a more open vegetation, a grassland with birch, lasted for longer. In north-western Britain and Ireland also, the juniper phase lasted much longer and tree birch only became important later. Birch was the first of the trees to spread and dominate the landscape. The herbs of open habitats of the Late Devensian stage became rarer and rarer.

2.6 A SURVIVING HERB FLORA

The glorious wild-flower meadowland of the Late Devensian and Pre-Boreal was eventually converted into a green monochrome wilderness of a very few species of trees stretching from shore to shore. Few species of the Pre-Boreal open grasslands could survive in the woodland shade. Even a species with wider tolerances, *Teucrium scorodonia* (wood sage), was affected. Its grassland ecotype was replaced by the woodland ecotype (Hutchinson, 1968). Open-habitat plants become etiolated in the woodland shade. Woodland plants have larger cotyledons and expand earlier. They also have a higher shoot/root ratio than open-habitat plants.

At the time of the maximum development of the wildwood the pollen of herbaceous plants reached a low of 10% of the total. This compares to today's landscape where non-tree pollen, including grass pollen, often exceeds 90% of the total. However, the landscape was not entirely unvaried woodland. If we could have flown over the wildwood at its period of maximum development, about 6000 years ago, we would have

seen variation in the colour and pile of the thick green tree canopy. Part of this variation would have been the changing dominance of tree species, but in other areas, splashes of colour would have indicated areas inimical to the growth of trees. Here, because of the climate and soil, or where there were steep slopes that allowed light to penetrate beneath a tree canopy, many of the herbs of the Late Devensian landscape survived.

Many of the refuges for Late Devensian species can still be identified today as some of our botanically most interesting areas. It is in these areas that some of our most beautiful and treasured wild-flowers survive, flowers like *Gentiana verna* (spring gentian), *Geranium sanguineum* (bloody cranesbill), *Primula farinosa* (bird's-eye primrose) and *Silene acaulis* (moss campion) (Figure 2.14).

The most important refuges are floristically rich because they allowed a range of different communities of flowers to survive in a range of open habitats. In the 106 hectares of Gaitbarrows National Nature Reserve in North Lancashire there are over 150 species of moss, 30 liverwort species, 25 fern species and nearly 450 species of flowering plants. This kind of floristic diversity is a measure of the breadth of habitats present, especially those where the growth of trees is limited; limestone pavements, marsh, fen and wet meadow habitats, as well as diverse woodland habitats. Similar habitats were present in the Late Devensian but, although floristically rich areas today have many of the same plant species as the Late Devensian landscape, the vegetation in most areas is quite different; the Late Devensian vegetation was developed on young soils at a time of rapidly fluctuating climate, while today's exists in a relatively stable climate on soils that have matured.

The most important similarity between floristically interesting areas is the presence of open plant communities with little shade. Many of them have thin limestone soils which inhibit the growth of several highly

(b)

(a)

(c)

Figure 2.14 (a) *Gentiana verna* (spring gentian); (b) *Geranium sanguineum* (bloody cranesbill); (c) *Primula farinosa* (bird's-eye primrose); (d) *Silene acaulis* (moss campion). (Copyright Jon B. Wilson.)

(d)

competitive shrubs and herbs which would otherwise dominate the vegetation and reduce floristic diversity. One of the reasons why ash woodlands have so diverse a field layer is that they are commonly developed on steep valley sides in hard limestone areas, where light penetrates the tree canopy. Refuges can be broadly grouped as:

1. areas above the tree line;
2. areas of exposed limestone or rugged topography;
3. wetlands; and
4. the coastlands.

However, this is in some respects an artificial classification because these kinds of refuges share many characteristics. Limestone is found at high altitude and at sea level. Wetlands are found on mountains and hills, in coastlands and in limestone areas. An uninterrupted vegetation succession can occur from a coastal saltmarsh to a freshwater wetland community. Topographical irregularities are found on mountains, at the coast and in limestone areas. The comparability of many refuges is pointed out by the shared presence of several species. For example, *Armeria maritima* (thrift) and *Plantago maritima* (sea plantain) are found both at the coast and on our mountains. Several Late-Devensian species are found today in lowland wetlands and on mountain cliffs or flushes. *Potentilla fruticosa* (shrubby cinquefoil) is found at 650 m above sea level in the Lake District and 15 m above sea level in the Burren (Elkington and Woodell, 1963).

2.6.1 ABOVE THE TREE LINE

Climates inimical to trees are found at high altitudes and at the highest latitudes in the British Isles, or where there is a high degree of exposure as at the coast. One important feature of vegetation is the tree line. It is difficult to detect in the British Isles because it has been greatly modified by grazing and burning. It is estimated to be at about 530 m

at present. There is active churning of the soil by ice formation above 600 m in the Scottish Highlands.

Some *Pinus sylvestris* (Scots pine) grows at 640 m above Loch an Eilean. It has the characteristic twisted shrubby form of the 'Krummholz' vegetation commonly found above the tree line in the Alps and elsewhere (Crawford, 1989). In these habitats trees are wind-pruned or liable to wind-throw in storms so that the canopy is opened. For example *Arctostaphylos alpinus* (mountain bearberry), one of the many species that takes advantage of the absence of trees, grows only above 600 m in the Highlands but also on coastal clifftops in the very north of Scotland where trees are wind-pruned. These areas share a very short growing season.

The tree line was much higher at the time of the climatic optimum about 5000 years ago, but the highest woodlands were patchy and the trees generally dwarfed. Trees that grow at high altitude, *Betula* (birch) and *Sorbus aucuparia* (rowan), do not cast a dense shade, and *Pinus sylvestris* (Scots pine) has a tendency to grow widely spaced as a result of grazing, so that open-habitat herbs can survive well below the tree line. Large areas of Scotland, like the Cairngorms and parts of Sutherland, and even parts of the Southern Uplands, were never forested (Lowe, 1993). Even on the relatively southern and low-altitude North Yorkshire Moors, pollen analyses from the high plateaus indicate that dense woodland may never have existed (Simmons and Cundhill, 1974).

Paradoxically, the very exposed slopes and summits of our highest hills and mountains today, far from providing a refuge of diverse herbs, more often than not have a range of species-poor communities which have little similarity to the herb-rich Late Devensian vegetation (Pearsall, 1971). A clue as to why this is so is provided by the history of the vegetation of north-east Scotland. A tall-herb flora survived here for a long time because a thick woodland never developed. Nevertheless it, too, was eliminated by the later spread of blanket peat (Peglar, 1979). The uplands have very high rainfall. Soils are leached and/ or peaty and very acid. The acid rankers and peaty soils support floristically poor upland grasslands, heathlands and moorlands. These kinds of vegetation are also largely the product of grazing. *Nardus stricta* (mat grass) is a common component and is often dominant. In some places shrubby species such as *Calluna vulgaris* (heather), *Vaccinium myrtilus* (bilberry), *V. vitis-idaea* (cowberry), *Empetrum nigrum* (crowberry) or *Arctostaphylos uva-ursi* (bearberry) are dominant.

There are rare pockets of floristic diversity. The Lake District, for example, has a good total list of herbaceous species but many species are very rare. Most hills in the Lake District are acidic and floristically poor.

Helvellyn and the Fairfield range are floristically the richest because they have pockets of calcareous rocks (Figure 2.15).

There are larger areas of calcareous rock in the Craven Pennines, which include Ingleborough and Penyghent. The distribution of plants on Ingleborough, which rises to 723 m, emphasizes the importance of the substrate rather than altitude. Beds of Carboniferous limestone alternate with Millstone grit and Yoredale shale and sandstone (Figure 1.17). Rare alpine plants are confined almost entirely to the belts of limestone on the western side (Raven and Walters, 1956). They include *Saxifraga oppositifolia*, *S. aizoides*, *S. hypnoides* (purple, yellow and mossy saxifrages), *Arabis hirsuta* (hoary rock-cress), *Potentilla crantzii* (alpine cinquefoil), *Asplenium trichomanes-ramosum* (green spleenwort) and *Poa alpina* (alpine meadow-grass).

Figure 2.15 View from Helvellyn.

Other high-altitude areas have different assemblages of rare herbaceous species (Ratcliffe, 1991). The southernmost location for several alpine species is the Brecon Beacons (Perkins, Evans and Ghillam, 1982) with *Saxifraga oppositifolia*, *S. hypnoides*, *Sedum rosea* (roseroot), *Thalictrum alpinum* (alpine meadow-rue), *Galium boreale* (northern bedstraw), *A. trichomanes-ramosum* and *Salix herbacea* (dwarf willow). The glory of Wales is, of course, Snowdonia, which has *Saxifraga nivalis* (alpine saxifrage) as well as *S. hypnoides*, *S. oppositifolia* and *S. cernua* (drooping saxifrage). *Lloydia serotina* (Snowdon lily) is exceedingly rare (Figure 2.16).

It has an extraordinary circumpolar distribution in the northern hemisphere (Woodhead, 1951) from the Swiss Alps, Carpathians, Caucasus, Eastern Himalayas and Siberia to the Rockies of North America. Since it has no special mechanism for

Figure 2.16 *Lloydia serotina* (Snowdon lily). (Copyright Jon B. Wilson.)

dispersal and fruiting is rather rare, at least in Wales, this may indicate a very ancient, perhaps even a relic Tertiary distribution. *Lloydia serotina* was discovered by Edward Lhwyd and sent to John Ray before 1690. Rare species such as *Lloydia* have become extinct in some localities because of the depradations of botanists. It is found very locally in only very few sites in Caernarvonshire.

Apart from Snowdon itself, there are several other interesting localities in North Wales. On Twll-Du (the Devil's Kitchen) in Cwm Idwal there are lime-rich rocks and base-rich flushes (Figure 3.8). On the cliffs themselves and on the scree and grassland below there is an enriched community of species, including *Saxifraga oppositifolia*, *Lloydia serotina*, *Oxyria digyna* and *Thalictrum alpinum* (alpine meadow-rue). There are woodland species, such as *Luzula sylvatica* (great woodrush) on ledges, representing a remnant of the 'tall-herb' community of the Late Devensian stage (Godwin, 1955). In places there are small ungrazed relict pockets of willow scrub dominated by *Salix lapponum* (downy willow), but with other dwarf willows present, and with *Vaccinium myrtilus* (bilberry). The community is confined to damp calcareous slopes and rock ledges, surviving where it is protected from grazing. The *Salix lapponum–Luzula sylvatica* scrub (Rodwell, 1991; W20) is found in the southern and central Highlands with outliers elsewhere in Scotland. *Salix lapponum* also grows on Helvellyn in the Lake District.

The limestone cliffs of Ben Bulben in County Sligo are probably botanically the most interesting of any mountain in Ireland. This is the only locality of *Arenaria ciliata* (fringed sandwort), represented here by an endemic subspecies (ssp. *hibernica*). *Salix phylicifolia* (tea-leaved willow) is present and *Silene acaulis* (moss campion) is common. Most high-altitude areas in Ireland, such as Macgillycuddy's Reeks, the Wicklow Mountains or the Mourne Mountains, are covered by species-poor blanket bog or heather moor.

The Highlands of Scotland are by far the most extensive high-altitude area in the British Isles but they are generally characterized by moorland and bog communities with low floristic diversity. The Cairngorms are rich in the total number of species, but the main rock type is acidic and many of the species are very rare and confined to the pockets of basic rock on its margins (Ratcliffe, 1981). The greatest floristic diversity in the Highlands is to the south in the Breadalbane, especially on Ben Lawers, which preserves some remarkable assemblages (National Trust for Scotland, 1986) (Table 2.5).

At present there is no woodland on Ben Lawers but there are rare tree remains of *Pinus sylvestris* (Scots pine) up to about 600–670 m. *Sorbus aucuparia* (rowan) and tree birch (*Betula*) have grown even higher in the past. On the southern slopes there is a relatively nutrient-rich and damp glacial drift. The lack of podzols here and the absence of a characteristic pine woodland field layer indicates that a birch–oakwood probably grew here, with *Pinus* only in drier spots. Floristic richness is associated here with the enrichment of the normally acid rocks of the Central Highlands by a band of mica schist (Caenlochan schist). The band of rock is rich in calcium, magnesium, sodium, potassium, iron, phosphate and sulphate. Most importantly, the mica schist breaks down into fine sand, leaving a moisture-retaining clay residue. This soft, erodable soil provides open areas for the rarities. Especially good refuges are the micaceous terraces and bogs flushed with lime-enriched ground water. There are also substantial screes and solifluction slopes. Rare plants grow on cliff ledges, in gullies, on dry grassy slopes and on the summit. Cliffs and edges provide a refuge from grazing.

Almost as important a refuge is the area in Perthshire, the corries and on the slopes at the head of Glen Isla, Glen Clova and Callater: Caenlochan, Corrie Fee and Corrie Kander. Many species are shared with the Breadalbane. Caenlochan and Glen Clova also have *Gentiana nivalis* (alpine gentian), *Astragalus alpinus* (alpine milk-vetch), *Cicerbita alpina* (alpine blue sowthistle) and *Lychnis alpina* (alpine catchfly).

Many different adaptations to growing in different climates can be observed. For example, many species from northern latitudes and high altitudes, where sexual reproduction is difficult, because of a lack of pollinators and a short flowering season, hedge their reproductive bets by reproducing asexually, producing stolons and rhizomes. A few, such as *Polygonum alpinum* (alpine bistort), have bulbils. The grasses *Poa alpina* (alpine meadow-grass) and *Festuca vivipara* (viviparous sheep's fescue) have spikelets which proliferate vegetatively to produce plantlets.

2.6.2 UNFRIENDLY SOILS

During the Late Devensian mature soils were eroded or overlain by young heterogeneous deposits. These young soils had many characteristics inimical to tree growth and favourable to diverse kinds of ruderals or wetland species. In time, as a result of an interaction between soil, climate and vegetation, soils have matured along with the maturing woodland. However, in some areas, especially in the uplands, on limestone, at the coast and in wetlands, soils inimical to trees have been maintained so that a diverse herbaceous flora has been preserved.

Some soils are inimical to tree growth simply because they are very thin. The importance of craggy ground as a provider of refuges has already been noted in the montane flora. Another example is provided by the flora of the Great Whin Sill, which runs through County Durham and into Northumberland up to Lindisfarne, a wall of basalt, the result of a great wall of lava

Table 2.5 Distribution of some plant species that can be found on Ben Lawers (many species can be found in several communities)

Scree and rocky slopes	Cliffs and ledges	Leached grassland and heath
Moss campion	Alpine saxifrage	Heather
Purple saxifrage	Roseroot	Bilberry
Mossy saxifrage	Mountain avens	Alpine Lady's mantle
Mountain scurvy grass	Wood cranesbill	Wild thyme
Alpine willow herb	Alpine cinquefoil	Mountain everlasting
Dwarf cudweed	Globeflower	Yellow rattle
Mountain sorrel	Alpine mouse-ear	Chickweed wintergreen
Mountain sandwort	Alpine saw-wort	Harebell
Alpine pearlwort	Meadow oat grass	Sheep's fescue
Alpine Lady-fern	Alpine meadow grass	Three-leaved rush
Newman's Lady-fern	Alpine fleabane	Cowberry
Holly fern	Black alpine sedge	
Wood anemone	Maidenhair spleenwort	
Wood sorrel	Green spleenwort	
Mossy cyphel		

Racomitrum heath	Snow beds	Flushed grasslands
Woolly hair moss	Starry saxifrage	Mountain pansy
Viviparous fescue	Sibbaldia	Alpine gentian
Spiked woodrush	Snow pearlwort	Alpine forget-me-not
Alpine bistort	Dwarf cudweed	Eyebrights
Alpine clubmoss	Dwarf willow	Fairy flax
Fir clubmoss		Cyphel
		Limestone bedstraw
		Devil's bit scabious

Mat-grass and peat bog	Springs	Flushes and mountain bogs
Mat grass	Yellow saxifrage	Scottish asphodel
Heath rush	Chickweed willow herb	Bristle sedge
Tormentil	Alpine willow herb	Scorched alpine sedge
Heath bedstraw	Alpine meadow rue	Russet sedge
Hare's-tail cottongrass	Starry saxifrage	Yellow sedge
Common cottongrass	Hairy stonecrop	Chestnut rush
Deergrass		Two-flowered rush
Cloudberry		
Crowberry		
Dwarf cornel		

injected up into Carboniferous limestone in the Permian Period. The rock is very hard and the slopes are mostly south-facing. In winter the shallow soils are wet, but summer drought prevents the establishment of trees (Richards, 1982). On the sandy steep northern slopes and cliffs a woodland of *Sorbus aucuparia* (rowan), *Betula* (birch) and *Fraxinus excelsior* (ash) is established, but not densely enough to prevent a range of ferns from growing.

On the dry, south-facing slopes the flora is

dominated by annuals such as *Senecio vulgaris* (groundsel), *Capsella bursa-pastoris* (shepherd's purse), *Sisymbrium officinale* (hedge mustard), *Veronica arvensis* (wall speedwell), and *Malva sylvestris* and *M. neglecta* (common and dwarf mallows), a group only remarkable for being very common in man-made habitats of rough and waste ground. Ruderals such as these were common in the Late Devensian vegetation. On the Whin Sill they form a natural community which may have existed as an unshaded reef in the sea of wildwood. In cracks and crevices there is a grassland of *Festuca ovina* (sheep's fescue) and *Deschampsia flexuosa* (wavy hair grass) and a range of herbs, with xeromorphic features that allow them to survive the summer drought. *Allium vineale* (wild onion), *A. oleraceum* (field garlic), *A. schoenoprasum* (chives), a rarity elsewhere, survive as bulbs; *Saxifraga granulata* (meadow saxifrage) produces bulbils; *Sedum acre, S. album* and *S. villosum* (biting, white and hairy stonecrops) are succulent; and *Erica cinerea* (bell heather) has hard, needle-like leaves which do not wilt in drought. One species particularly associated with the Great Whin Sill is the Late Devensian relict *Juniperus communis* (juniper). *Saxifraga oppositifolia* (purple saxifrage) survived for many thousands of years here, although it has since become extinct.

Calcareous soils are inimical to rapidly growing competitive species because of low nutrient availability. The main limestone rock types in the British Isles are the Carboniferous limestone, the Permian Magnesian limestone and the Cretaceous chalk, but almost every geological age has produced its own limestone. Limestone is widespread in the British Isles but often it is overlain by glacial deposits or peat. There are very large areas of Carboniferous limestone in Ireland overlain by glacial till. Most limestone areas were once covered by wildwood and developed a calcareous brown earth. Only later, when they were cleared, erosion occurred so that a rendzina developed,

supporting a chalk and limestone grassland. Such limestone areas, although floristically diverse now, did not provide a refuge for herbaceous species of open habitats at the time of the maximum development of the woodland.

Some limestone areas, however, did provide refuges. These were areas of bare limestone plucked clean by glacial action or erosion, or cliff ledges, steep slopes and screes so unstable that any developing soil was constantly eroded away. Upland landscapes are especially exposed to erosion. Scree slopes are common features of our landscape on the slopes of our hills but also in the limestone dales. Some rock strata, like Caenlochan schist and sugar-limestone erode very easily to produce a coarse sand. In the upper parts of river courses river banks, gorges and terraces are particularly susceptible to erosion. All these processes constantly expose new areas for colonization by herbs.

Alternatively, some soils are very thin, restricting tree root growth. Extensive 'karst' landscapes of exposed hard limestone have little soil for tree roots. Limestone erodes by solution leaving little residue, so there is little potential for a mineral soil development. Patterns of fissures are created, the grykes, separating the slabs of clints. Limestone pavements are well developed in the Burren and parts of the Pennines. 'Karst' landscapes also have many topographical irregularities, steep slopes and cliffs which provided refuges.

A bare limestone surface can be colonized (Avery, 1990). Crustose lichens colour the surface. Weathering of the limestone surface is speeded by the biological activity of the lichens, and shallow depressions are made which become occupied by prostrate mosses. As depressions develop further they are taken over by cushion mosses. Partially decomposed organic matter, especially the droppings of small arthropods, mites, springtails and insect larvae make a soil. With soil formation a thin grassland of *Festuca ovina* (sheep's fescue), with *Sesleria caerulea* (blue moor-grass) in northern England, develops. With a stabilization of the soil, and the closure of the turf, *Festuca rubra* (red fescue) becomes more important. Surface leaching then allows more phosphate to be available as the pH drops to slightly acidic. With the development of a brown earth *Agrostis capillaris* (common bent) can become dominant. Humus accumulation and incipient podzolization sees the elimination of calcicoles and the invasion of *Deschampsia flexuosa* (wavy hair grass). Erosion restores the bare limestone.

2.6.3 LIMESTONE LANDSCAPES

One kind of limestone landscape in which elements of the Late Devensian flora have been preserved is the steep-sided coombs, gorges, cliffs and crags of the Mendips in Somerset. In many of these there is an extensive development of limestone scree, or else the soils are a particularly thin rendzina. In the Mendips there is *Sedum album* (white stonecrop), a garden escapee in many places but native here, and *Lithospermum purpurocaeruleum* (purple gromwell). In different areas there are many whitebeams, species which set seed asexually and include several endemic variants: *Sorbus anglica, S. bristoliensis, S. wilmottiana, S. eminens, S. porrigentiformis, S. vexans* and *S. subcuneata*. There are many ferns, such as *Cryptogramma crispa* (parsley fern). Among the plants growing in Cheddar Gorge are *Dianthus gratianopolitanus* (Cheddar pink), *Saxifraga hypnoides* (mossy saxifrage), *Orchis morio* (green-winged orchid), *Thalictrum minus* (lesser meadow-rue) and *Mecanopsis cambrica* (Welsh poppy). *Arabis scabra* (Bristol rock-cress), *Hornungia petraea* (Hutchinsia), *Veronica spicata* (spiked speedwell), *Orobanche hederae* (ivy broom-rape) and *Carex humilis* (dwarf sedge) grow in the Avon Gorge. In west Gloucestershire *Allium sphaerocephalon* (round-headed leek) can be found.

The dales of Derbyshire and Yorkshire have similar habitats, containing some areas of rich limestone flora, survivors of the tall-herb vegetation of the Late Devensian stage (Grime, 1963). The pollen record from Lismore Fields near Buxton in Derbyshire demonstrates the survival of areas of open vegetation in the local area through the period of maximum woodland development (Wiltshire and Edwards, 1993). The ash woodlands and south-facing grasslands of the dales have a rich flora: *Trollius europaeus* (globe flower), *Hornungia petraea, Viola lutea* (mountain pansy), *Potentilla neumanniana* (spring cinquefoil), *Silene nutans* (Nottingham catchfly), *Rubus saxatilis* (stone bramble) and *Draba muralis* (wall whitlowgrass). Lathkill Dale is a notable site for *Polemonium caeruleum* (Jacob's ladder). This is a plant of screes and grassy glades in open upland woodland. Monk's Dale has *Geranium sanguineum* (bloody cranesbill) and *Parnassia palustris* (grass of Parnassus) (Scurfield, 1959).

In northern Britain the colder, wetter conditions generally favour peat formation, leaching and podzolization. Floristically rich areas are associated either with peculiar limestone characteristics or where there are open wetland habitats, flushes, stream margins and fens which provided refuges (Wheeler, 1980). Upper Teesdale is one such area (Clapham, 1978). It provides a link between the montane vegetation described in the previous section and other limestone areas, because altitude has had a role in the maintenance of its floristic richness. The main areas of richness today are parts of Cronkley Fell and Widdybank Fell, which lie at 546 m and 510 m, respectively. These were below the possible tree line of 760 m. Nevertheless *Dryas octopetala* survived on Widdybank Fell at the time of maximum woodland development. The population in Littondale, north Yorkshire must have survived on limestone crags or unstable river banks (Elkington, 1971). The vegetation history of Upper Teesdale has been worked out in detail (Turner,

1978). Twigs of *Betula* (birch), *Salix* (willow), *Juniperus communis* (juniper) and *Populus tremula* (aspen) are preserved in the peat. At lower levels, where the Cow Green reservoir now lies, the first woodland to arise in the Holocene (Postglacial) was at first dominated by *Pinus sylvestris* (Scots pine) and *Corylus avellana* (hazel) and later by *Quercus* (oak), *Ulmus* (elm) and *Alnus glutinosa* (alder).

However, there are scattered fossil records of rare herbs of open habitats. Rare plants, like *Betula nana* (dwarf birch), may have survived in areas above 760 m on the fells above Moorhouse, adjacent to the present centre of diversity, or in the open woodlands at the highest altitudes on Widdybank and Cronkley fells. The uplands were more favourable for many herbs in the first half of the Holocene, not just because of a better climate but because blanket peat was not widespread. Several species, such as *Carex ericetorum* (rare spring-sedge), would have been happy in the early pine woodland but may have found the later closed deciduous woodland too shady. Some species did not survive. *Polemonium caeruleum* (Jacob's ladder) was once present but became extinct in Upper Teesdale. *Onobrychis viciifolia* (sainfoin) has also been recorded and is now extinct here. These species may have suffered more from the later spread of blanket peat and more competitive acid-soil grasslands after the decline of the woodland, rather than from the spread of the wildwood itself.

The Whin Sill runs into Upper Teesdale, and this is by far its most botanically exciting stretch. Here it metamorphosed limestone into sugar-limestone which erodes freely into a coarse calcareous sand. It is on the eroding sugar-limestone, on the flushes and calcareous grasslands associated with them, that most rarities are found today (Figure 2.17, see plate section).

This is an analogous situation to the base-rich flushes on Ben Lawers and so it is no surprise that Upper Teesdale has a similar range of species. One rare speciality found in

both places, but hardly anywhere else, is *Myosotis alpestris* (alpine forget-me-not).

The montane flora in Upper Teesdale includes *Dryas octopetala* (mountain avens), *Saxifraga stellaris* (starry saxifrage), *Armeria maritima* (thrift), *Gentiana verna* (spring gentian) and *Helianthemum canum* (hoary rock-rose), all with a fossil record from the time of maximum woodland development (Turner, 1978). *Helianthemum canum*, which is one of the Upper Teesdale sugar-limestone specialities (Figure 2.17), also grows on Scout Scar near Kendal in Westmorland, on the Burren in Ireland and on the Great Ormes Head and the Gower coast, but the Teesdale plants are a different subspecies from the others, *Helianthemum canum* subsp. *levigatum*. It is shorter, with smaller, more hairless leaves, and fewer flowers in each inflorescence. Some other surviving Teesdale specialities have no fossil pollen record, either because they were very rare, like *Primula farinosa* (bird's-eye primrose) or else their pollen is indistinguishable from more common related species. The pollen of *Potentilla fruticosa* (shrubby cinquefoil), for example, cannot be distinguished from that of *Potentilla erecta* (tormentil) or *Potentilla anserina* (silverweed).

Potentilla fruticosa is also found in western Ireland (Elkington and Woodell, 1963). It is a species with a widespread distribution in the northern hemisphere and a wide ecological tolerance. Before the spread of the wildwood it was more widespread in the British Isles but now it is restricted to small areas in Teesdale, Wasdale and Ennerdale in the Lake District, and to the west of Ireland, especially in parts of the Burren. One thing its sites have in common is that they are all damp and open. In the Lake District it is rooted in rock ledges and crevices. In Teesdale it grows on the banks of the River Tees, where it is liable to flooding, and also in a flush above (Figure 2.18).

Almost every geological age has produced some limestone, and limestone floras are widely scattered from north to south throughout the British Isles. In north-west Scotland the Cambrian limestone runs from Durness through Inchnadamph to the limestone pavement of the Rassal ashwood at Kishorn. The Durness limestone flora has many of the same plants as Ben Lawers. In the south many interesting limestone coastal floras are located on coastal cliffs and promontories, like that on the Devonian limestone of Berry Head in South Devon or on the Carboniferous limestone of the Gower. These were all significant in providing open-habitat refuges from the wildwood for open-habitat Early Holocene herbs. In contrast, the extensive areas of chalk in the south were once covered with dense woodland.

2.6.4 LIMESTONE PAVEMENTS

Upper Teesdale shares a range of species with the Burren including *Gentiana verna* (spring gentian), *Dryas octopetala* (mountain avens), *Helianthemum canum* (hoary rock-rose)

and *Potentilla fruticosa* (shrubby cinquefoil). The Burren variants of *Potentilla fruticosa* and *Helianthemum canum* are genetically distinct from those in England (Proctor, 1957; Elkington and Woodell, 1963). The Burren variant of *H. canum* has some similarities to a variant from the Pyrenees called *Helianthemum canum* subsp. *piloselloides*, which has larger and less pubescent leaves than normal. The limestone grassland in which *Gentiana verna* (spring gentian) is found in Ireland and Teesdale share many other species (Elkington, 1963).

The Burren is the most impressive area of 'karst' landscape, with the largest continuous areas of limestone pavement in the British Isles (Figure 2.19, see plate section).

In spring and early summer, on a sunny day, it is a paradise for botanists (Dickinson, Pearson and Webb, 1964). The seemingly bare slabs of limestone rise from sea level to about 300 m. Ice-smoothed rocks and other signs of recent glaciation extend up to 180 m.

Figure 2.18 *Potentilla fruticosa* on the banks of the Tees.

In pockets on the limestone, in thin grassland and especially in the cracks in the limestone, called scailps here, there is a rich flora of species of open habitats (Ivimey-Cook and Proctor, 1966; Nelson and Whalsh, 1991). However, the pollen record from the eastern part indicates that over much of the area there was once a woodland of *Pinus sylvestris* (Scots pine), *Corylus avellana* (hazel) and *Taxus baccata* (yew) which gradually gave way to a mixed oak woodland of *Quercus* (oak), *Fraxinus excelsior* (ash), *Corylus avellana* (hazel), *Ulmus* (elm) and other trees. That woodland was only cleared in historical times. Today on parts of the limestone a dense and shady hazel scrub has developed. In some areas the woodland was growing on substantial deposits of stony boulder clay. Nevertheless species of open habitats did survive in the Burren and spread later when the woodland was cleared.

The exposure to high winds at the coast, carrying salt, may have restricted the growth of scrub and woodland, especially at Black Head. *Corylus*, for example, suffers from die-back in severely exposed situations. The most exposed north-western part of the Burren is the most treeless part today, and here there is a floristically rich *Dryas*-heath in thin soils over the limestone, which contains many other rarities. Other characteristic species of the Burren are *Asperula cynanchia* (squinancy wort), *Teucrium scorodonia* (wood sage), probably the open-habitat ecotype, *Antennaria dioica* (mountain everlasting) and, on its margins, *Geranium sanguineum* (bloody cranesbill). The Burren flora includes many orchids: *Anacamptis pyramidalis* (pyramidal orchid), *Platanthera chlorantha*, *P. bifolia* (greater and lesser butterfly-orchids), *Gymnadenia conopsea* (fragrant orchid), *Ophrys* (bee and fly orchids), *Epipactis atrorubens*, *E. palustris* (dark-red and marsh helleborine), *Listera ovata* (twayblade), *Neotinea maculata* (dense-flowered orchid), *Spiranthes spiralis* (autumn lady's tresses), *Orchis mascula* (early purple orchid) and *Dactylorhiza incarnata*

(marsh orchid). The spotted orchid occurs in two forms, the normal pink and a white one called *Dactylorhiza fuchsii* forma *okellyi* named for a local character, botanical collector and guide of the end of the nineteenth century called Patrick Kelly (Figure 2.20).

Twenty-four different species of fern are recorded, including *Phyllitis scolopendrium* (hart's tongue), *Asplenium trichomanes*, *A. adiantum-nigrum*, *A. marinum*, *A. ruta-muraria* (maidenhair and black and sea spleenworts and wall rue), *Adiantum capillus-veneris* (maidenhair fern), *Ceterach officinarum* (rusty-back), *Cystopteris fragilis* (brittle bladder-fern), *Polystichum aculeatum* (hard shield fern) and *Polypodium cambricum* (southern polypody) (Figure 2.21).

The list of species is long and even includes many calcifuges such as *Calluna vulgaris* (heather) which inhabit hummocks of peaty soil that overly the limestone in places. Some species, normally calcifuge, like *Empetrum nigrum* (crowberry), *Arctostaphylos uva-ursi* (bearberry) and *Hypericum pulchrum* (slender St. John's wort), can be found intimately mixed with *Dryas octopetala*.

Britain has no limestone landscape that can quite challenge that of the Burren, but there are extensive areas of limestone pavement in north-western England. The development of the limestone landscape in the Craven Pennines and around Malham, has been studied extensively (Clayton, 1981). Limestone pavement areas were deeply stripped by glacial action, 'plucked' of their loose and fragile top layers down to relatively solid and unfissured, unweathered lower strata. Striae, the scratch marks of glacial action, have been preserved in a few situations where the limestone was later protected from erosion by a calcareous glacial till. However, in most places the new limestone surface has been eroded, creating grikes, deep fissures, even where it has been covered by glacial till. Indeed, the process has been most rapid under leached or acid glacial till, so that the buried limestone surfaces of Derbyshire are

Figure 2.20 Orchids of the Burren: (a) *Dactylorhiza fuchsii* var. *fuchsii* and var. *okellyi*; (b) *Ananacamptis pyramidalis*; (c) *Gymnadenia conopsea*. ((b) and (c) Copyright Jon B. Wilson.)

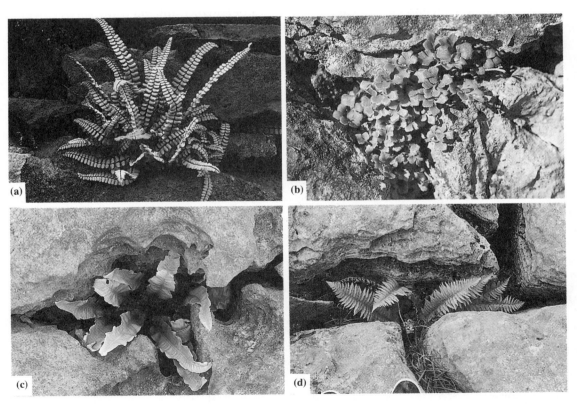

Figure 2.21 Ferns of the limestone: (a) *Asplenium trichomanes*; (b) *A. ruta-muraraia*; (c) *Phyllites scolopendrium*; (d) *Polystichum aculeatum*. ((b) Copyright Jon B. Wilson.)

more irregular, with more closely spaced grikes, than the exposed ones in north-west Yorkshire, which sometimes have massive clints, slabs of relatively uneroded limstone between grykes.

Throughout the Early Holocene epoch the pollen diagram for Malham Tarn shows that there was the usual woodland succession from *Juniperus*, through *Pinus* and *Corylus*, to *Quercus*- and *Ulmus*-dominated woodlands. The presence of the acid leaf litter of *Pinus* and *Juniperus* may have speeded the process of erosion. In time the eroded surface became more fully exposed, but there is no evidence of rapid erosion following woodland clearance. The presence of many woodland herbs, such as *Sanicula europaea* (sanicle) and *Mercurialis perennis* (dog's mercury), and *Phyllitis scolopendrium* (hart's tongue) in the grikes today might be taken as evidence for the survival of a relict woodland underflora. However, the 'tall-herb' flora of the pre-woodland Holocene landscape included 'woodland' herbs, as does the existing high-altitude flora of cliffs and screes. The grikes do provide similar conditions of shade and humidity to woodland, though probably more important today is the measure of protection they afford from grazing (Silvertown, 1983).

The grikes provide a refuge for species of a much earlier Late Devensian and Early Holocene flora. There is fossil pollen of species of open habitats, such as *Helianthemum* (rock-rose) and *Polemonium caeruleum* (Jacob's ladder), from the time of maximum woodland development, which may indicate that the woodlands were thin and patchy. Perhaps the growth of trees was limited by the high-altitude climate as well as by the absence of soil. Much of the Craven area lies above 400 m in altitude. There are also present today many open-habitat species for which bare limestone pavement is not necessarily a suitable habitat, including *Dryas octopetala* (mountain avens), *Saxifraga oppositi-folia* (purple saxifrage), *Primula farinosa*

(bird's-eye primrose), *Polygala amarella* (dwarf milkwort), *Bartsia alpina* (alpine bartsia) and *Carex capillaris* (hair sedge).

The magnificent lowland limestone pavement of Gaitbarrows, which is only 20 m above sea level and 3 km from the coast of Arnside in North Lancashire, provides an interesting comparison to the upland limestone pavements of the Pennines. The development of the vegetation is recorded in the pollen preserved in the sediments of the adjacent marl lake, called Hawes Water (Oldfield, 1960). Here, as in parts of the Burren, *Corylus* remained a very important species throughout the Early Holocene, only declining somewhat with the spread of *Ulmus* and *Quercus*. Later, *Tilia cordata* (small-leaved lime) rose to prominence. *Corylus* and *Tilia* are still important species today in the woodland in the area surrounding the pavement at Gaitbarrows (Figure 2.22).

Both these species cast a deep shade, and indeed, with the development of the woodland in this area, the pollen record shows a marked decline in herbs of open dry-land habitats.

The development of woodland cannot have been restricted by the climate in this lowland, mild-climate site. However, even here there is scattered fossil pollen showing the survival of *Artemisia* (mugwort), *Succisa pratensis* (devil's bit scabious), *Valeriana* (valerian), *Rumex* (sorrel), *Helianthemum* (rock-rose), and *Chenopodium* (goosefoot). Looking today, it is clear that although woodland can colonize the pavement, its ability to do so depends in part on the structure of the pavement. Open areas are present where there are very large unfissured clints. The survival of little-eroded very large clints is itself an indication that there has not been a covering soil. Only where grikes are close together can a tree can find purchase and stability, by growing and rooting across several grikes. Several species probably survived in open areas on the limestone: *Geranium sanguineum* (bloody cranesbill),

Figure 2.22 Gaitbarrows.

Epipactis atrorubens (dark-red helleborine), *Polygonatum odoratum* (angular Solomon's seal), *Campanula rotundifolia* (harebell), *Filipendula vulgaris* (dropwort), *Galium sterneri* (limestone bedstraw), *Minuartia verna* (spring sandwort), *Sedum acre, S. album* (biting and white stonecrop), *Briza media* (quaking grass), *Asplenium ruta-muraria* and *A. trichomanes* (wall rue and maidenhair spleenwort).

2.6.5 THE WETLAND REFUGE

It is likely that many open-habitat species of limestone areas survived 6000 years ago, not on the limestone itself, but at the margins of associated wetlands. In the Burren there are the temporary lakes, the turloughs. Turlough literally means 'dry lake', a name used because at times the lakes consist mainly of wide, grassy hollows. When it rains heavily, water, draining from the limestone nearby, wells up to fill the lakes. There can be a 9 m change in water level in the turlough. In winter the turloughs can remain filled for weeks. Growing on the bed of the turloughs is a sward of *Potentilla anserina* (silverweed), *Hydrocotyle vulgaris* (marsh pennywort) and *Carex nigra* (common sedge) and the moss *Drepanocladus sendtneri*. The upper limit of flooding is marked by the presence of *Potentilla fruticosa* (shrubby cinquefoil). Regular flooding prevents the colonization of trees. There is a transition between the turloughs and a string of permanent lakes (Dickinson, Pearson and Webb, 1964; Nelson and Whalsh, 1991).

Wetlands were once very widespread (Figure 2.23). The Somerset Levels (Figure 2.24) are one of the few surviving areas regularly inundated.

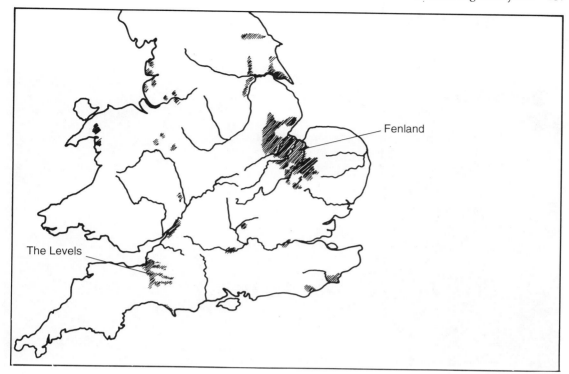

Figure 2.23 The distribution of major wetland areas in England and Wales. (From Purseglove, 1989.)

At the margins of more permanent bodies of water there are other wetland communities. Those developed under the influence of the calcareous ground water include *Schoenus nigricans* (black bog-rush), *Cirsium dissectum* (meadow thistle) and *Carex panicea* (carnation sedge). In wetter areas *Leontodon autumnalis* (autumn hawkbit) and the moss *Scorpidium scorpiodes* grow. Similar kinds of communities are widespread in Britain around springs or seepage lines from calcareous bedrock or overlying glacial drift with a high calcareous content. Common species are *Succisa pratensis* (devil's bit scabious), *Potentilla erecta* (tormentil), *Molinia caerulea* (purple moorgrass) and the mosses *Campyllium stellatum* and *Calliergon cuspidatum*.

This is only one of many kinds of mire communities in Britain (Figure 2.25).

The National Vegetation Classification (Rodwell, 1992a) records 38 distinct communities and many more subcommunities. A large part of the diversity of mire communities is related to the height of the water table, how much the water table fluctuates, and water quality. The diversity is also related to the kind of management or grazing regime these mires experience. Changes in the hydrological conditions, as a consequence of deforestation, can have a profound effect on species composition. Reduced soil aeration and water-holding capacity lead to accelerated surface run-off. Snow melt is quicker, leading to a brief spring flood stage. Lake-margin communities have the potential for rapid change in species assemblages. Nevertheless, many mire communities are some of our most natural kinds of vegetation, showing clear zonations. Mire vegetation is widespread, and although it has been much

Figure 2.24 A view of the Somerset Levels at time of flood (courtesy Ian Hodge).

reduced in recent years because of drainage, 6000 years ago it provided a very significant refuge from the wildwood for shade-intolerant species (Daniels, 1978).

2.6.6 AQUATIC PLANTS

Water quality can vary from dystrophic (0–2 p.p.m. $CaCO_3$ and pH < 6.00) through oligotrophic (0–10 p.p.m. $CaCO_3$ and pH 6–7), mesotrophic (10–30 p.p.m. $CaCO_3$) to eutrophic (>30 p.p.m. $CaCO_3$, pH > 7.0). Oligotrophic waters tend to be clear, with a thin population of aquatic algae, because they are nutrient poor and unproductive. In contrast, eutrophic waters can become a green soup of algae.

Other important characteristics for determining vegetation development are the speed of water flow, the depth of the water

and the degree of exposure. These can influence the amount of mixing of the water and therefore the distribution of nutrients and oxygen. In still or stagnant eutrophic waters light does not penetrate far. For aquatic margin communities, the degree of exposure to wave action is critical.

Aquatic plants with long linear leaves, long flexible stems and petioles, and with dissected or linear leaves are found in faster-flowing water. Commonly, aquatic plants readily produce strong adventitious roots and rhizomes, which anchor them to the substrate. This ability, combined with a propensity for the plant to fragment, makes vegetative reproduction especially common in aquatic plants. Fragments of stem or rhizome are carried by water flow and root readily where they come to rest. It is partly because they can reproduce clonally so

Figure 2.25 Aquatic and wetland vegetation: (a) see plate section; (b) view of mature fen system at Sutton Broad, Norfolk. (Copyright Jon B. Wilson.)

readily that hybridization is particularly common in aquatic plants. Hybrids often have reduced sexual fertility but in aquatic plants even sterile hybrids can survive for a long time and gain a wide distribution with vegetative fragments spread by flowing water.

Different species of pondweed (*Potamogeton* spp.) are distributed depending on the flow rate, the base status and nutrient conditions of the water. The water-crowfoots (Ranunculaceae), relatives of the buttercups, are also important. Both the water-crowfoots and pondweeds produce different-shaped submerged and floating leaves. There is a startling change from a linear leaf in the pondweed and a finely dissected one in water-crowfoots in submerged conditions to

a broad and entire floating leaf. Both submerged and floating leaves can be found on the same plant, but different species vary in their ability to produce each kind of leaf.

There is a clear ecological separation of different *Potamogeton* species. *Potamogeton pectinatus* (fennel pondweed), with linear leaves, is found in eutrophic waters. *Potamogeton crispus* (curled pondweed) is present in slightly less eutrophic waters. Aquatic plants have a very thin cuticle, allowing the uptake of inorganic nutrients, but some, like *P. crispus*, have well-developed stems so that they can obtain nutrients from the silt they are rooted in. In stagnant waters *P. natans* (broad-leaved pondweed) can be present. *Potamogeton coloratus* (fen pondweed) favours the most calcareous waters, and *P. polygonifolius* (bog pondweed) acid waters.

One problem that aquatic plants face is the lower level of light in aquatic environments. At the surface, 5–25% of light is reflected, and more is reflected by silt particles or plankton in suspension in the water. In still, eutrophic water this can be made worse by the presence of a floating layer of *Lemna* (duckweed) and by floating filamentous or thallose algae such as *Enteromorpha*.

A particularly beautiful aquatic species in some fen waters is *Hottonia palustris* (water-violet) (Figure 2.26). Like an aquatic primrose, it flowers early, thereby avoiding the shady conditions later in the season when the canopy of other aquatics has fully developed. Other aquatic species are shade tolerant and some are able to photo-synthetize even in only 1% of available light.

In stagnant waters low oxygen levels can be a problem to aquatic plants. Plants with surface or emergent leaves, like the water-lilies or rushes and reeds, can conduct oxygen to their roots via an aerenchyma tissue rich in air passages. Lack of oxygen is especially damaging in the seed or seedling stage. Some plants, such as *Phragmites australis* (common reed), can tolerate low oxygen levels but only grow when oxygen is present (Crawford, 1989). *Scirpus lacustris* (common club-rush) seems to be able to grow in anoxic conditions but actually only extends the stems from buds preformed in more oxygen-rich conditions. *Glyceria maxima* (reed sweet-grass) dies within 7–15 days in anoxic conditions. Alternatively, carbon dioxide can be in short supply. Diffusion rates in water are 10 times slower than in air. Some aquatics such as *Littorella uniflora* (shoreweed) have 'aquatic acid metabolism' which allows them to garner scarce carbon dioxide. In calcareous water some aquatics can utilize the bicarbonate ions which are readily available as a carbon source.

2.6.7 MARL LAKES

Marl lakes have a high pH and very high levels of calcium carbonate ($CaCO_3$ >100 p.p.m.). Malham Tarn is the only upland marl lake in Britain. Marl is precipit-ated calcium carbonate, produced in calcium-enriched waters on the removal of carbon dioxide by photosynthesizing organisms, mainly aquatic algae. The pH of the water is between 8.0 and 8.5. The algae are particularly diverse. A stonewort, *Chara delicatula*, is the most important. There are extensive beds of aquatic vegetation including pondweeds, especially *Potamogeton lucens* (shining pond-weed), *Myriophyllum verticillatum* (whorled water-milfoil) and the moss *Fontinalis anti-pyretica*. With their high level of dissolved calcium carbonate, marl waters have a very low nutrient status because nutrients such as phosphate are precipitated out of solution (in the latter case as calcium phosphate) along with the marl. This reduces the potential productivity of free-floating algae so that the water is particularly clear, resembling acid (oligotrophic) waters in this respect.

Another feature Malham Tarn shares with upland acidic lakes is the lack of a marginal reedbed, except for a small patch near the inflow to the fen (Proctor, 1974). Young *Phragmites australis* (common reed) is frost sensitive. In the warmer conditions of the Atlantic stage there was a well-developed reedbed. *Carex rostrata* (bottle sedge) colonized the marl at times when the tarn was shallower (Piggott and Piggott, 1959). Today it is found in sheltered bays around the tarn.

Carex rostrata, *C. elata* (bottle and tufted sedge) and *Phragmites australis* are present around the lowland marl lake of Hawes Water near Gaitbarrows. The marl enriches the soil of the exposed parts of the lake margins encouraging a calcicolous community. In a small, drier patch on the margin of Hawes Water there is a rich wet assemblage which includes *Primula farinosa*

(a)

(b)

Figure 2.26 Aquatic and marsh plants: (a) *Iris pseudacorus* (yellow iris); (b) *Potamogeton polygonifolius* (bog pondweed); (c) *Hottonia palustris* (water-violet); (d) *Lobelia dortmanna* (water-lobelia). ((a) and (d) Copyright Jon B. Wilson.)

(d)

(c)

(bird's-eye primrose), *Parnassia palustris* (grass of Parnassus) and *Gymnadenia conopsea* (fragrant orchid).

Marl lakes are a rare feature of the British Isles today but they were more widespread in the past. In Ireland they are more common than in Britain and several large lakes have relatively high base status. These include Loughs Owel and Derraveagh in West Meath and lakes of the Corrib complex in Connemara (O'Connell, Ryan and MacGowran, 1987). In the Fenland of East Anglia there used to be numerous marl lakes, the meres, some of which were very large. Holme Fen near Peterborough developed at the margin of one called Whittle Mere, which covered 607 hectares. The largest was Redmere near Lakenheath in Norfolk. Only a few survive today, but their former number and extent is indicated by the marl-enriched soils they left behind. The areas used agriculturally are marked by lime-induced manganese deficiency which can reduce crop yield if not rectified by spraying. The drainage of the meres is described later.

2.6.8 THE HYDROSERE

There is a natural succession from open-water to dry-land communities, called the hydrosere. A complete hydrosere is really a fiction but zonations are commonly encountered which represent parts of the hydrosere (Hardy, 1939; Sinker, 1962; Walker, 1970). Deep water is occupied by *Chara* (stonewort) and *Myriophyllum* (water-milfoil). In the early stages of the hydrosere the *Potamogeton* species (pondweeds) are commonly important. In the past 100 years various introduced water weeds, such as *Elodea canadensis* (Canadian pondweed), have become important. In shallow water, plant detritus builds up and rooted aquatics trap silt so that the soil surface rises closer and closer to the surface of the water.

An interesting group of amphibious plants grows on the shores of lakes in places where the more vigorous reeds and grasses are restricted because of wave exposure or liability to large changes in water level. Shore plants can grow submerged or on the wet soils of the exposed shore. They include *Littorella uniflora* (shoreweed), *Hypericum elodes* (marsh St. John's wort), *Corrigiola littoralis* (strapwort), *Limosella* spp. (mud-wort), *Elatine* spp. (waterwort) and *Lobelia dortmanna* (water-lobelia) (Figure 2.26d). These plants do not flower normally when submerged. *Lobelia dortmanna* produces closed self-pollinating flowers in submerged conditions and open flowers when exposed. There are also a number of spore-producing aquatics, such as *Isoetes* (quillwort) and *Equisetum palustre* and *E. fluviatile* (marsh- and water-horsetail).

In stable conditions, a community of free-floating aquatic plants gradually gives way to one dominated by rooted aquatics such as *Typha latifolia* (bulrush), *Phragmites australis* (common reed) and sedges (Cyperaceae). Within the influence of fluctuating ground-water levels, especially where the water is base rich, a richer reed-swamp and fen develops. Forming a boundary at the edge of the fen, a highly competitive and species-poor reed- or sedge-bed is often found. Most commonly *Phragmites australis* (common reed) forms reed beds in the fairly nutrient-rich conditions now common in the lowlands. Commonly, *Phragmites* is dominant, forming an almost single-species stand. *Phragmites* is highly competitive. It has a dense network of horizontal and vertical rhizomes and erect stems. A dense horizontal mat of roots arises from the upper parts of the rhizomes (Haslam, 1972); this often forms the fibrous part of *Phragmites* peat.

Phragmites can grow in deeper water than sedge communities, but in many slightly calcareous but otherwise nutrient-poor fen systems, with pH 6.5–8.0 and low in nitrate-nitrogen and phosphate, *Cladium mariscus* (great fen-sedge) dominates the water margin, producing large plants, and a dense sedge-bed. Other vegetation may be suppressed by *Cladium* leaf litter, which is particularly resistant to decomposition. *Carex paniculata* (greater tussock-sedge) is particularly important in fairly base-rich and fertile conditions. In land-locked hollows with a fluctuating water table *Carex elata* (tufted sedge) may be dominant. In poor, acid fens, with pH 5.5–7.3, *Carex rostrata* (bottle sedge), *Potentilla palustris* (marsh cinquefoil), *Menyanthes trifoliata* (bogbean) and *Eriophorum* spp. (cotton grass) are common.

As alluvium and leaf litter accumulates, a fen flora can develop. Fens do not develop just beside open water. They can arise in several different situations and have different floras, depending upon the situation. In hollows where drainage is impeded a 'topogenous fen' can develop. In a small, enclosed hollow a 'basin mire' can develop. Along a soakway or seepage zone, where there is lateral movement of water, a 'soligenous fen' can develop. These exist on hill slopes and valley sides. In the valley bottom a 'valley mire' develops. Where the valley is broad, in the poorly drained flood plain of a river, a 'flood-plain mire' develops. In many situations fens are only a transient vegetation. Rooted reeds and sedges raise the ground surface further so that a wet woodland, carr, develops. Elements of the herb flora may survive within the shade of the carr (described below).

Phragmites australis (common reed) is not very shade tolerant and is soon eliminated from the woodland as the canopy closes, surviving longest where *Alnus* is not strongly developed and the canopy is mainly patchy *Betula pubescens* (downy birch) and *Salix cinerea* (grey willow) (Rodwell, 1991; W1, W2, W4). The tussock sedges are slightly more shade tolerant than *Phragmites* so that below the alder canopy the fen community can survive in modified form. However, as the woodland develops many important species of the open fen, such as *Cladium mariscus*

(great fen-sedge), *Glyceria maxima* (reed sweet-grass) and *Juncus subnodulosus* and *J. acutiflorus* (blunt-flowered and sharp-flowered rush), are eliminated. Others, such as *Thelypteris palustris* (marsh fern) and *Calamagrostis canescens* (purple small-reed) grow on in the fen woodland as abundantly as in the open fen (Rodwell, 1991; W5). Some tall herbs are especially likely to survive, along with *Carex paniculata* (greater tussock-sedge) or the more shade-tolerant sedge, *Carex acutiformis* (lesser pond-sedge).

The closing of the tree canopy can be accompanied by a rich development of fern tussocks of *Athyrium filix-femina* (lady fern), *Dryopteris dilatata* (broad buckler-fern), *Osmunda regalis* (royal fern) and *Thelypteris palustris*. Shrubby species such as *Rubus fruticosa* agg. (brambles), *Lonicera periclymenum* (honeysuckle) and *Rosa canina* agg. (dogrose) can form a tangled underscrub shading out the herbs. *Galium palustre* (common marsh-bedstraw) and *Solanum dulcamara* (bittersweet) or, in more fertile conditions, *Galium aparine* (cleavers) and *Convolvulus arvensis* (common bindweed), scramble over the vegetation. There are geographical differences in fen communities. Wicken Fen has little *Alnus glutinosa*, probably because of the continuous cutting it has experienced, but in the Broads *A. glutinosa* is the most important tree.

Fen woodlands are generally developed on a very unstable peat surface. The weight of the woodland can compress the surface so that the height of the fen surface above ground water does not increase despite the accumulation of peat. There may even be a degeneration of the fen surface so that it floods and *Alnus glutinosa* and tussock ferns die. Where *Alnus glutinosa* trees are rooted in sedge tussocks these can be unstable, rolling over, tilting the trees aside, so that the canopy is opened up for the colonization of light-demanding species. However, if the surface is not too wet, *Salix cinerea* (grey willow), *Betula pubescens* (downy birch) and

Frangula alnus (alder buckthorn) can take advantage by colonizing the gaps.

As important for the survival of a rich herb flora, the full fen succession has been modified by centuries of different kinds of management, described in the next chapter, maintaining the open conditions required by many fen herbs.

2.6.9 FROM FEN TO BOG

The topogenous peatland of a fen, which is irrigated by ground water, may be replaced by a different kind of open community, a rain-fed (ombrogenous) bog, as the ground surface rises away from the influence of ground water. The mosses *Drepanocladus*, *Bryum* and *Scorpidium* are replaced by the bog moss, *Sphagnum*. Soil conditions become more acid. A transition fen may be developed. In some fens a superficial layer of peat and *Sphagnum* (bog moss) between sedge tussocks is developed. Small sedges root in the *Sphagnum* layer and many acid-loving species establish. Because the peat is light, it floats on the water surface and keeps the roots of small plants away from the influence of ground water. The whole community can physically rise with a rising water table so that it does not drown.

Three major types of peat soils can be identified: fen mires, raised bogs and blanket bogs. Fen mires are influenced by ground water and have higher mineral nutrient levels. They are minerotrophic, either relatively highly or moderately mineral rich, eutrophic or mesotrophic. In raised bogs the peat is rain-fed and mineral deficient, ombrotrophic. *Sphagnum* (bog moss) species dominate. Blanket bogs are similar, but develop on top of mineral soils. They have a vegetation succession associated with the accumulation of fen peat and the rise of the peat surface away from the influence of ground water so that it becomes ombrotrophic. *Phragmites australis* (common reed) and sedges (Cyperaceae) give way to water-tolerant

shrubs and trees. A rise in the fen surface of only 25 cm at Askham Bog, near York has led to a succession away from a fen mire with a rich ground flora to an acid bog woodland of *Betula* and *Quercus* with *Sphagnum* and *Molinia caerulea* (purple moor-grass) dominant in the herb layer.

This kind of succession is seen in pollen diagrams from many fen/bog sites (Poore, 1956; Godwin, 1984). *Phragmites* peat is frequently recorded under *Sphagnum* and *Eriophorum* peat, as at Tregaron Bog (Godwin and Newton, 1938b). At Shapwick Meare in the Somerset Levels the early succession was from reed-swamp to birch fen-wood and then as the humus layer built up further away from the influence of ground water there was a transition from a rheotrophic (rock-loving) or topogenous (ground-forming) fen to an ombrotrophic (rain-loving) or ombrogenous (rain-forming) bog (Godwin, 1941). An alder carr, rich in *Lysimachia vulgaris* (yellow loosestrife) gave way to poor fen communities with *Carex rostrata* (bottle sedge) but including *Equisetum* (water-horsetail) and *Juncus acutiflorus* (sharp-flowered rush) or *Molinia caerulea* (purple moor-grass) and thence to more acid-loving open communities including *Eriophorum* spp. (cotton grass) or *Sphagnum* pools and hummocks.

Borings in Tarn Moss at Malham show a layer of marl at the lowest level. Above that there is a layer of fen peat. At a level deposited 8000 years ago there was a rapid rise of *Alnus* (alder) pollen. Almost at the same time there was the rise of *Sphagnum*, showing the beginning of development of a raised bog. Later the *Alnus glutinosa* was eliminated, even though, as shown by a layer of humified peat, conditions were relatively dry. Above this layer there is a thick, almost unhumified, peat rich in *Sphagnum imbricatum*, a hummock moss. Near the surface, the peat is more humified again, and becomes richer in the remnants of *Eriophorum vaginatum* (hare's-tail cotton-grass). At Wybunbury Moss there is a layer of fen mud containing fragments of *Phragmites australis* (common reed) and *Menyanthes trifoliata* (bogbean) (Green and Pearson, 1968b). Wood fragments are common, showing the development of a wet woodland containing *Alnus glutinosa* (alder), *Salix* spp. (willow) but also *Pinus sylvestris* (Scots pine), *Corylus avellana* (hazel), *Betula* (birch), *Malus sylvestris* (apple), and *Cornus sanguinea* (dogwood). The present woodland is a tangle of *Lonicera periclymenum* (honeysuckle) and *Rubus fruticosa* (brambles). The presence of *Eriophorum* sp. (cotton grass) and *Vaccinium oxycoccus* (cranberry) may show the transition to more ombrotrophic bog conditions.

Even on surviving fens, on a raised part, acid-loving plants like *Myrica gale* (bog myrtle), which normally grow in areas with pH 5.0–6.0, can be found. The northern part of Woodwalton Fen has a remnant acid bog vegetation. This represents the transition from a soil supplied by base-enriched ground waters to one supplied by neutral/acidic rain, low in nutrients. In the Norfolk Broads at Buckenham Broad the species of bryophytes present show the long-term survival of a mainly neutral fen, which fluctuated between the slightly base-rich fen, indicated by the mosses *Drepanocladus revolvens*, *Acrocladium giganteum*, *Campyllium stellatum*, *Campto-thecium nitens* and *Cratoneuron commutatum*, and the more acid bog, containing the bog mosses *Sphagnum palustre*, *S. teres* and *S. acutifolium*. In some of the Broadland valleys there are lenses of reed and sedge peat in the mixed brushwood peat, showing an alternation of carr with more open fen and reed swamp communities.

The fen to bog succession did not develop fully in many cases. Borings from the lowland mosses around Gaitbarrows and Hawes Water show no transition to *Sphagnum* bog. The influence of calcareous ground water was maintained in valley mires along the line of a calcareous stream. In some valleys in the New Forest, alder carr is maintained in a central strip with parallel bands of more and

more acid vegetation to either side (Rodwell, 1992a). Similarly, near the margin of a bog where ground waters seep in from adjacent mineral-rich high ground, floristically enriched communities survived which included *Alnus glutinosa* (alder), *Cladium mariscus* (great fen-sedge), *Carex paniculata*, *C. diandra* (tussock sedges), *Solanum dulcamara* (bittersweet) and *Lycopus europaeus* (gypsywort). In addition, the development of bogs was halted or reversed either because of subsidence or because of changes in water level, caused by variation in precipitation, drainage and, in coastal lowlands, by changes in sea level.

Nevertheless, the development of raised bogs, like the development of fen-woodlands, greatly decreased the areas that could serve as refuges for herbs of open habitats. The timing of the development of raised bogs was not the same at all sites, so that many fens survived for many thousands of years, providing a continuous refuge for shade-intolerant plants.

2.6.10 BOGS

Species-poor acid bog communities, dominated by *Sphagnum* spp. (bog moss), can develop almost anywhere in the British climate (Daniels, 1978). Again and again, it is possible to see the results of a transition from a fen to a bog. The latter is fed only by rain and is rich in hydrogen ions and sulphate ions. Primary peat forms in basins or depressions. In time the primary peat acts as a reservoir for the development of secondary peat which lifts the bog surface away from the basin. Finally, a tertiary peat may develop above the influence of ground water. The tertiary peat grows, forming a raised bog, until the surface dries out.

The undulating Postglacial landscape was rich in ill-drained hollows which were the foci for fen and then raised bog development. Some hollows were once occupied by huge blocks of ice left by the retreating ice sheet.

Others developed in valleys dammed by moraine. Lowland central Ireland, covered by calcareous glacial till, was particularly rich in such sites. Similarly, many fens and bogs developed in depressions in the glacial drift that covered the Cheshire, North Staffordshire and North Shropshire plain. Here in particular some sites were subsidence hollows produced by the solution of deep pockets of salt deposits.

The continued subsidence of these hollows at Wybunbury Moss and at Chartley Moss led to the peculiar situation where the bog collapsed, breaking horizontally into two parts so that the top layer floated on a lake (Green and Pearson, 1968a). At Chartley Moss the lake, which is 10 m deep, breaks the surface and there are lenses of water within a more continuous peat. At Wybunbury the light, unhumified upper peat floats like a raft on the lake, which is mostly hidden. The denser, well-humified lower fen peat, which arose early in the succession from fen to bog, lines the lake bottom. The bog surface continues to grow actively. *Sphagnum recurvum* is the most important *Sphagnum* species. Other plants are the normal ones found in acid bogs. Chartley Moss is the southernmost limit of *Andromeda polifolia* (bog rosemary). It is possible, though a little dangerous, to 'walk on water' across the moss surface, but there are no rights of access because of the danger and to preserve the delicate bog structure. Rooted in drier areas of the floating bog are rather stunted *Betula* (birch) and *Pinus sylvestris* (Scots pine) trees. Jumping up and down causes waves to roll across the bog so that the tree trunks wave backwards and forwards as a wave passes under them.

Raised bogs grow especially fast in wetter periods. In the wetter conditions of western Britain a direct transition from reed swamp through poor reed/sedge fen to *Sphagnum*-dominated mire can occur. The important species in this kind of succession is often *Carex rostrata* (bottle sedge). Other species are

Potentilla palustris (marsh cinquefoil), *Carex nigra* (common sedge), *Eriophorum angustifolium* (common cotton-grass) and *Succisa pratensis* (devil's bit scabious). The most important *Sphagnum* species present are *Sphagnum squarrosum* and *S. recurvum*. A direct succession from fen to bog without the intervening carr stage is recorded in a sequence at Goldcliffe in Gwent from 5750 ^{14}C years ago.

Above the bottom layer of fen mud at Wybunbury Moss, there are thick layers of peat, alternating highly humified and slightly humified. These different kinds of peat were formed in different conditions, either in pools or on *Sphagnum* hummocks, or at different times, in very wet or drier times. Unhumified peat is formed in the anaerobic conditions of dystrophic or oligotrophic *Sphagnum* pools. It is largely undecomposed and light brown in colour. Humified peat, which is dark brown, is formed on *Sphagnum* hummocks where the greater aeration, because of the drier conditions, allows part decomposition of the peat. Two layers can be discerned in the peat. The lower layer, or catotelm, is the compressed peat which acts as a reservoir of captured rainwater. The upper layer, which includes the living surface of *Sphagnum*, is called the acrotelm. It is only 10–50 cm thick but it is the layer in which other plants root.

In many areas acid peat communities and richer calcareous communities may form a mosaic vegetation. Calcicole and calcifuge plants can be found growing side by side: calcicoles on mineral soil or where there is calcareous ground water and calcifuges on peat. As the bog grows, the flow of water becomes canalized and diverted away from the peat basin. The bog becomes more heavily influenced by rainfall, either directly or from seepage. The mire surface becomes untouched by moving water as the raised bog accumulates. A rich fen commonly has a slightly basic pH of about 7.5 and a raised bog a highly acidic pH of about 3.8. With increasing acidity the species composition of the bog changes.

There are variations in bog vegetation which have been ascribed to a characteristic cycling of the vegetation as the *Sphagnum* hummocks build and decay (Clymo and Hayward, 1982). It has been suggested that in some conditions the *Sphagnum* hollows are converted to hummocks by the gradual build up of peat, while the greater decomposition of peat on the hummocks converts these into hollows. It is unlikely that the alternation between hollows and hummocks is an inherent part of bog development. If it does happen, it may take a very long time, over a time span in which it is difficult to see whether there is a regular cyclic change, or whether the change is influenced by fluctuations in rainfall or changes in drainage patterns.

On the margins of the bog there is better drainage so that a more compact humified peat is formed. In time the bog rises to form a dome, rising perhaps 6 m from the surrounding mineral soil surface, often marked by concentric rings of hollows and hummocks. In the wet hollows *S. cuspidatum* is dominant. *Sphagnum tenellum* and *S. papillosum* replace it as the hummock builds. Eventually these are replaced by *Leucobryum*, *Rhacomitrium* and *Cladonia* on top of the hummock. There is also a succession of flowering plants from *Narthecium ossifragum* (bog asphodel), *Eriophorum* spp. (cotton grass), through *Erica tetralix* (cross-leaved heath) and *Trichophorum cespitosum* (deergrass) to *Calluna vulgaris* (heather) and *Vaccinium myrtilus* (bilberry) on top of the hummock. The quality of the ground water may influence the development of the bog. The flow of water determines the nutrient status and oxygenation of the water. Ground water may contain calcium and bicarbonate ions which will counter the effect of the hydrogen ions produced by biological activity. The hydrogen ions react with bicarbonate to give carbon dioxide and water or, more simply, they may be flushed away in the flow.

The different conditions in which a bog is forming are indicated by the different species of *Sphagnum* present. In pools in the hollows and forming soft, wet lawns is *Sphagnum cuspidatum* (Rodwell, 1992a; M1), along with *S. auriculatum* (Rodwell, 1992a; M2) in the very oceanic western and south-western parts of Britain. There is a transition to other species of *Sphagnum*, through *S. tenellum*, *S. pulchrum*, *S. papillosum* and then to *S. magellanicum*, *S. capillifolium*, *S. rubellum* and *S. imbricatum*, which prefer the drier, better aerated and exposed hummocks. Dryness here is relative because, although the hummocks can grow up to 0.5 m high, bog moss draws up water.

Sphagnum peat can hold 5–600% water by weight. The leaves and stems of *Sphagnum* have hyaline cells; cells with pores open to the outside, and empty and dead at maturity, which act as water reservoirs. The structure of *Sphagnum* also turns each plant into a wick, drawing up water. The stems have tightly overlapping leaves and some branches hang down close to the stem aiding this effect. On tops of the hummocks other mosses, such as *Racomitrium lanuginosum*, and *Cladonia* lichens are frequent.

2.6.11 THE ADAPTATIONS OF BOG PLANTS

Sphagnum-dominated bog communities are very species-poor (Figures 2.27, 2.28).

The conditions for vascular plant growth on a raised bog are very difficult. They are very acid because of the accumulation of acidic by-products of plant metabolism, because rainfall is slightly acidic and because of the action of *Sphagnum* itself (Clymo and Hayward, 1982). *Sphagnum* exchanges hydrogen ions for the positive ions of mineral nutrients, thereby acidifying the water percolating through the bog. There are very low nitrates and phosphates, with little input from rain and a lot locked away because of the slow breakdown of peat. *Andromeda polifolia* (bog rosemary) and *Rhynchospora alba*

Figure 2.27 *Sphagnum* bog: (a) Thursley Common, Surrey; (b) Wybunbury Moss, Cheshire. ((a) Copyright Jon B. Wilson.)

Figure 2.28 Bog plants: (a) *Narthecium ossifragum* (bog asphodel); (b) *Eriophorum angustifolium* (cotton grass); (c) *Anagallis tenella* (bog pimpernel); (d) *Drosera anglica* (great sundew). (Copyright Jon B. Wilson.)

(white beak-sedge) grow in the most nutrient-poor conditions. *Eriophorum vaginatum* (hare's-tail cotton-grass), which grows on deep peat, has a remarkable ability to concentrate mineral nutrients, especially phosphorus (Wein, 1973). Phosphorus, as phosphates, and potassium are conserved by being translocated back from the leaves before leaf fall. *Myrica gale* (bog myrtle) is unusual in this community in having nitrogen-fixing nodules on its superficial roots. Insectivorous plants, such as the sundew (*Drosera* spp.), butterwort (*Pinguicula* spp.) and bladderwort (*Utricularia* spp.) are common. *Drosera rotundifolia* (round-leaved sundew) favours the high ridges, *D. longifolia* (great sundew), the low ridges, and *D. intermedia* (oblong-leaved sundew), the hollows.

Eriophorum angustifolium (common cotton-grass) is interesting because it is one of the few species that can grow on both acid and base-rich soils (*Festuca ovina* is another) (Phillips, 1954). *Eriophorum angustifolium* shows enhanced growth in the presence of calcium, and at higher temperatures it behaves more as a calcicole. It spreads very effectively by its rhizomes, colonizing bare peat and mud pools. *Eriophorum angustifolium* was probably a very significant element in the development of mires in the warm, wet Atlantic stage of the Postglacial. *Eriophorum* peat is widely recorded in bog and mire stratigraphies. Mires have developed directly on mineral soils in shallow depressions and wide valleys and even on flat or gently sloping ground. The widespread development of blanket mires is often associated with disturbance of a natural woodland by man. With the lack of transpiration by trees, trampling by animals created muddy pools providing foci for colonization by *Eriophorum angustifolium*. The development of blanket mires is described in the following section.

Dystrophic pools are found in the peaty situations of *Sphagnum* bogs. The water is brown with a suspension of plant detritus. Low light levels inhibit photosynthesis and breakdown of the detritus is inhibited by the lack of nutrients and oxygen. There are slightly higher levels of oxygen at the surface, where turbulence mixes the water. Here breakdown of detritus allows some release of nutrients, which in the summer surface algae take advantage of, augmenting oxygen levels. Therefore, in the summer the breakdown of organic material is enhanced, giving rise to a fine residual sediment called 'Dy'. One of the few plants to occupy these waters is *Utricularia* spp. (bladderwort), which survives the low nutrient conditions by trapping and digesting minute water animals in its bladder-like traps. In well-oxygenated water, peat is formed slowly and is heavy. It sinks, allowing water to flow over it. Where the water is stagnant or oxygen-poor the peat forms rapidly. It is light and floats on the surface. A mat of peat may encroach over the surface of the water, eventually covering the surface. It may become thick enough to support a well-developed living plant community.

If there is some lateral movement of water, through runoff from surrounding mineral soils, bringing more eutrophic conditions, bog-pool communities (Rodwell, 1992a; M1 and M2) are replaced by a bog community with a carpet of sedges, especially *Carex rostrata* (bottle sedge) but also *C. nigra* and *C. curta* (common and white sedge). Other plants are *Juncus effusus* and *J. acutiflorus* (soft and sharp-flowered rush), *Eriophorum angustifolium* (common cotton-grass) and *Agrostis canina* and *A. stolonifera* (velvet and creeping bent) and *Molinia caerulea* (purple moor-grass). The mosses are *Sphagnum cuspidatum*, *S. auriculatum* and *S. palustre* and also the robust *Polytrichum commune* (Rodwell, 1992a; M4). In more nutrient-rich conditions this gives way to a community in which *Sphagnum squarrosum* is the dominant bog moss. *Sphagnum* has pale-green tufts of the moss *Aulocomnium palustre* poking out of it. There is a slightly richer herb community, including *Potentilla palustris* (marsh cinquefoil) and even *Succisa pratensis* (devil's bit scabious) (Rodwell, 1992a; M5). In the most nutrient-rich conditions along seepage lines and channels there is a richer herb flora, with a mixture of short sedges or taller rushes and a range of other species. *Sphagnum recurvum* is more abundant.

Rooting conditions in a bog are very water-logged and anaerobic. Some species, such as *Juncus effusus* (soft rush), have air channels in their roots to oxygenate the growing root-tip. Plants of the pools, such as *Eriophorum* spp. (cotton-grass) and *Rhynchospora alba* (white beak-sedge), have a well-developed aerenchyma which allows them to grow. Some grasses and sedges have two kinds of roots: superficial unlignified ones are produced as conditions dry in spring, but die back in winter. In wetter conditions, generally in western Britain, *Andromeda polifolia* (bog rosemary) and *Narthecium ossifragum* (bog asphodel) are more common. A lateral move-ment of water may bring more aerated condi-tions. *Narthecium ossifragum* is tolerant of the strongly reducing conditions but grows best where the oxygen concentration is 0.6–7.1 p.p.m., preferring flushed habitats (Summerfield, 1974).

The shrubby species *Calluna vulgaris* (heather), *Erica tetralix* (cross-leaved heath), and *Vaccinium oxycoccus* (cranberry) produce superficial roots and can survive on the hummocks. In drier conditions *Calluna vulgaris* is the dominant higher plant. For example, in a small raised bog there is usually a lagg, a drainage stream, which forms a boundary. Drainage is improved along the channel and a line of *Calluna* often takes advantage of this. On the slopes above the lagg, trees such as *Betula* (birch) and *Pinus sylvestris* (Scots pine) can grow because of more aerated conditions. The lowland heaths and upland bogs share many other calcifuges. There is a continuous spectrum of types from dry to wet peat communities. A wide range of blanket peat communities with different altitudinal and geographical distributions can be recognized (Rodwell, 1992a).

2.6.12 COASTAL REFUGES

Although saltmarshes have low diversity, other coastlands have provided one our richest refuges for herbs of open habitats. Even today the coastlands provide perhaps the largest, and least modified, most natural vegetation in the British Isles. Even areas that have become highly developed may have patches of botanical importance. In some cases the preservation or modification of coastal dune systems for golf courses has preserved the refuge. Part of the value of the coast as a refuge is the diversity of habitats it offers. One-third of Britain's species grow in the vicinity of the coast.

In some respect the coastlands combine all the features that encourage the preservation of open herb habitats; exposure, magnified by the harmful effects of salt spray; wetlands in saltmarshes and dune slacks; rocky areas and thin soils, including extensive limestone areas; topographical irregularity and steep slopes on sea cliffs disrupting the vegetation canopy and allowing penetration of light; and, most importantly, continual disturbance through erosion and deposition of soils. The coast also provides a refuge for species at their northern limit in Britain because in south-western Britain the coast is relatively free of frost. Seemingly paradoxically, the coastlands also provide a refuge for Late Devensian relicts such as *Armeria maritima* (thrift), *Plantago maritima* and *P. coronopus* (sea and buck's-horn plantain) and *Artemisia* (mugwort). *Hippophae rhamnoides* (sea buck-thorn), widespread inland in the Late Devensian stage, is now only native on the calcareous dunes of eastern Britain (Pearson and Rogers, 1962). On the continent it is still found inland, on the shingle banks of mountain streams.

The coast is a refuge for many rarities. *Brassica oleracea* var. *oleracea* (wild cabbage) is an interesting coastal rarity found on cliffs in southern England. It is probably the only ancestor of a crop plant which is native to

Britain. Several of our endemic variants are found in coastal communities.

Many of the most floristically rich coastal areas are also limestone areas. An interesting comparison can be made between various western coastal limestone outcrops (Table 2.6).

They include Berry Head in south Devon, the largest outcrop of Devonian limestone. Brean Down and Steepholm in Somerset, are outliers of the Mendips. The Carboniferous limestone is also extensively exposed on the South Gower coast and also at Great Orme's Head in Anglesey. The presence of the rare south-west European plant *Aster linosyris* (goldilocks aster) links the Gower, Great Orme's Head, Brean Down and Berry Head. Other rare species are shared in part and each site has its own specialities.

One rare species is *Hornungia petraea* (Hutchinsia), found on the Gower coast, both on cliffs and fixed dunes, and on Great

Orme's Head (Ratcliffe, 1959). It is a small crucifer with white or purplish petals and, unusually for a crucifer, pinnate leaves. The fact that it grows on the coast is really only an accident. It is more widely distributed in the limestone areas of the Derbyshire and Yorkshire Dales and in the Mendips where it grows on rock ledges or in scree. It finds similar conditions of skeletal limestone soils at Great Orme's Head and on the Gower coast and also in calcareous dunes in a few other localities around the coast of Wales. It seems to require sunny, well-drained positions which are seasonally very dry.

The rock-roses (*Helianthemum* spp.) which were widespread in the late-glacial stage have interesting coastal distributions (Proctor, 1957) (Figure 2.29).

Helianthemum canum (hoary rock-rose), but a different variant from the one in Upper Teesdale, is present on Great Orme's Head and the Gower coast only because it finds the

Table 2.6 The distribution of some rare calcicole species on western coastal limestone

Species	Berry Head	Brean Down	Gower coast	Great Orme's Head
Hutchinsia			+	+
Small restharrow	+		+	
Goldilocks aster	+	+	+	+
Honewort	+	+		
Autumn squill	+			
Spring squill			+	+
Somerset hair grass		+		
Dwarf sedge		+		
Bloody cranesbill			+	+
Small hare's-ear	+			
Yellow Whitlow grass			+	
Hoary rock-rose			+	+
White rock-rose	+	+		
Spring cinquefoil		+	+	+
Wild cotoneaster				+
Spotted cats-ear				+
White stonecrop	+			
Rock stonecrop	+	+		+
Spiked speedwell			+	+

Figure 2.29 Rock-roses: (a) *Helianthemum canum*; (b) *H. appeninum*; (c) *H. nummularium*. ((c) Copyright Jon B. Wilson.)

open habitats it requires there. It grows only above the area of greatest abundance of halophytes, the plants best adapted to the salt winds of the coast, like *Armeria maritima* (thrift). *Helianthemum appeninum* (white rockrose) is confined to the areas of Brean Down and Berry Head in Britain. It is not limited to its localities by the climate nor by a requirement for salt; in Europe it is chiefly a montane species, and it thrives in cultivation outside its normal range. Brean Down and Berry Head provide two patches of lowland limestone grassland, refuges from the woodland. It has been introduced successfully to the Pennard cliffs of the Gower. The related *Tuberaria guttata* (spotted rock-rose) is another coastal rarity, found in a few places in North Wales and in the far west of Ireland (Proctor, 1960). It grows in peaty coastal heath.

The spiked speedwell, a different subspecies from the one in the Breckland, called *Veronica spicata* subsp. *hybrida*, is on the cliffs of the Gower and also Strumble Head, Stanner Rocks and the Avon Gorge, and has scattered occurrences elsewhere. *Ononis reclinata* (small rest-harrow) is found on the Gower and further west in South Wales, on Berry Head and in the Channel Isles.

Not all the specialities of coastal sites are long resident there. One speciality of the Gower is *Draba aizoides* (yellow whitlowgrass) which is otherwise found on the European mountains from the Pyrenees to the Tatra mountains of Slovakia. It grows up to 3400 m in the Alps, and so might potentially have survived the Late Devensian, certainly the Loch Lomond stadial, in Britain. However, its absence from other potential sites today within the British Isles indicates that its colonization of the Gower is by long-distance dispersal by birds, perhaps only relatively recently. It was discovered on the Gower in 1795 (Kay and Harrison, 1970). A similar disjunct pattern of distribution is that of *Arabis scabra* (Bristol rock-cress) on the limestone of the Avon Gorge and the Alps.

The Lizard is one of the most interesting coastal areas (Byfield, 1991). This is the southernmost part of the British mainland. Frosts are very rare and the growing season long. The temperature rises above 6°C almost every day of the year, which is enough for growth. The winters are warm and moist but the summers dry; average annual rainfall is only 880 mm. This is the area with the most Mediterranean-type climate in the country and its rarities include many with the characteristic life-forms of the Mediterranean vegetation: annuals, including several rare species of clover, and geophytes, like *Scilla verna* and *S. autumnalis* (spring and autumn squill). A very important influence on vegetation in the Lizard is the presence of serpentine rock, but very few of the specialities of the Lizard are confined to serpentine soils (Proctor and Woodell, 1971).

A major influence is the exposure of the coast. The coastal influence reaches far inland, as shown by the presence of *Scilla verna*, *Plantago maritima* (sea plantain) and *Silene uniflora* (sea campion) in the centre of the Lizard, 5 km from the sea. Maritime species are more abundant in positions exposed to the south-western gales. The Lizard experiences 30 gale days each year, so trees are rare. Trees and shrubs are more common on the more sheltered east coast of the Lizard. Occasional spring or summer gales are probably more damaging than winter gales because of the presence of tender parts of the plant. Plants projecting above the general level of the vegetation are much more affected. The windward side of trees and shrubs is pruned back by the injury caused by the salt-laden wind. A yearly total of sodium deposited from sea salt has been calculated to range from 44 kg/ha to 8650 kg/ha (Malloch, 1972).

On the Lizard there are two major coastal heath communities, an *Erica vagans* (Cornish heath) and *Ulex europaeus* (gorse) dominated community in more sheltered places (Figure 2.30), and a *Festuca ovina* (sheep's fescue) and *Calluna vulgaris* (heather) dominated community in more exposed places.

The latter is also present around much of the coast of western and northern Britain, and along the eastern and northern coasts of

Figure 2.30 The Lizard peninsula: (a) view of the coastal heath, Goonhilly Downs; (b) *Erica vagans* (Cornish Heath). (Copyright Jon B. Wilson.)

Ireland, everywhere except on limestone. It is dominated by *Calluna vulgaris*, *Festuca ovina*, *Plantago maritima* (sea plantain), *Scilla verna* (spring squill), *Lotus corniculatus* (bird's-foot trefoil), and *Thymus polytricus* (wild thyme) (Rodwell, 1992a; H7). The heath marks the inland limit of maritime vegetation. In it *Calluna* grows stunted and congested.

Below it in a more exposed situation is a maritime grassland where *Festuca rubra* (red fescue) and *Holcus lanatus* (Yorkshire fog) are the dominant grasses (Rodwell, 1992b; MC9). In less-exposed situations on thin soils, and in ungrazed situations, the grassland is dominated by *F. rubra* but with *Silene maritima* (sea campion), *Plantago maritima* and *P. coronopus* (sea and buck's-horn plantain). *Sedum anglicum* (English stonecrop) and *Asplenium marinum* (sea spleenwort) can be found. In more sheltered areas a denser grass sward can develop, with *Agrostis capillaris* and *A. stolonifera* (common and creeping bent) and *Holcus lanatus* (Yorkshire fog). In North Cornwall there is *Agrostis setacea* (bristle bent). On the Lizard, in the thin grasslands and erosion pans, *Hernieria ciliolata* (fringed rupturewort), *Isoetes histrix* (land quillwort), *Juncus capitatus* (dwarf rush), *Moenchia erecta* (upright chickweed) and *Orchis morio* (green-winged orchid) (Hepburn, 1943) grow. A widespread community occurs where there are nesting birds and the soil is enriched by guano; *Rumex acetosa* and *R. acetosella* (common and sheep's sorrel) are found with *Poa annua* (annual meadow-grass) and *Stellaria media* (chickweed).

Several common species have coastal ecotypes, variants genetically adapted to the exposed conditions. On the Lizard there are dwarfed ecotypes of *Leucanthemum vulgare* (ox-eye daisy), *Stachys officinalis* (betony), *Serratula tinctoria* (saw-wort), prostrate ecotypes of *Asparagus officinalis* (asparagus), *Cytisus scoparius* (broom) and *Schoenus nigricans* (black bog-rush) and hairy varieties of heather (*Calluna vulgaris* var. *hirsuta*) and common sorrel (*Rumex acetosa* var. *hirtulus*)

(Malloch, 1971). Another hairy plant of the Lizard and a few other coastal areas in the west, and very rarely inland, is *Genista pilosa* (hairy greenweed). At the edge of cliffs, in the eroded soils present and in the most exposed situation, *Armeria maritima* (thrift) is the most common species. *Sagina maritima* (sea pearlwort) is also very resistant to coastal exposure. Also found in this most exposed zone in many different places around the coast are *Crithmum maritimum* (rock samphire), *Limonium binervosum* (rock sea-lavender) and *Spergularia rupicola* (rock sea-spurrey) which all have the ability to root in rock crevices.

A cliff vegetation will often include woodland herbs, such as ferns of various species, *Hyacinthoides non-scripta* (bluebell), *Angelica sylvestris* (wild angelica) and so on, especially where, in sheltered conditions, a sward of grasses such as *Festuca rubra* (red fescue) has developed. Succession towards woodland is not uncommon on sheltered parts of the coast, but it is always liable to be disrupted by cliff erosion. Between Axmouth and Lyme Regis, in the easternmost corner of south Devon, many landslides have occurred in an area with a wide range of rock types, some lime-rich, dating from the Triassic to the Cretaceous. This has created a diverse environment, rich in potential for the development of a paramaritime vegetation. There are exposed slopes with wind-pruned shrubs. There are wet areas where ground water drains down. There are shady places rich in ferns and woodland herbs. One large landslide happened in the winter of 1839, detaching 6 ha of land to create what is called Goat Island. In the chasm which was created there has been a succession towards an ash woodland, one of the best examples of this kind of succession in Britain. From old photographs it is possible to see that the succession really got going only in this century.

2.6.13 UNSTABLE HABITATS

The most important feature of the coast as a

refuge is the continual disturbance that occurs. Changes in sea level have eliminated communities and also exposed new ground for colonization. Mapping of some coastal areas shows remarkable changes in outline. Spurn Point, the south-eastern corner of Yorkshire, has a range of temporary habitats – sea and river shores, saltmarsh and sand dunes – under imminent danger of disappearing. The loss of land, of a freshwater pond and a marsh, and the gain of tidal pools have all been described in recent years. From a small area 340 species of flowering plants have been recorded (Crackles, 1990). Some do not persist but newcomers are arriving all the time, 36 since 1978 at the last count.

The changing coastal outline of part of south-east England is illustrated in Figure 2.31.

Coastal successions have been constantly disrupted. Some changes have occurred overnight following particularly bad storms.

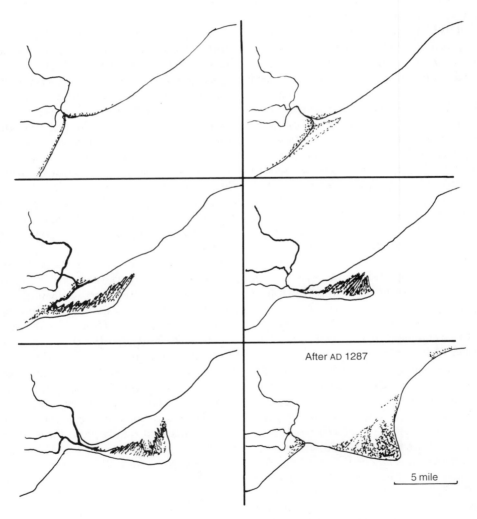

Figure 2.31 Stages in the development of Dungeness (adapted from Firth, 1984).

New open habitats have been supplied regularly and frequently. Coastal communities are rich in species which have become agricultural weeds, and ruderals which take advantage of the 'waste' places. In the strandline in sheltered conditions there is a soil enriched and mulched by the rubbish left by the sea. Here in sheltered places above saltmarshes can be found species of *Atriplex*, *Rumex* and *Polygonum* (orache, dock and knotweed). Other weedy species are the mayweeds (*Matricaria* spp.) and campions (*Silene maritima*, *S. dioica* and *S. latifolia*); there are also *Crambe maritima* (sea kale), *Salsola kali* (saltwort) and *Honckenya peploides* (sea sandwort).

Regosols are lithomorphic soils formed on unconsolidated pebble beaches and also sand dunes. They are the most unstable of habitats. Shingle beaches are widespread around our coast (Chapman, 1976). Where they fringe the coast or form part of a spit they are regularly washed by storm waves and splashed by sea spray and their vegetation is very limited and ephemeral. At the upper margins, on beaches which are stable in spring and summer, weeds such as *Galium aparine* (cleavers), *Senecio vulgaris* (groundsel) and *Atriplex* spp. (orache) establish (Scott, 1963a). More stable conditions are found on shingle bars or barriers. Scolt Head Island in north Norfolk is a particularly well-developed barrier island, which has developed from a series of spits like the adjacent Blakeney Point. With its well-developed saltmarsh on the landward side, it is over 1 km wide in parts.

The broadest areas of stable shingle are found where the wave approach is mainly from two directions. Dungeness is the largest area of shingle in the British Isles (Figure 2.32).

Here the main direction of storm waves are from the south-west or from the east down the Channel. It has developed mainly in the past 2000–3000 years, especially in a period of great storms 750 years ago (Firth, 1984). There was at first a sand barrier and gravel

Figure 2.31 Dungeness

bar protecting a saltmarsh. An apron of shelly sand provided a platform for storm-beach gravel. The shingle area started as a spit reaching to the east across the bay containing Romney marsh. The River Rother at first flowed behind the spit but breached it at the Rye end in the thirteenth century, so that the cuspate structure of Dungeness was created. A similar process, but at an earlier stage, can be seen at Orfordness in Suffolk. Dungeness is still growing east at a rate of 2–3 m/year. Every 5–8 years a new shingle ridge is formed, thrown up by a storm.

The shingle area of Dungeness covers 2035 ha but there are associated areas of dunes, salt- and freshwater marsh and open water. This all adds up to a remarkably diverse range of habitats, helping to account for the 600 species of plant recorded there (Scott, 1965). As well as exposure and instability, key features of shingle beaches are the quantity and quality of the matrix between the pebbles and the availability of water. The free-draining soils can be a problem for seedling establishment. Deep down there is a reservoir of fresh water from rainfall, floating on the denser sea water, rising and falling with the tide. However, since this is usually 2 m or more below the surface, the tap roots of even large plants may not be able to reach it. More important for their survival is the nature of the shingle matrix, the finer material between pebbles. On beaches without matrix, little grows except lichens and a few tolerant species like *Lathyrus japonicus* (sea pea) (Scott, 1963a).

Where the matrix is sandy, species that also grow on sand dunes are found, such as *Glaucium flavum* (yellow horned-poppy) (Scott, 1963b) (Figure 2.33a). It persists in the stabilized ground but only colonizes open areas. At Dungeness it avoids the lows, which are liable to waterlogging (Scott, 1963b).

From the strand-line debris an organic matrix can develop, retaining moisture and boosting nutrient levels. There are species particularly associated with the strand-line, such as *Atriplex glabriuscula* (Babington's orache). On beaches like Blakeney Point and Chesil Beach, *Suaeda vera* (shrubby sea-blite) starts as a drift-line plant, establishing with long tap roots (Figure 2.33d) (Chapman, 1947b). As plants are overcome by shingle, adventitious roots are sent out. Older buried parts of the plant die, but the plant can appear to have climbed up the beach as it produces new shoots higher up. On more stable beaches like Dungeness, especially on the landward slopes of the ridges, called the 'fulls', *Crambe maritima* (sea kale), *Silene uniflora* (sea campion), *Trifolium scabrum* (rough clover), and *Rumex crispus* (curled dock) establish. *Crambe maritima* has a relatively weak tap root but a broad, flat root system which can extend as much as 2 m out from the plant (Scott and Ranwell, 1976). If buried, it produces a ring of side shoots. It has a large store of starch which allows regrowth after burial.

'Lows' between the ridges have a coarser shingle and the community is species-poor. Some shingles have a silt/clay matrix. They share species with saltmarsh like *Atriplex portulacoides* (sea purslane), *Glaux maritima* (sea milkwort) and *Seriphidium maritimum* (sea wormwood). In Norfolk, the mainly Mediterranean species, *Limonium bellidifolium* (matted sea-lavender), reaches into the upper saltmarsh.

On Dungeness the leaf litter produced by a procumbent variant of *Cytisus scoparius* (broom) is especially important in the succession towards a well-developed acid heath community dominated by *Ulex europaeus* (gorse) and even to woodland (Ferry, Waters and Jury, 1989). On the fulls, along with *Cytisus*, more common in areas away from the sea, is *Prunus spinosa* (blackthorn), in a prostrate vegetation rich in lichens and bryophytes. The development of the *Cytisus* vegetation can be cyclical, as plants decline and die after 15–20 years. Alternatively, once stabilized, an acid heathland with *Festuca rubra* and *F. filiformis* (red and fine-leaved

Figure 2.33 Plants of the shingle: (a) *Glaucium flavum*; (b) *Crambe maritima*; (c) *Calystegia soldanella*; (d) *Suaeda vera*. ((a)–(c) Copyright Jon B. Wilson.)

sheep's fescues) and *Arrhenatherum elatius* (false oat-grass), later invaded by *Ulex europaeus* (gorse), develops. Uncommon elsewhere are *Silene nutans* (Nottingham catchfly) and *Crepis foetida* (stinking hawk's-beard). On the south-western Holmstone Beach of Dungeness a climax scrub of *Ilex aquifolium* (holly), *Sambucus nigra* (Elder), *Taxus baccata* (yew), *Ulex europaeus* (gorse), *Cytisus scoparius* (broom) and *Prunus spinosa* (blackthorn) has developed.

The Open Pits on Dungeness are areas of open-water and freshwater marsh which have developed from brackish pools. Here a willow carr is developed in some places. The largest has a floating fringe of *Typha latifolia* (bulrush), *Potentilla palustris* (marsh cinquefoil), *Iris pseudacorus* (yellow iris) and *Rumex hydrolapathum* (water-dock). In places the hydrosere has continued towards incipient bog formation with *Sphagnum*.

2.6.14 DUNE SYSTEMS

The most important coastal habitat, where exposure has helped to maintain a rich open herb community, is the dune system; such systems are some of the most floristically rich communities. Many hundreds of native vascular plants are found in the dune systems of the British Isles, and more than that number of introduced exotics have also found a home there. There are notable dune systems at Newborough Warren in Anglesey (Ranwell, 1960a, b), Kenfig Burrows in West Glamorgan (Orr, 1912), Whitford Burrows on the Gower (Ghillam, 1977), Braunton Burrows in Devon (Willis *et al.*, 1959), along the Northumberland Coast and the Sands of Forvie in Scotland.

The Holocene development of the largest coastal accumulation in Ireland, the Magilligan Foreland in County Londonderry, has been radiocarbon dated. The development of the foreland dates back to the accumulation of beach ridges 7000–6500 years ago (Wilson and Farrington, 1989). There

have been many different phases of flooding, soil development and human occupation, sand-dune encroachment from different directions, and peat development in wet areas between ridges.

The greatest extent of dunes is the Culbin Sands on the Moray Firth. Here bare, mobile sand dunes can be found (Steers, 1937).

A difference can be observed between dune systems exposed to westerly gales, where the dunes are mobile and transient, and those where the dunes eventually stabilize. Sand dunes are not confined to the coast but were present inland in the Breckland, which in some respects represented a coastal refuge inland (section 2.6.15).

Within dune systems there are many kinds of open habitat: slopes in the sun or shade, sheltered or exposed to salt-laden wind. There are dune plateaus, wind blowouts and hollows. There are wide variations in the accretion rates of sand and in the availability of water. On the wind-exposed foredune, where the sand is just above tidal limits, accretion rates are slow (Figure 2.34).

A strand-line of rubbish, colonized at first by annual weeds, can aid colonization by foredune plants. *Elytrigia juncea* (sand couch) propagates readily from seed. The seedling root elongates very rapidly down to moist sand, and lateral roots spread out just below the surface. Short lateral rhizomes establish a rosette of shoots binding the sand and then much longer rhizomes colonize surrounding areas, their tips turning up for expansion in the following spring. *Leymus arenarius* (Lyme grass) adjusts to burial by sand by producing horizontal and oblique rhizomes (Bond, 1952). The foredune can build at about 30 cm a year. In time *Leymus arenarius* and *Elytrigia juncea* are limited by unavailability of water and by competition from *Ammophila arenaria* (marram).

The yellow dune is dominated by *Ammophila arenaria* but there are still areas of bare sand. *Ammophila* stabilizes the dune; it has unlimited potential for horizontal and vertical

Figure 2.34 Dune systems: (a) view of part of the fore and yellow dune at Newborough Warren, Anglesey; (b) view of the mature dune system at Kenfig Burrows. ((b) Copyright Jon B. Wilson.)

growth, being able to tolerate burial to a depth of up to 1 m in a year (Huiskes, 1979). Initially it has a tufted growth form compensating for moderate accretion of sand by leaf growth. With greater depth of burial, vertical shoots are produced from axillary buds. Rhizomes have a sharp point to push through the sand. Adventitious roots bind the sand. As long as sand accretes, growth continues. Lime-rich dunes of tremendous height have developed at Braunton Burrows. However, if accretion slows, the rhizomes elongate so rapidly that the production of adventitious roots becomes inhibited as the tip of the shoot is pushed into the dry surface sand.

Greater stability allows the colonization of *Eryngium maritimum* (sea holly), *Calystegia soldanella* (sea bindweed) and *Euphorbia paralias* (sea spurge) (Figure 2.35).

These species have deep roots. The seedlings of *Calystegia soldanella* (sea bindweed) produce a root 5–6 cm long in 3–4 days, even before the seed leaves have emerged from the seed coat. By 14 days the root is down in permanently moist sand 10–15 cm deep. The yellow dune is converted to a vegetation-covered grey dune (Figure 2.34b). *Carex arenaria* (sand sedge) spreads into the yellow dune by producing rhizomes. As the dune becomes stabilized it is colonized by mosses, *Bryum* species, especially *B. algovicum*, *Tortula muralis* and lichens, *Cladonia*. The mosses have some limited ability to elongate to compensate for a rising sand level. In these communities *Ammophila* grows as a scattered and sparse element (Orr, 1912).

The capillary rise of water from the water table is only about 40 cm even in fine sand, so that the water table may be many metres below the surface. Water from rain and dew is especially important. In the thin soil diurnal temperature variation can be high so that there is a heavy dew fall, up to 0.9 ml per 100 ml of sand. Water percolates down into dune hollows called 'slacks'. At Newborough

Warren where bare, wet sand is exposed on the seaward-facing slope it is colonized by *Sagina nodosa* (knotted pearlwort) and *Juncus articulatus* (jointed rush), and then *Agrostis stolonifera* (creeping bent) hummocks form (Ranwell, 1960a, b). As the sand accretes again and the surface dries *Salix repens* (creeping willow) colonizes the slack along with *Salix purpurea* (purple willow) to form a scrubby vegetation. *Hippophae rhamnoides* (sea buckthorn) may be present, enriching the soil because of its nitrogen-fixing nodules (Pearson and Rogers, 1962). *Carex arenaria* and *C. flacca* (sand and glaucous sedge) can be found.

In the slacks at Newborough *Epipactis leptochila* (narrow-lipped helleborine), *Parnassia palustris* (grass of Parnassus) and other orchids grow. The dune-slack community is a temporary formation because eventually it is buried by the sand of a migrating dune. Peripheral dune communities have the seepage of ground water into them. The short, rabbit-grazed turf of a mature dune system can be very diverse. Several southern European species, such as *Mathiola sinuata* (sea stock) and *Lagurus ovata* (hare's-tail) are found in the dunes behind (St. Ouen's Bay in Jersey).

Dune systems can be base rich with shell fragments. On the north coast of Sutherland at Bettyhill calcareous sand is blown into the surrounding bog, enriching the communities there. In the damp sand there is *Carex maritima* (curved sedge), *Saxifraga aizoides* (yellow saxifrage), *Saxifraga oppositifolia* (purple saxifrage), *Epipactis atrorubens* (dark-red helleborine), *Dryas octopetala* (mountain avens), *Trollius europaeus* (globe flower), *Oxytropis halleri* (purple oxytropis), *Ajuga pyramidalis* (pyramidal bugle), and the endemic *Primula scotica* (Scottish primrose). Machair is a particular kind of calcareous dune grassland found on north-western coasts (Gimmingham, Gemmell and Greig-Smith, 1948; Ranwell, 1974), most extensively developed in the Outer Isles, especially on

(a)

(b)

(c)

(d)

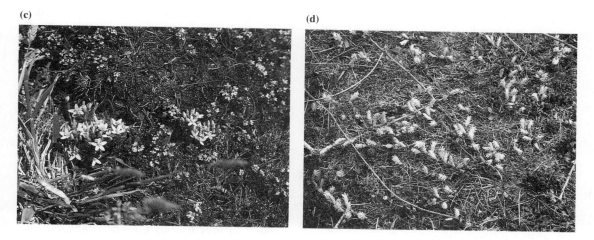

Figure 2.35 Sand-dune plants: (a) *Euphorbia paralias*; (b) *Eryngium maritimum*; (c) *Centaurea erythraea* and *Thymus polytrichus*; (d) *Salix repens*. ((b) Copyright Jon B. Wilson.)

South Uist. There is a grassland of *Festuca rubra* (red fescue), *Holcus lanatus* (Yorkshire fog), *Agrostis stolonifera* (creeping bent) and *Cynosurus cristatus* (crested dog's-tail) with a broad range of common grassland herbs such as *Bellis perennis* (daisy), *Ranunculus bulbosus* (bulbous buttercup), *Achillea millefolium* (yarrow), *Lotus corniculatus* (bird's-foot trefoil), *Thymus polytrichus* (wild thyme), *Senecio vulgaris* (groudsel), *Plantago lanceolata* (ribwort plantain), *Galium verum* (lady's bed-straw), *Anthyllis vulneraria* (kidney vetch), *Taraxacum officinale* agg. (dandelion), *Prunella vulgaris* (selfheal), *Centaurea nigra* (common knapweed), *Carex flacca* (glaucous sedge) and *Euphrasia* spp. (eyebrights).

Alternatively, leaching can make the surface layers of dune plateaus acidic so that a kind of dune heath develops with *Festuca rubra* (red fescue) and a range of small annuals like *Sagina maritima* (sea pearlwort), *Stellaria media* (common chickweed) and perennials, including *Trifolium* spp. (clovers) and *Potentilla anserina* (silverweed). At Newborough on moderately lime-rich sands, there is a clear zonation from strand-line through dune communities to a *Calluna vulgaris* heath and on to scrub. The dune system at the Culbin Sands and the Sands of Forvie are lime deficient. Stable dunes first have an open, lichen-rich *Empetrum nigrum* (crowberry) heath, with many sand-binding plants like *Carex arenaria* (sand sedge) and *Ammophila arenaria* (marram), but also herbs and mosses. As the *Empetrum* community closes, the grasses and other herbs and the mosses and lichens are out-shaded. Finally, a heathland dominated by *Calluna vulgaris* with *Campanula rotundifolia* (harebell), *Festuca ovina* (sheep's fescue) and *Agrostis capillaris* (common bent) or a kind of acid grassland with *Nardus stricta* (mat grass) or *Molinia caerulea* (purple moor-grass) arises.

2.6.15 THE BRECKLAND

The Breckland is a part of East Anglia, an area of about 700 km^2 between Newmarket and Norwich. It is a unique area; several Breckland species do not grow, or are very rare, elsewhere in the British Isles, for example *Artemisia campestris* (field wormwood) (Figure 2.36), *Muscari neglectum* (grape hyacinth), *Silene otites* and *S. conica* (Spanish and sand catchfly) and several species of *Veronica* (speedwell) (Figure 2.37) (Trist, 1979).

It may seem strange to include the Breckland next to a section on coastal vegetation, but, in the relatively recent past, travellers have likened the Breckland to part of the coast. Even into the seventeenth century there were extensive areas of travelling sands in the Breckland. In 1668 the village of Santon Downham was nearly overwhelmed by a sand storm.

The sandy, wind-blown soils have some similarities to those of coastal dune systems. Soils are either calcareous (light-coloured) or non-calcareous (dark-coloured). Wind-blown sand, sometimes 1–2 m deep, is widespread. These are glacial sands and loamy sands which are very free draining. The non-calcareous soils are very low in potassium and are often podzolized. A number of sand-dune species are present, including *Carex arenaria* (sand sedge). It produces two kinds of roots: sinkers which can reach down 2 m, and short, fine roots near the surface which are produced each year (Noble, 1982). Rhizome buds can survive for more than 7 years after deep burial by sand, allowing later regeneration if the dune moves on. It is recorded from the Middle Devensian Glacial Stage at Barnwell station, Cambridgeshire, growing in a landscape through which mammoth and reindeer grazed. More common at the coast, it may have survived the spread of woodland in the Postglacial, growing in sandy glades within the wild-wood, which, in the Breckland, was probably a thin cover of *Pinus sylvestris*.

The interaction between plants, soils, climate and humanity is particularly well marked in the Breckland. This is the area of

Figure 2.36 Distribution of *Artemisia campestris* (field wormwood). (Map redrawn from Hultén, 1971.)

Britain with the most continental climate; the sandy soils tend to exaggerate the aridity and temperature fluctuations. It is a landscape area defined by the interaction of soil and climate. It has the lowest rainfall in the British Isles (600 mm/yr) with the lowest rainfall in the growing season of February to June. It has the most extreme range of temperature (mean absolute monthly minimum and maximum: for January, −6.7°C and 12.4°C; for July, 6.6°C and 28.5°C). It is especially liable to suffer late spring frosts. In 1966 there were 11 frosts in April, 16 in May and 6 in June. A temperature of −5.5°C was recorded on 1 June 1962.

The Breckland rarities include species that are more common in central and eastern Europe. These so-called 'steppe' plants owe their presence in the Breckland to a stage in the colonization of Britain when a dry, continental climate reached across Europe and into Britain, at a time when the North Sea was grassland dissected by many rivers. A Breckland rarity, *Thymus serpyllum* (Breckland thyme), is a relic of this time, not found elsewhere in Britain. It is widespread in northern and central Eurasia as far as China. In Europe it is more or less restricted to well-drained sandy soils, leached glacial drift or sand dunes (Pigott, 1955). It is particularly sensitive to competition from other species. Other Breckland species are ruderals of low competitive ability which prefer the open ground of the mobile soils.

(a)

Figure 2.37 Breckland rarities: (a) *Veronica spicata* (spiked speedwell); (b) *Silene conica* (Spanish catchfly); (c) *S. otites* (sand catchfly); (d) *Muscari neglectum* (grape hyacinth). ((a)–(b) Copyright Jon B. Wilson.)

(c)

(b)

(d)

In areas free of superficial deposits, the pattern of Breckland soils is complex. In the Glacial stage permafrost developed. Frost-heave destroyed the vegetation, creating bare patches. Without the insulating blanket of vegetation the bare areas melted in summer but in autumn were subject to the injection of chalky material as they froze, raising chalky ridges. Lower areas between the ridges filled with peat and sand. Patterns of polygons with a radius of 5 m, or, on slopes, stripes about 7 m apart, were created. Today *Calluna vulgaris* (heather) alternates with chalk pasture plants, or the grasses *Dactylis glomerata* (cocksfoot) alternates with *Agrostis capillaris* (common bent). The varying performance of a species like *Pteridium aquilinum* (bracken) also marks out the mosaic of the soils beneath (Watt, Perrin and West, 1966).

2.6.16 CHANGING SEA LEVELS

Throughout the early Holocene there were dramatic changes of sea level, an overall rise in sea level, leading to the flooding of the English Channel. Even afterwards, sea level has fluctuated so that lowland coastal areas have become liable to flooding because of raised water tables.

The first phase of flooding happened as the ice melted after the Glacial stage. There were two influences: the greater quantity of water in the oceans, a eustatic change, and the reduced weight of ice on the land, an isostatic change. In the Early Holocene coastal areas in Scotland and Northern Ireland were inundated. In time the land that had been weighed down with ice rebounded isostatically, and by about 5000 years ago these areas had drained again.

Meanwhile, further south, rising sea levels, rising at between 0.6 and 1.5 m/century at one period, caused a backing up of the rivers as the North Sea Plain was being flooded. Fen peat began to form in the channels of some present-day inland sites as the flow rates of rivers decreased (Godwin, 1940). In the East Anglian Fenland, there was a forest of *Pinus sylvestris* (pine) with some *Quercus* (oak), *Ulmus* (elm) and *Corylus avellana* (hazel). At channel margins there were sedges, grasses and ferns. At Shippea Hill, mesolithic man left a layer of microliths 7600 years BP, which is overlain by a peat, indicating the development of wetland conditions, while the typical mature woodland of *Tilia cordata* (small-leaved lime), *Quercus* (oak) and *Ulmus* (elm) was becoming established on the higher ground in the vicinity. The presence of *Alnus glutinosa* (alder) indicates the wetness of the site.

Peat began to be formed generally about 7000 years BP (Godwin, 1940). Eventually in the period 6600–6000 years ago the rise in sea level caused the rivers to burst their banks and the extensive flooding killed the trees. Brackish conditions began to prevail. In the Fenland there was a large lake. Peat formation ceased and fen clay was deposited. Inland, the fen clay has a buttery texture. Fine clay particles settled in lagoons only intermittently invaded by sea water. It would have been a wonderful scene, with flotillas of pelicans languidly floating across the sparkling water. Nearer the present-day coastline there was tidal flow. The sediment was siltier and there was a vast saltmarsh.

The rise in sea level was not a steady, regular process. There were surges dated to different times. For example, eustatic rise of the sea was modified by down-warping of the land in the east in southern England. The complexity of the changes in sea level that could occur is shown in a sequence of coastal plant communities beside the Bristol Channel on the Gwent levels. Fen communities developed from saltmarsh about 6700 years ago (Godwin, 1941) as the sea receded. A short regression to reed swamp occurred after 600 years as the sea level rose again, giving way to an open saw-sedge fen and fen grassland, which lasted another 300 years. From this a bog developed 5750 years ago lasting until 3400 years ago, when the bog was flooded by the sea.

Fenlands remained an important element in our lowland vegetation, until recent drainage. In the East Anglian Fenland there were several phases of flooding, marine flooding near the coast and in estuaries, freshwater further inland. At Wiggenhall St. German, 4 miles south-west of King's Lynn, there are four peat beds separated by fen clays and marine silts. At Wicken Fen, oak woodland was succeeded by a wet fen 4200 years ago, killing and preserving massive 'bog' oaks in the peat (Godwin, Clowes and Huntley, 1974). There was another major phase of marine incursions about 2000 years ago in the Iron Age and Roman period. The end of each marine phase was marked by a succession away from saltmarsh plants such as *Armeria maritima* (thrift), *Plantago maritima* (sea plantain) and *Juncus maritimus* (sea rush), and plants common in the waste places above the tide-line, such as *Artemisia* spp. (mugwort and wormwood), to plants of brackish conditions, such as *Phragmites australis* (common reed), and then to those of freshwater swamps, *Typha latifolia* (bulrush) and *Nymphaea* spp. (water-lily). This was followed by the normal development of freshwater fen and carr (fen woodland).

2.6.17 FROM SEA TO LAND

Phragmites australis (common reed) is particularly important in the succession from saltmarsh to fen because it slightly salt-tolerant. It is recorded in the transition from saltmarsh to fen peat in many sites (Godwin, 1984). The succession from *Phragmites* is directly to *Sphagnum* bog in some places, as over the submerged forest at Borth and Ynyslas, Cardiganshire (Godwin and Newton, 1938a).

The seedlings of *Phragmites* show an enhanced germination in slightly brackish water. It forms a band behind many saltmarshes, and here its performance is clearly related to the degree of freshwater runoff diluting the saline soils of the upper saltmarsh. Other plants found in this zone are *Juncus maritimus* (sea rush), *Beta vulgaris* ssp. *maritima* (sea beet) and even the very common grasses *Agrostis stolonifera* (creeping bent) and *Festuca rubra* (red fescue).

Some saltmarsh species have a wide tolerance of different salinities. They include *Spergularia* spp. (sea spurrey), which requires fresh water to germinate. *Aster tripolium* (sea aster) also has a wide tolerance of salinity but performs best in brackish conditions. Highly salt-tolerant species include *Atriplex portulacoides* (sea purslane) and *Limonium* spp. (sea lavenders) which have salt-excreting glands. *Salicornia* spp. (glasswort) are the most halophytic species of all and are restricted to saline conditions. Their seeds germinate even in salt solutions more concentrated than sea water. Tolerant or halophytic species can contain high concentrations of proline, glycine betaine, sorbitol, mannitol or pinitol. These do not damage cell activities but their presence restricts the uptake of salt. A consequence of this is that many coastal plants, such as *Salicornia*, are succulent.

The saltmarsh provides several contrasting habitats and there is a strong zonation of species (Figure 2.38).

There is an important difference between the lower, or submergence, marsh, which is submerged almost every day of the year and may be exposed for 10 days or more at a time. A characteristic plant of the submergence marsh is *Spartina* (cord grass) (Figure 2.39). Summer drought limits the establishment marsh is *Spartina* (cord grass) (Figure 2.39).

Summer drought limits the establishment of its seedlings. Emergence marsh plants often have dense, woody root-stocks or compact, short, rhizomatous growth. Within the transition from lower to upper marsh other features can modify the conditions of exposure and salinity. Creeks have banks slightly raised and are better drained than the surrounding marsh. *Atriplex portulacoides* (sea purslane) can be dominant in the middle

Figure 1.62 View of Cairngorm (copyright Jon B. Wilson).

Figure 2.17 View of sugar limestone flora on Cronkley fell, *Helianthemum canum* in flower.

Figure 2.19 The Burren.

Figure 2.25 Aquatic and wetland vegetation: view of one of the lodes at Wicken Fen, *Phragmites australis* fringing.

Figure 2.38 Salt marsh systems: *Limonium vulgare* (common sea lavender) in the middle marsh, Negden, Isle of Sheppey (copyright Jon B. Wilson).

Figure 3.13 Heathland and moorland: Chobham Common, Surrey (copyright Jon B. Wilson).

Figure 3.8 Grasslands: (chalk downland, Devil's Kneadingtrough, Kent)

Figure 3.8 (Upland grassland, Cwm Idwal, north Wales).

Figure 2.38 Saltmarsh systems: (a) *Salicornia* (glasswort) in the lower marsh at Pegwell Bay, Kent; (b) see plate section. (Copyright Jon B. Wilson.)

marsh but lower down in the marsh it is concentrated on the creek banks (Chapman, 1950).

In the emergence marsh some of the highest salinities can be found in salt-pans where sea water is concentrated by evaporation. About the only plant growing in these is *Salicornia*. The pans also suffer the greatest, and most rapid fluctuation in salinity; in rainstorms the salt can be washed away. The saltmarsh habitat is particularly harsh. Tidal flow and wave action can damage the plant surface, allowing the uptake of salt. Saltmarsh plans have strong, leathery or hard leaves. In *Salicornia* the leaves are tiny scales and the plant is reduced to a branched, succulent stem. A large proportion of saltmarsh plants are perennial, sprawling, tufted or bushy herbs. They have well-developed rhizome and rooting systems. Most die back in winter, storing useful metabolites in a rootstock and getting rid of excess salt in their dying leaves.

Figure 2.39 *Spartina anglica* (common cord-grass).

Perennial species are very important in saltmarshes because of the way they stabilize the coastal mud, and calm the water so that sedimentation occurs. At the forefront of these, low down on the marsh, is *Spartina anglica* which spreads very vigorously, binding the mud with deep, strong anchor roots (Thompson, 1990). It can cope with rapid accretion of sediment by rapid root and shoot growth. The smaller grass, *Puccinellia maritima* (common saltmarsh-grass), is less tolerant of high accretion rates and water-logging but colonizes areas of bare mud by producing long stolons. The annual species, *Salicornia europaea*, *S. dolichostachya* and *Suaeda maritima* (annual sea-blite), are seasonally important, colonizing bare areas of mud (Chapman, 1947a). *Salicornia* has a related perennial species called *Sarcocornia perennis* (perennial glasswort) which is important in the middle of some marshes in south-eastern England.

Saltmarshes form where tidal water counters the flow of rivers. Mud banks build up in the area below a marsh, called the slob lands. When they are high enough to be exposed for several days at the neap tide part of the tidal cycle, the mud banks are colonized by plants. In the past *Zostera* (eelgrass) was very important in this process, colonizing the area below low-water mark, but from 1931–34 populations were largely wiped out by disease (Tutin, 1942). One place that they are still important is in the lagoon behind Chesil Beach. Here all three native species can be found, along with *Ruppia maritima* (beaked tasselweed). They provide food for water-fowl.

Waterlogging of the soil is as important a factor in plant distribution on saltmarshes as in other wetlands. Alluvial soils are an important subcategory of gley soils. They are periodically inundated unless they are protected by embankments. The soil remains fluid (unripened) and completely reduced unless embankment and drainage is carried out. With drainage a physical ripening occurs from the surface downwards as aeration increases. The burrowing of animals and, in summer, the activity of diatoms can aerate the mud. Soils are often fine silts and clays which drain poorly at low tide and are poorly aerated. The presence of sodium ions in salt water restricts aggregation of silt particles, reducing pore size in the soil. Anaerobic mud

is rich in toxic black iron sulphides. In the reducing conditions present toxic ferrous and manganous ions predominate over ferric and manganic ions and toxic sulphide ions are present (Pearson and Havill, 1989). The annuals *Salicornia* (glasswort) evade the more waterlogged conditions of winter. *Puccinellia* (saltmarsh grass), *Spartina* (cord grass) and *Aster tripolium* (sea aster) have a high toler-ance to anaerobic soils because of the aerenchyma, tissue with air canals, in their roots.

There is very marked variation in the behaviour and appearance of individuals of species in different ecological zones of the saltmarsh. *Aster tripolium* behaves as a short-lived perennial on the lower marsh and an annual on the upper marsh. Many distinct *Salicornia* variants have been described as species (Ingrouille and Pearson, 1987; Ingrouille, Pearson and Havill, 1990). *Puccinellia maritima* has a wide range of genetic variants and also responds to grazing by growing as a prostrate, small-leaved, rapidly tillering variant (Gray and Scott, 1977). Stem and leaf characters of *Plantago maritima* (sea plantain) vary with the change in salt concentration across a marsh (Gregor, 1930, 1938). Plants with upright flowering scapes are concentrated in the upper marsh.

There are very extensive saltmarshes around the British coast. Those of the north Norfolk coast are particularly spectacular, a haze of blue when *Limonium vulgare* (common sea-lavender) is in flower. Potentially salt-marshes might have provided an important refuge for herbaceous open-habitat species at the time of the maximum development of the wildwood, because the large tidal range ex-perienced and exposure to waves prevents the growth of trees. However, the environment is so harsh that few species are found here.

2.7 THE SPREAD OF THE WILDWOOD

Between 10 000 years ago and 5000 years ago dense, shady wildwood arose over most of the British Isles (Birks, Deacon and Peglar, 1975). Surviving woodlands are only a threadbare reminder of that natural wildwood and yet the best clues of what it was like. For example, there is a rough south-east to north-west cline in existing woodland types. This mirrors the likely abundance of different tree species in the wildwood, the direction of colonization of the British Isles and the timing of the rise to abundance of important tree species, earliest in the south-east and latest in the north-west (Huntley and Birks, 1983). Species that survived the cold stages in refuges far away in south-eastern Europe, being adapted to southern continental climates, were the last to arrive and failed to colonize the north.

The British Isles have a poor tree and shrub flora compared to the continent (Table 2.7).

Notable species of previous interglacials, such as *Picea abies* (Norway spruce) and *Abies alba* (silver fir) failed to colonize in the Holocene.

Table 2.7 Native trees and shrubs

Native woody species in order of rising to prominence in the Holocene	Other native woody species
Juniper	Bay willow
Downy birch	Bird cherry
Silver birch	Goat willow
Aspen	Holly
Rowan	Hawthorn
Scots pine	Crack willow
Hazel	Black poplar
Wych elm	Yew
Pedunculate oak	Whitebeam
Sessile oak	Midland thorn
Small-leaved lime	Crab apple
Alder	Wild cherry
Ash	Strawberry tree
Field maple	White willow
Beech	Wild service tree
Hornbeam	Large-leaved lime
	Box
	Plymouth pear

Refuges for each native tree have been identified and rates of migration calculated. From this kind of analysis *Corylus avellana* (hazel) and *Pinus sylvestris* (Scots pine) must have migrated at 1500 m/year between 10 500 and 10 000 BP, and *Ulmus* (elm), *Quercus* (oak), *Tilia cordata* (small-leaved lime) and *Fagus sylvatica* (beech) migrated at between 50 and 1000 m/year, with *Ulmus* the fastest and *Fagus* slowest (Huntley and Birks, 1983). The time of arrival in Britain is complicated by the distance different species had to migrate. *Quercus* and *Ulmus* only had to migrate from a Late Devensian refuge in the Bay of Biscay which they shared with *Corylus*. *Pinus sylvestris* had refuges in Iberia and western Britain. *Tilia* and *Fagus* had to come from refuges in south and south-east Europe, in Italy and the Balkans.

Mapping pollen abundance is useful for showing what were the important tree species at different ages, but calculated rates of migration only relate very poorly to what we know about the biology of the different species. Some of the earliest to establish, *Ulmus* and *Pinus*, have light, wind-dispersed seeds but *Corylus* and *Quercus* do not. *Fagus* must have arrived in Britain before it was cut off by the sea, which is at least 4000 years before the pollen maps show it as an important woodland tree.

It would be wrong to imagine the development of the British vegetation in the Holocene as if waves of different woodland types surged across Europe to colonize Britain from the south-east, with hazel and pine woodland then oak and elm woodlands, and finally lime and beech woodlands (Figure 2.40).

Some woodland herbs are known from the Late Devensian, including *Anemone nemorosa* (wood anemone) and *Mercurialis perennis* (dog's mercury) which are known from Scotland (Godwin, 1984; Shirreffs, 1985). There are notable absences from the pollen record of the Early Holocene of important woodland herbs, such as *Hyacinthoides*

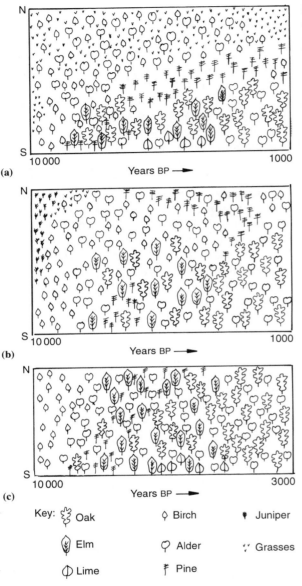

(a)

(b)

(c)

Key: Oak   Birch   Juniper

Elm   Alder   Grasses

Lime   Pine

Figure 2.40 Regional development of woodland: (a) eastern Britain; (b) western Britain; (c) central Britain. (After Bennett, 1986).

non-scripta (bluebell) and *Primula vulgaris* (primrose). Nevertheless, an entire flora and fauna had colonized Britain by about 8300 years ago, and most without leaving any early trace. By this time the English Channel had

flooded. The colonization of Ireland was even earlier (see section 2.3.4).

An alternative model is that in the Early Holocene there was a much more heterogeneous vegetation. Superimposed on a changing pattern of regional variation (Bennett, 1988), there was complex pattern of local variation. Long-distance dispersal or perglacial survival provided foci within Britain for the subsequent expansion of particular species. Some species had many such foci and so their patchy distribution soon coalesced. *Betula* and *Juniperus* both did this. Other species were very slow to expand, either because of inherently low rates of spread or because they had to wait for some change in the climate or soil to give them a competitive advantage. For many centuries woodland composition was in flux.

Two kinds of factors have determined the structure of our native vegetation: first, there are the essential determinants of a plant community and, secondly, there is chance, climatic and edaphic factors and the biological characteristics of species. The accidental events and influences which are not necessarily predictable may have been the most influential. The important biological characteristics include the genetically determined relative reproductive and competitive abilities and ecological ranges of different plant species. The relative importance of essential and chance determinants of our vegetation has changed over time. In the Early Holocene the chance processes of migration were very important in vegetation change. Later essential determinants became more important. However, the essential determinants of the vegetation have not been constant. They have changed, either regularly or haphazardly, anyway. For example, the soil was maturing and the development of some kinds of vegetation was delayed until soils had matured to a state that could support them. In turn, the direction and rate of soil development depended upon the vegetation growing on it. Some kinds of vegetation maintained one kind of soil in an area where the climatic regime would have determined another.

2.7.1 THE SPREAD OF HAZEL

The Boreal period starts at different dates throughout the British Isles. It is indicated by the rise of *Corylus avellana* (hazel) to dominance. The early spread of hazel is a good indicator of the rapidly ameliorating climates. The Boreal was a period of dryness, certainly in large parts of eastern England. At first a large part of the North Sea Plain was dry land and so East Anglia was many miles from the sea. It was warm. *Emys orbicularis*, the freshwater turtle, could be found in East Anglia. *Betula–Corylus avellana* woodland was established by 9000 years BP at Llyn Gwernan in Wales and Loch Maree in Scotland. The rise of *Corylus* was rapid, perhaps from refuges in England (Figure 2.41) (Deacon, 1974) but its survival through the Windermere stadial stage is rather unlikely (Huntley, 1993).

It does not grow anywhere today where the mean annual temperature is less than about 2.5°C. The climate in the Early Holocene was very seasonal, which may

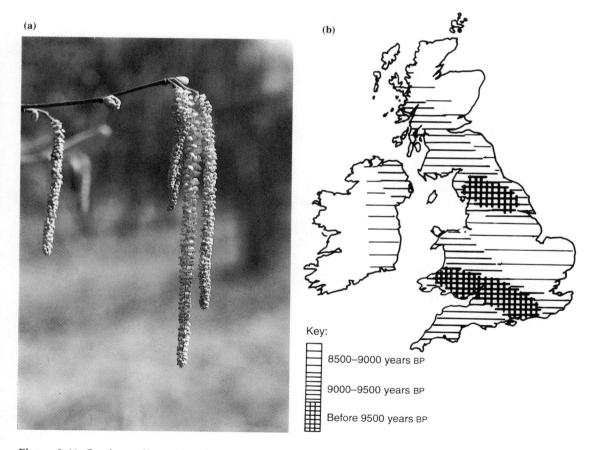

Figure 2.41 *Corylus avellana*: (a) catkins; (b) distribution in different ages of the Early Holocene (Deacon, 1974). ((a) Copyright Jon B. Wilson.)

have favoured *Corylus* over other species. The behaviour of *Corylus* was different in each interglacial. Perhaps each had a unique climate. *Corylus* became abundant earlier and was a more important part of the vegetation in each successive interglacial, and possibly this was the result of progressive evolutionary adaptation.

Corylus avellana became the dominant tree species because it spread faster than oak or elm, and because it was adapted to the fresh soils present at the time. However, other species, such as *Ulmus* (elm) and *Fraxinus* (ash) would also have been favoured. Hazel's own seedlings are shade intolerant. We know it today as an understorey shrub or as coppice but in the Boreal it was growing as a canopy tree. It will grow up to 6 m high (Rackham, 1980). When it grows like this, it flowers very profusely and this may explain the superabundance of *Corylus* pollen in the Boreal. *Corylus avellana* pollen was 17 times more abundant in Ireland and 4 times more abundant in England than all other tree species put together. The rise of hazel woodlands is accompanied by a decline in the herb flora.

The abundance of *Corylus avellana* in the Boreal is still something of a mystery. Perhaps humans had something to do with it. The human culture had changed from the palaeolithic of the Devensian, one which exploited open habitats, to the mesolithic, which, in some forms, exploited the woodland. An intriguing idea is that the activities of mesolithic man favoured hazel. Its nuts were an important part of the diet, but it seems unlikely that it would have been planted. Hazel coppices easily and responds to fire by growing vigorously from the base, though Rackham (1980) doubted that it is especially fire-tolerant. If mesolithic man was using fire to chase game towards his spears, or to maintain open grassland communities where game congregated, this would have encouraged hazel. At several sites, for example at Cothill Fen, charcoal is present in the Boreal deposits at times when the vegetation became more open (Day, 1991). In the dry climates of the Boreal stage, fires would have spread easily and wildfires may have been a common occurrence.

2.7.2 THE DEVELOPMENT OF SOILS AND WOODLAND SUCCESSION

Pennington (1986) has noted a lag between the time of arrival of tree species and their subsequent expansion in the Late Devensian and Early Holocene. The former is most closely related to the time of a change in the climate, recorded by changes in the fossil insect fauna, and the distance the species had to migrate. The timing and degree of later expansion of woodland is related to edaphic factors, especially the condition of the soil. The lag in plant response varied between 500 and 1500 years. For example, in north-west Scotland, although summer temperatures rose rapidly 10 000 years ago, as they did everywhere else in Britain, birch woodland only expanded between 9500 and 9000 years ago, by which time soils were already acid.

Theoretically, given the essential characteristics of the plant species and the nature of the habitat, it might be possible to predict what communities, dominated by which few species, would arise inexorably by vegetation succession. This might be complex because there are local as well as regional differences, so that relative competitive abilities are altered on different soils and climates. For example, local differences in vegetation history and a general lack of synchronicity have been noted between five sites within the Isle of Skye, where they have been related to local habitat differences: slope, aspect, soil type and degree of exposure (Birks, 1973).

The importance of differences in soil in determining the timing and nature of woodland successions is emphasized by a comparison of two pollen stratigraphies from sites only 3 km apart in East Anglia (Bennett, 1983a, b). Here the most important difference

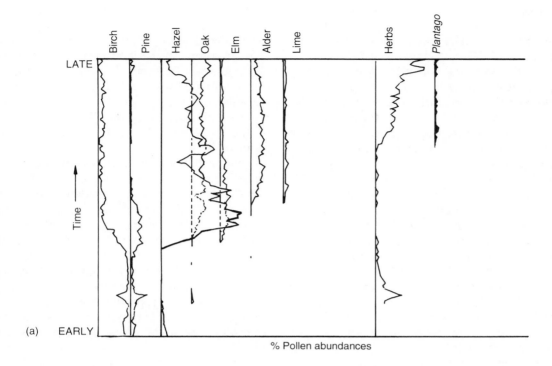

(a)

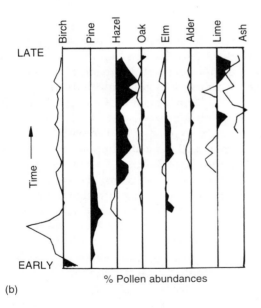

(b)

Figure 2.42 The Holocene succession: (a) at Hockham Mere; (b) pollen percentage difference between Hockham Mere (shaded) and Stow Bedon (unshaded). (After Bennett, 1983a,b.)

is that Hockham Mere records the vegetation on the Breckland sands and Bedon Mere records that on chalky boulder clay (Figure 2.42).

Pinus sylvestris (Scots pine) was an important part of the woodland around Hockham Mere on the Breckland sands from about 9500 years ago but it never established on the richer soils around Bedon Mere. At Hockham Mere there are two phases of oak expansion. The first, from 9300 to 8900 years ago, was probably *Quercus robur* (pedunculate oak). Its doubling time was 66 years, but it remained only a minor component of the *Corylus avellana–Pinus sylvatica* woodland.

The second phase of oak expansion, from 8600 to 8200 years ago, shows a slower rise, with a doubling time of 173 years. This was perhaps of *Quercus petraea* (sessile oak). By this time *Pinus sylvestris* was in decline, perhaps out-competed by the spread of *Q. petraea* on to the poorer soils where it had survived the first expansion of oak (Bennett, 1986). *Pinus sylvestris* largely disappeared by 7500 years ago, although it may have remained an important tree in the valleys and on raised bogs. The arrival of *Alnus glutinosa* (alder) later led to a decline of Scots pine in these habitats too. It survived in Scotland only on base-poor rocks of the north and west. On the better and thicker soils, developed on glacial drift, in lowland and eastern Scotland, oak woodland established (Pennington, 1986). The colonization of oak around Bedon Mere contrasts with that at Hockham Mere; the doubling time is fast and continuous, so that it has been suggested that it was due to *Q. robur* spreading at the expense of *Populus tremula* (aspen) on the more fertile soils.

2.7.3 THE CLIMATE IN THE FIRST HALF OF THE POSTGLACIAL

In the few thousands of years of the climatic optimum, from 8000 to 5000 years ago, in the Atlantic period, the so-called climax communities of our natural vegetation mosaic, wildwood of various sorts, were established (Figure 2.43).

Mean temperatures and rainfall 6500 years ago had reached their Holocene maximum. Compared to the first half of this century, summer and winter temperatures were 1°C higher. Rainfall in the uplands was 120% and in the lowlands 110% of present average rainfall.

The distribution and relative abundance of three contrasting woodland species, *Hedera helix* (ivy), *Viscum album* (mistletoe) and *Ilex aquifolium* (holly) have been used to help determine past climatic regimes in the wildwood (Iversen, 1944). They are helpful because of their contrasting biology. Ivy is one of our very few woody climbers, and is only absent from the most wet and acid soils. Mistletoe is an epiphytic hemiparasite. It grows on a wide range of trees but is especially prevalent on apple. Holly is an understorey tree which is intolerant of very base-rich and wet soils. It produces a mor humus and grows readily on acid soils, including podzols. The distribution of these species in north-western Europe is closely delimited by temperature, measured by a combination of the mean temperature of the warmest and coldest months (Figure 2.44).

In Britain each had a different distribution in the Atlantic period. *Ilex* did not reach as far north, especially in the east. It is limited to areas with a mean temperature for the coldest month of −0.5°C or warmer. The warmer the winter, the less warm the summer has to be for it to reproduce effectively. *Hedera* did not spread as far north, especially in the west. It has a similar warmth requirement to *Ilex* but can withstand a colder mean coldest month of −2.0°C. *Viscum* extended further north than both *Ilex* and *Hedera* in the east but was absent from its present range in the west. This was probably related to its need for high summer temperatures. It has a greater cold tolerance than the others, with a coldest month limiting-temperature of −8°C.

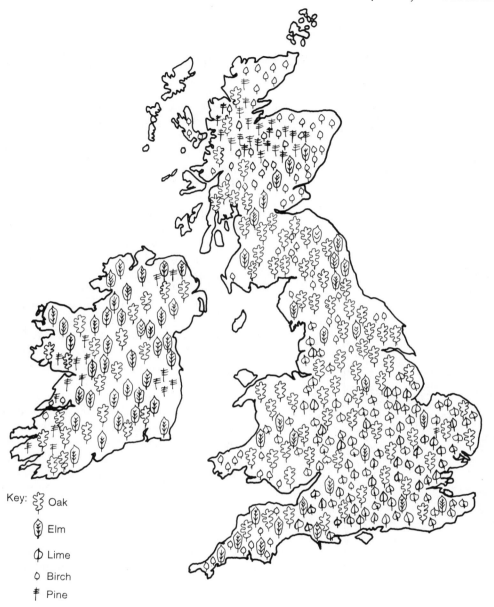

Key:
- Oak
- Elm
- Lime
- Birch
- Pine

Figure 2.43 Distribution of woodland types in wildwood 5000 BP.

Together, these past distributions indicate that climatic gradients at the end of the Atlantic stage were probably steeper than at present. There may have been warmer summers in the east but colder summers in the west. This helps to explain the dominance of oak or elm in the north and west. After the Atlantic stage, in the succeeding Sub-Boreal stage, which started about 5000 years ago, there was a more extreme

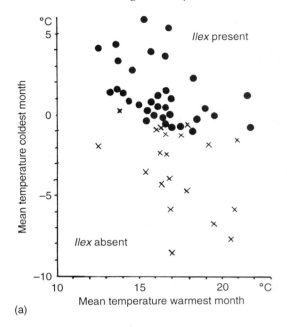

(a)

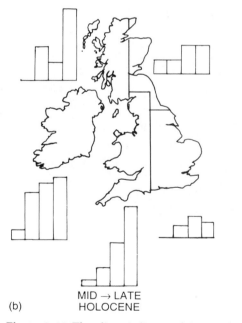

(b)

Figure 2.44 The climatic limits of *Ilex aquifolium*: (a) distribution of fertile and sterile populations in relation to mean temperature for warmest and coldest month; (b) abundance in successive Holocene stages in different parts of the British Isles (from Iverson, 1944 and Godwin, 1984).

continental-type climate, but this was felt mainly in south-eastern England. Winters were colder and there was less rain. Ivy declined for a while but it rose again in abundance in the succeeding Sub-Atlantic stage, from about 2500 years ago, when the winters may have become milder and there was almost certainly more rainfall.

The climatic optimum of warm, wet conditions followed the drowning of the North Sea Plain. The surrounding sea isolated and insulated the British Isles in a warm, wet blanket. For the time being, the presence of woodland cushioned the soils from the increased wetness of the climate. Transpiration of the trees dried the soil despite higher rainfall. Runoff into rivers may actually have decreased in some areas in the Atlantic period. However, when the trees were cleared there was a potential for a very rapid maturing of the soils, especially in the uplands, because of increased leaching. Acid and podzolized soils were to become widespread as more and more woodland was cleared. Different plant communities, adapted to the new soils, heathlands and moorlands arose. Woodland found it more difficult to re-establish on the impoverished acid soils. This was part of a continuing process of the contraction of the range of plants of open calcareous habitats, first because of the spread of the wildwood, and then later the spread of leached acid soils. Not only were the soils no longer suitable, but the new communities were too competitive. The rich herb flora of the Early Postglacial was restricted more and more to a few kinds of refuge.

The change to the warmer, wetter conditions of the Atlantic period is marked by the increased importance of the wetland tree *Alnus glutinosa* (alder) from about 7000 years ago. Its late establishment was due in part to its slow migration into north-western Britain from the south-east, perhaps limited by the steeper climatic gradients present. It is first recorded in the Thames estuary 8600 years ago, reaching Tregaron Bog in west Wales 7200 years ago and Loch Garten in northern Scotland 5800 years ago. It took advantage of rising water tables due to the rising sea level and flooding estuaries so that its distribution expanded earlier in the lowlands than the uplands.

The essential characteristics of *A. glutinosa* made it particularly suitable for the warm, wet climatic conditions of the Atlantic period. Catkins are formed in July but pollination does not occur until the following spring. The female catkins are exposed throughout winter and may be damaged by frost. The seeds take a long time to mature from pollination in February until the end of September. High humidity is required for germination in spring but its seedlings are very susceptible to drought and cold in early spring. Today, *A. glutinosa* is very widespread in the British Isles but it is confined to habitats that offer the necessary conditions for seedling establishment; river banks and other wetlands. It can survive flooding because its roots have a spongy 'aerenchyma' tissue, which allows oxygen to diffuse down into them, and it can also produce adventitious roots from the trunk.

The rise of *Alnus* provides an interesting example of how plant communities have developed through the interaction of essential and chance processes. It was particularly suited to the climatic conditions of the Atlantic period.

2.7.4 CLIMATE, SOIL AND WOODLAND TYPES

A major distinction of present woodland communities relates to different climatic regimes found across the British Isles. Similar patterns of variation were present in the wildwood. In this way past and present woodlands are in some ways comparable. As today, 5000 years ago there were cold northern upland and subalpine woodlands, cool and wet north-western submontane woodlands, and warm and dry south-eastern lowland woodlands. Although our surviving woodlands are much modified and may not be remnants of the primeval wildwood, we can understand a lot about the ecology of the wildwood by studying existing woodlands. For example, there is a fundamental distinction between wet and dry communities, the former dominated by *Alnus glutinosa* (alder), *Betula* (birch) and *Salix* (willows), and the latter by a range of other tree species.

Today it is possible to make a distinction between woodland communities on different soil types. Woodland communities can be categorized into three major types:

1. those on a rendzina or brown calcareous earth;
2. those on a brown earth of low base status;
3. those on a ranker, brown podzolic soil or podzol.

The main kinds of woodland are shown in Table 2.8.

The distinction between woodland types may not have been as clear in the wildwood as it became later, partly because of the haphazard nature of colonization and establishment. Most importantly, soils were immature and drainage patterns were not settled. Neutral or calcareous soils were widespread but changing the whole time as they became leached and organic material was incorporated into them.

Brown soils have developed under the broad-leaved deciduous woodland of much of the British Isles. They are present over many different rock types and now cover about 45% of England and Wales, 40% of Ireland and 10–15% of Scotland. They are more than 30 cm deep, and are moderately well drained and well aerated. They have a mull humus; leaf litter is rapidly decomposed and incorporated in the soil. They are neither very calcareous nor very siliceous. They gain their colour from the presence of ferric oxide. Normally there is an A-horizon which is characteristically dark grey–brown and is

Table 2.8 Distribution of dry-land woodland types according to *British Plant Communities* (Rodwell, 1991)

	Rendzinas and brown calcareous earths	Brown earths of low base status	Rankers, brown podzolic soils and podzols
Cool and wet north-western submontane zone	W9, ash/rowan/dog's mercury woodland	W11, *Q. petraea*/birch/wood sorrel woodland	W17, *Q. petraea*/birch/*Dicranum* woodland
Warm and dry south-eastern lowland zone	W8, ash/maple/dog's mercury woodland	W10, *Q. robur*/bracken/brambles woodland	W16, *Quercus*/birch/wavy hair grass woodland
Zone of natural beech dominance	W12, beech/dog's mercury woodland	W14, beech/brambles woodland	W15, beech/wavy hair grass woodland
Locally in southern Britain	W13, yew woodland		

slightly acid. It merges into a yellow–reddish brown B-horizon. There are many brown soil variations, depending in part upon the nature of the mineral material in which they developed, either sandy, calcareous, alluvial or colluvial, and also on the climatic regime. In areas of high rainfall, more than 1000 mm/year, and especially under mixed deciduous/coniferous forests, an acid brown earth develops. A luvic brown soil has a Bt-horizon where clay has been washed down from more superficial layers (Avery, 1990). The majority of brown earths have been cleared for agriculture because they have good fertility and a good range of mineral nutrients. Agricultural usage, especially ploughing, destroys their normal stratification.

2.7.5 HERBIVORES AND HUMANS IN THE WILDWOOD

Accidental events and chance have kept our vegetation in nearly constant flux. The 'chance' processes include different patterns and rates of migration, fluctuations in climate, interactions with herbivores, human activity and episodes of disease. In today's vegetation chance grazing is important in maintaining open plant communities and preventing the regeneration of a closed woodland, but grazing does not seem to have been a significant factor in the wildwood. The spread of shady trees such as oak and elm was inexorable. The loss of open conditions was marked by the extinction of reindeer, Irish elk and arctic fox in the British Isles. Horse became much rarer. Red deer, roe deer, elk, auroch and wild boar lived on, in or on the margins of the woodlands. There was a dearth of large herbivores in the wildwood. In some places wild boar rooted for acorns. Rare aurochs and red or roe deer ventured into the woodland in search of browse, but large areas of woodland were required to feed a few animals. It has been suggested that up to 2 km² of woodland could provide browse for only 20–30 cattle, but they probably lived as diffuse populations and, although they may have had a significant local effect, they did not prevent the spread of the wildwood. The Atlantic period saw the establishment of a mature shady woodland in which the dominant trees suppressed other vegetation.

There were some open areas which may have supported a denser population of animals. A few mesolithic people camped in open areas on river banks and lake sides which provided game, fowl and fish. Animals were hunted with bow and arrow and dogs. Elk waded through the marshes of the lowlands. Beaver dammed streams. In the uplands the woodlands of pine, birch and hazel were more open. On the highest plateaus of the North Yorkshire Moors substantial levels of *Calluna vulgaris* (heather) pollen and *Pteridium aquilinum* (bracken) spores have been recorded (Simmons and Cundhill, 1974). The woodland may have formed part of a mosaic scrubby heathland through which herds of deer ran. Mesolithic man moved into the uplands seasonally for hunting when the lowlands became too wet in winter.

It is striking that most mesolithic sites are outside or on the margins of the woodland. Nevertheless mesolithic man had a significant effect on vegetation development in some places, influential in the development of heathland and moorland. However, to mesolithic people the wildwood must have seemed a dense impenetrable barrier. Where the wildwood was slightly more open they established pathways. Some, like the Ridgeway in the Chilterns, and the Pilgrim's Way in Kent and Surrey, which runs just below the chalk scarp of the North Downs, used natural features as sign posts. It was the later neolithic culture that fully entered the wood, using it, changing it, clearing it, dominating it and ending the dominion of trees.

2.8 THE WILDWOOD, PRIMARY AND ANCIENT WOODLANDS

Despite human activity during the Atlantic period, a dense and shady mature wildwood covering much of the British Isles developed between 10 000 and 8000 years ago. Some existing woodlands may have had a direct continuity with this wildwood, but they are rare. They are called primary woodlands, in contrast to the secondary woodlands that have arisen in sites after clear-felling. The composition of primary woodlands has changed over the millennia in response to changing climate and soils. The succession towards one dominant climax species has not been permanent. The disappearance of *Tilia cordata* (small-leaved lime) from the woodlands it once dominated, in much of south-eastern England, is an example of the scale of the changes that have taken place. The vegetation succession continues today as *Fagus sylvatica* (beech) and *Carpinus betulus* (hornbeam) show signs of preventing the regeneration of oak in many woodlands. Indeed, in some cases there may have been a cyclical process whereby one tree species became dominant for a time only to be replaced by another, before reasserting its dominance later. Most importantly, in recent millennia primary woodlands have been modified by humans, managed as a source of timber and firewood, and changed by planting with trees from elsewhere (Rackham, 1980, 1976), Even if there are any surviving areas of primary woodland, they are different from the wildwood.

Another distinction between kinds of woodlands, that between ancient and modern woodlands, has been made. Ancient woodlands are very old woodlands, perhaps more than 300 years old. They may or may not be primary woodlands, but because they are ancient they have had time to accumulate a diverse range of underflora herbs and shrubs. A number of species have been suggested as indicators of antiquity, *Tilia cordata* is one. Others are *Convallaria majalis* (lily-of-the-valley), now naturalized as a garden escapee in many places, *Dryopteris aemula* (hay-scented buckler-fern), *Primula elatior* (oxlip), *Paris quadrifolia* (herb Paris), *Crataegus laevigata* (Midland hawthorn) and *Sorbus torminalis* (wild service-tree). Good indicators of antiquity, are, like *Primula elatior* (oxlip), slow to spread, and rarely colonize secondary woodlands, even over quite short distances. Lists of indicators of the antiquity of woodlands have been constructed for different small geographical areas (see Peterken, 1974).

The National Vegetation Classification identifies 19 different main kinds of woodlands in Britain today, 13 'dry' and 6 'wet' (Rodwell, 1991). These have been distinguished not on the basis of indicator species but on the total assemblage of plants. Many of the kinds of woodlands identified are represented by both ancient and recent woodlands, and even by relatively recent secondary woodlands. The floristic differences between them are best understood in terms of the soils and climate in which they are found and not on their antiquity. In many cases, even though humans have influenced the physiognomy of the woodlands, for example through coppicing, it seems that they can still be readily classified as one of a few major woodland community types (Table 2.8). This also means that variation in present woodlands may provide a reasonable simulacrum of the variation in the wildwood, with the difference that the wildwood was growing on younger soils and in slightly different climatic conditions.

2.8.1 JUNIPER WOODLANDS

Juniper woodlands were some of the earliest to develop in the Early Postglacial. The start of the Pre-Boreal is marked by a sudden abundance of juniper pollen as it responded to the sudden warmth.

Lowland juniper woodlands, like those of the Early Postglacial are not known today. *Juniperus communis* is a rare or occasional element of several other woodland types of very different kinds, including lowland woodlands, but only as an undershrub. However, *Juniperus communis* (juniper) can be the dominant woody species of a woodland found in drier areas, or free-draining areas, at high altitudes, especially in the eastern Highlands of Scotland (Rodwell, 1991; W19). Perhaps here and in its scattered occurrences elsewhere, in Upper Teesdale, the Lake District and the Southern Uplands of Scotland, these juniper-dominated communities

represent a ghost of the Pre-Boreal and Boreal juniper woodlands. A striking feature of the *J. communis* woodlands is the great diversity of growth forms exhibited (Gilbert, 1980). Some plants are tall and columnar, others pyramidal and others decumbent or prostrate. These are genetic variants. The presence of genetic diversity may be evidence for the antiquity and the relictual status of these *J. communis* woodlands (Gilbert, 1980).

The soil under juniper has an only mildly acid humus which is incorporated into the soil relatively rapidly. The flora of juniper woodlands can be very like that of pine woodlands, with a similar range of mosses. *Vaccinium myrtilus* (bilberry) and *Galium saxatile* (heath bedstraw) are prominent. Less-demanding calcifuges, such as *Luzula pilosa* (hairy wood-rush) and *Agrostis vinealis* or *A. capillaris* (brown and common bent), may be present. *Oxalis acetosella* (wood sorrel), *Viola riviniana* (common dog-violet), *Deschampsia flexuosa* (wavy hair grass) and *Anemone nemorosa* (wood anemone) are common and may be abundant. Sometimes the associated fern flora is diverse, with several shelter-demanding and slow-colonizing species.

2.8.2 BIRCH WOODLANDS

Birch has the advantage of producing many light, wind-dispersed seeds. It has a rapid growth rate and a short generation time of only 10–12 years. It is a 'ruderal' tree, by far the best colonizer of our native tree species. It rapidly took advantage of the Postglacial landscape. It is rather intolerant of shade but quick to take advantage of any opening. Soon it is eliminated by the competition of taller, shadier trees. *Betula pubescens* (downy birch) is a colonizer, with *Calluna vulgaris* (heather), of bare ground left by retreating glaciers in the Scandinavian mountains. It may have been the more important species of the two native tree birches in periods of cold, though in central Scotland it is replaced by *Betula pendula. Betula pubescens* favours wetter conditions than *B. pendula*. This ecological difference has reproductively isolated the two species because they grow in different areas. However, the two species hybridize very readily to obscure differences between them, and recently reproductive isolation has been weakened because of the disturbed habitats created by humans.

The birch woodlands of the Early Holocene were probably much richer in species than those commonly found today. They were growing in a better climate on younger, more calcareous soils. In today's landscape birch is much more familiar as a colonist of clear-felled or disturbed woods. Birch invades clearings. It is said that any tree planting in an ancient woodland turns into birch (Rackham, 1980). Although birch leaf litter produces a moder humus, so that relatively high pH levels can be maintained, birch cannot convert a mor humus to a mull humus. When birch has colonized areas where other trees have been felled the under-flora which was previously present is normally retained.

Perhaps those present-day meadow-birch woods of Scandinavia, which occur on sloping calcareous to slightly acid ground, can give us a better idea of our native Pre-Boreal birch woods. The herb layer includes *Trollius europaeus* (globe flower), *Aconitum napellus* (monk's hood), *Corydalis*, *Filipendula ulmaria* (meadowsweet), *Oxalis acetosella* (wood sorrel), *Polygonatum* (Solomon's seal), *Geranium sanguineum* (cranesbill), *Milium effusum* (wood millet), *Myosotis* (forget-me-not) and ferns.

Herb-rich birch woodlands are found today in a few places in north Scotland, for example at Inverpolly and Strath Lungard, Ross and Cromarty. They have a relatively rich under-flora, considering the latitude, with several species of grass and herbs, which survive because birch does not cast a deep shade. Other woody species are present but not in sufficient quantity to cast a deep shade.

There is a mixture of plants that will grow on acid or basic soils.

2.8.3 PINE WOODLANDS

In East Anglia *Pinus sylvestris* (Scots pine) was a component of the hazel woodland. It later became the dominant species over much of the British Isles for a while. In the Loch Lomond stadial a pine woodland had probably survived somewhere in western Britain, perhaps in an area now drowned by rising sea level, as well as in southern England. In the Pre-Boreal it appears to have established in two distinct areas from these refuges (Figure 2.45).

Today two kinds of pine can be recognized in Britain. That of native pine woodlands in Scotland is *Pinus sylvestris* subsp. *scotica*, and that from England, probably introduced, is *Pinus sylvestris* subsp. *sylvatica*. *Pinus sylvestris* subsp. *scotica* remains pyramidal for a long time as it grows, and at maturity becomes rounded in outline. It has shorter leaves and cones. *Pinus sylvestris* subsp. *sylvatica* has a flat-topped canopy. The difference between variants may date from the separate geographical origin of the two races in the Early Holocene.

In the Early Holocene, Scots pine was favoured in a similar way to hazel. Pine is relatively shade-sensitive and, interestingly, in a pine woodland when a tree dies it may leave a gap too small for it to regenerate in (Prentice and Lehmans, 1990). The development of pine-rich woodland in the Early Holocene is closely associated with a history of fire. Pine is said to be fire-adapted; in four sites studied in Co. Mayo pine survived the longest on the site with an almost continuous record of charcoal, a site which also had coarse, free-draining soil (Bradshaw, 1993). The pine forest that developed in the Boreal was growing on fresh, nutrient-rich soils. Today, native British pine woodlands are largely confined to marginal soils in Highland Scotland (Rodwell, 1991; W18). The soils here

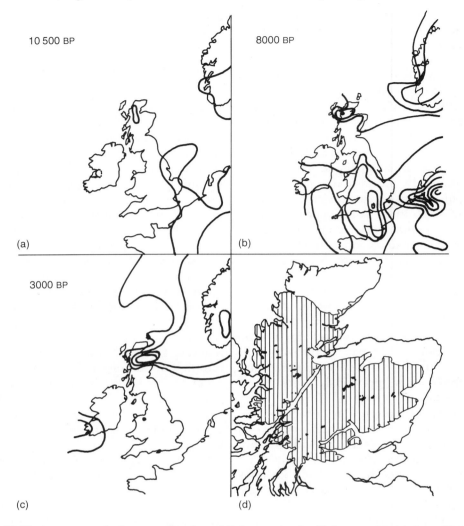

Figure 2.45 The progress of pine woodland establishment in the Holocene: (a)–(c) from Huntley and Birks (1983); (d) likely former distribution of the Caledonian pine woodland with surviving patches indicated (from Bunce and Jeffers, 1977).

have a long history of leaching and podzolization and are unlike those of the Boreal (Birks, 1970, 1972a, b). Perhaps some present-day pine woods of Scandinavia are more akin to these Boreal pine woodlands. Some are on rich brown earths and others are on rendzinas. There are several kinds. Most have *Empetrum nigrum* (crowberry) and some have *Juniperus communis* (juniper). Those on

rendzinas have a relatively rich field layer of calcicoles, such as *Epipactis atrorubens* (dark-red helleborine), *Melica nutans* (mountain melick), *Trifolium medium* (zigzag clover) and *Geranium sanguineum* (bloody cranesbill) along with the bilberries (*Vaccinium* spp.), heathers and heaths (*Calluna* and *Erica*). The moss *Leucobryum* is common. One important difference to the pine woodland of Scotland is the presence of *Picea abies* (Norway spruce). It was also a late arrival in Scandinavia but did not manage to colonize the British Isles before the Channel was flooded.

As broad-leaved deciduous species, especially oak, spread, the importance of *Pinus sylvestris* declined in most parts of Britain. Pine covered a significant proportion of Ireland 7500 years BP, except on the limesone-derived soils of central and northern Ireland. It was first ousted from wetter soils by the spread of *Alnus glutinosa* (alder) and further by the spread of blanket peat, confining it to small areas of western Ireland by 5000 BP (Bradshaw and Browne, 1987). In the cold northern uplands on podzols, especially the Highlands of Scotland, *Pinus sylvestris* remained the dominant tree species. Today it is still the potential dominant in the Highlands of Scotland, although only small patches of the Caledonian pine-woodland remain (Bunce and Jeffers, 1977). Although generally species-poor, the surviving patches of Caledonian forest, such as Rothiemurchus, are a glorious part of our landscape (Figure 2.46).

A character, which greatly adds to their beauty, is their openness; a mountain parkland through which the mountain background can be glimpsed, because the pine canopy generally occupies less than 70% of the view.

Pine woodlands do not have a rich herb flora but do contain a few interesting and beautiful plants. *Linnaea borealis* (twinflower) is named after Linnaeus, the founder of modern botany (Figure 2.47). He was a genius and very well aware of it but very easily slighted. He whinged about *Linnaea* that it was 'a plant in Lapland, low,

Figure 2.46 The Caledonian pine forest at Rothiemurchus.

Figure 2.47 Herbs of the pine forest: (a) *Goodyera repens*; (b) *Linnaea borealis*. (Copyright Jon B. Wilson.)

insignificant, forgotten, flowering for a short time: it is named after Linnaeus, who resembles it' (Lindroth, 1983). The orchid *Goodyera repens* (creeping lady's-tresses) is one beautiful resident of pine woodlands. It is more common in drier, eastern woodlands.

The pine woodlands also contain a range of interesting herbs that are limited by climatic conditions, such as the wintergreens: chickweed, intermediate, round-leaved, serrated and one-flowered wintergreens. These species come from several different genera, *Pyrola*, *Orthilia* and *Moneses* (Figure 2.48).

They share a common name because of the way they survive and grow, though slowly, at low temperatures, staying green in the winter to conserve their hard-gained growth.

At higher temperatures they respire faster than they can photosynthesize so that there is a net loss of energy. These species may be the first to decline if there is climatic warming because of the greenhouse effect.

The pine forest underflora is a mosaic of communities, the character of pieces of the mosaic determined by the degree of canopy closure and by edaphic factors such as the wetness of soils and their nutrient status. The more open kinds of pine woodland have a well-developed shrub layer and there is a thick carpet of mosses.

A geographical pattern of variation has been detected in the Caledonian pine forest, with different areas having different proportions of the various pieces of the pine-woodland mosaic (Bunce and Jeffers, 1977).

Figure 2.48 The wintergreens: (a) *Trientalis europaea* (chickweed wintergreen); (b) *Pyrola minor* (common wintergreen); (c) *Moneses uniflora* (one-flowered wintergreen); (d) *Orthilia secunda* (serrated wintergreen). (Copyright Jon B. Wilson.)

In the eastern Highlands, in the drier conditions, mineral podzols predominate and *Vaccinium myrtilus* (bilberry) or *Erica cinerea/Goodyera repens* woodland variants are more common. In wetter areas, especially in the south-western Highlands and on the westward-facing higher-altitude slopes of the central Highlands, peaty podzols predominate. Here an *Erica tetralix/Sphagnum* variant is more important. In the very wet parts of Wester Ross in the north-western Highlands there is a distinct community lacking *Erica tetralix* (cross-leaved heath) and *Molinia caerulea* (purple moor-grass) but having a very rich moss community including, *Scapania gracilis*, *Diplophyllum albicans* and *Thuidum tamariscinum* and the liverwort *Anastrepta orcadensis*.

2.8.4 WOODLANDS IN THE COLD

Paradoxically, Pre-Boreal and Boreal woodlands, which experienced a much better climate and soils that were young and base rich, survive in a much modified form only as fragments of upland woodlands. However, upland woodlands preserve only a ghost of the Early Holocene woodlands. The later spread of blanket peat and the intensive use of the uplands for grazing has greatly reduced their extent and greatly changed their composition. Only tiny, relatively insignificant, patches survive. Changes in climate since the climatic optimum of 6000 years ago, and the changes in soil that have accompanied them, have greatly reduced the potential for the development of upland woodlands. The growth of these different species at altitude is limited in different ways. In northern England up to 580 m, the growth of *Betula pubescens* is directly related to the accumulated air temperature above 6°C in that season. In contrast, growth of *Pinus* in one season is related to temperatures in the previous season, because the nodes from which needles are produced are formed in the previous July and August.

A reduction in temperature reduces the ability of trees to produce viable seed. *Sorbus aucuparia* (rowan) seed decreases in weight and viability with altitude. In addition, at the highest altitudes the shortness of the growing season prevents young seedling shoots from maturing and hardening before winter. The most northerly surviving natural woodland in the British Isles is one of *Populus tremula* (aspen) growing at Berriedale on Hoy in the Orkneys (Crawford, 1989). It has survived only because of the ability of *P. tremula* to reproduce by suckers. These trees have never been observed to produce seed. The Orkney aspens have also been protected from grazing by their bitter taste.

A retreat of woodlands from uplands after the first half of the Holocene is highlighted by the survival of remnant woods and by fossil tree stumps preserved in peat at high altitude. Pine woodlands grew at 790 m in some places 7000 years ago but today the highest remnant is at Creag Fhiaclach, Glen Feshie above Loch an Eilean, where it grows at 640 m. Here some plants adopt the dwarf, twisted Krummholz form familiar in many tree species in exposed high-altitude situations. The Krummholz growth form is not commonly seen in the British Isles, perhaps because here upper altitudinal limits are defined as much by grazing and soils as by climatic exposure. Today the upper altitudinal limit of several pine woodlands is only about 530 m and here trees have the normal upright form. Another estimate is that the altitude limit of tree birch in northern England is now 580 m. There was widespread birch woodland over the fells of northern England up to 760 m 8000 years ago.

In part, upland woodlands are now limited by changes in the soil and the spread of blanket peat. Changes in upland soils may have been initiated and encouraged by mankind's activities. Upland woodland ecosystems are in a fragile state at their ecological limits. In some areas the retreat of woodlands from their altitudinal limits or from areas edaphically marginal may have occurred without any interference from mankind. After climatic cooling and increased wetness in the later Holocene podzols and an acid mor humus developed. The blanket peat of upland moorlands replaced woodland. Today even if the climate ameliorated and climatic or edaphic factors were to change, the re-establishment of upland woodland would now be prevented by the higher levels of grazing encouraged by humans.

Today, apart from *Juniperus communis* and the dwarf willows, the highest-altitude trees are remnant *Betula pubescens* (downy birch) and *Sorbus aucuparia* (rowan) trees. The relationship of these upland birch and rowan woodlands to the previous pine woodlands is shown by their very similar shrub and field layer. *Betula pubescens* has a high-altitude 'Krummholz'-like variant, a shrubby sub-species with small leaves, called *B. pubescens* subsp. *tortuosa*. *Sorbus aucuparia* maintains an upright form even at the highest altitudes. Here it is excluded by grazing and the presence of blanket bog from most areas, and is confined to rock outcrops.

2.8.5 NORTHERN AND UPLAND OAK WOODS

In the north and west of Britain 6000 years ago there was an oak- or hazel-dominated woodland. Now seminatural woodlands are dominated by *Quercus* (oak) with *Betula* (birch) or ash (*Fraxinus excelsior*) and field maple (*Acer campestre*). *Corylus avellana* (hazel) is only less important now because it is most frequently seen only as an understorey or coppice plant below other tree species. As a coppice or understorey plant it flowers less profusely than as a canopy tree.

There are two native species of oak, *Quercus robur* (pedunculate oak) and *Q. petraea* (sessile oak). In some woodlands the oaks are very variable. This is a good indication of the ancient status of the woodland. Planted oak woods are usually much more homogeneous, perhaps because the acorns from only one or a few favoured trees were sown. Oak variability may also reflect a long history of hybridization between the two species of oak. The two species hybridize to a considerable degree, so that in some woodlands the distinction between the species is blurred. The lobing of the leaf, the length of the petiole (leaf stalk) and the length of the peduncle (acorn stalk), are all key characters for identifying the species, but these can form a continuous spectrum of variation (Figure 2.49).

The length of the peduncle is the most useful character, though it is not always readily available because oak has the habit, like many tree species, of having occasional 'mast' years when they fruit abundantly. *Quercus robur* has long peduncles, and also a more or less hairless undersurface to the leaves, in contrast to the sessile acorns and stellate pubescence of *Q. petraea*.

There is a considerable overlap in the ecological range of these two species, but *Q. robur* favours richer, more basic soils and is a more effective colonist than *Q. petraea*. Its seedlings are less tolerant of shade than *Q. petraea*. It also occurs more generally at higher altitude and is the most common oak in the far north. For these reasons it seems likely that *Q. robur* was the early colonist of many sites. It is a frequent pioneer species of abandoned land. *Quercus* is not the most competitive of our native trees. It is relatively slow growing as a young tree, but can withstand a certain amount of shade, maximizing the reception of light with its multistoried canopy (Figure 2.49c).

Quercus and *Ulmus* remained the dominant trees in the wildwood in northern and western Britain, but in most of England they were displaced as the dominants by the spread of *Tilia cordata* (small-leaved lime). The surviving northern and upland oak woodlands have a more direct continuity with the wildwood than lowland oakwoods. However, like woodlands everywhere in Britain, they are much reduced today from

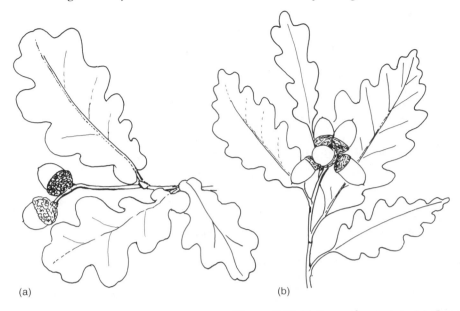

(a)

(b)

(c)

Figure 2.49 Native oak species: (a) *Quercus robur* (pedunculate oak); (b) *Q. petraea* (sessile oak); (c) Prince's Covert Wood, Claygate, Surrey.

their former extent and have been managed for thousands of years. These deciduous woodlands have survived because they have been an important part of the rural economy. Different kinds of management and other kinds of interference have obscured potential differences in natural woodland types. In other cases they have encouraged the development of distinct types (Rodwell, 1991).

On the poorest soils, on the rankers and acid podzolic soils which are common in upland sites, the field layer is normally rather thin but there is normally a rich carpet of mosses (Rodwell, 1991; W17). Different subcommunity types have different moss species, related to a pronounced change in the relative oceanicity of the climate, from west to east. A subcommunity, with the prominent mosses *Isothecium myosuroides* and *Diplophyllum albicans*, is found where there are more than 200 wet days each year. In the east, where there are 180 wet days or fewer, the cover of mosses is lower and *Rhytidiadelphus triquetrus* and *Pseudoscleropodium purum* are more common. On less acid soils, on acid brown earths, there is less *Vaccinium myrtilus* and *Calluna vulgaris*, and also less *Deschampsia flexuosa*, while a field layer of other herbs is more important (Rodwell, 1991; W11). A range of grasses is important and broad-leaved herbs are much more common on the richer soils; *Galium saxatile* (heath bedstraw) and *Potentilla erecta* (tormentil) indicate leached soil profiles. *Pteridium aquilinum* (bracken) can be extensive, growing vigorously on the deeper, well-drained soils. Where there is little grazing the community may be rich in ferns such as *Pteridium aquilinum* (bracken) and *Dryopteris* species, *D. filix-mas*, *D. dilatata* and *D. remota* (male fern, broad-buckler fern and scaly male fern), and in brambles (*Rubus fruticosa* agg.).

Oxalis acetosella (wood sorrel) is more or less ubiquitous in northern oak woodlands. It is also found in other habitats, as a woodland remnant in rough pasture, on grassy banks and also in deep, shaded rock fissures where light is low and moisture high, even to relatively high altitude. *Oxalis acetosella* is particularly successful because it is very shade tolerant. Genetic differences for adaptation to shade between variants in different localities are present (Packham, 1978). More importantly, *O. acetosella* shows a marked developmental plasticity in response to different levels of light. Leaflets are wider, thinner and have a higher chlorophyll content in the shade. Inside the leaf, the cells are funnel-shaped to trap light. In heavy shade relatively more shoots are produced than roots. In addition in lowland situations, and in the shade, young shoots are produced earlier and old leaves senesce later, allowing the plant to remain green in winter to photosynthesize after tree-canopy leaf fall. *Oxalis acetosella* has a slow growth rate but also a low respiration rate, so that little photosynthate is wasted. It has a light compensation point, the intensity of light above which there is a net gain of energy, of only 300–500 lux (Packham and Willis, 1977). This compares to 2000 lux in *Deschampsia flexuosa* (wavy hair grass) which is dependent on sun flecks for photosynthesis. Another adaptation of *Oxalis acetosella* is its unpalatability to potential herbivores because it contains oxalic acid.

A woodland subcommunity which is found in wetter hollows and plateaus is characterized by the presence of *Anemone nemorosa* (wood anemone) (Figure 2.50). It can push its way up through a mat of leaf litter because the young shoot is crosier-headed to protect the flower bud, and it can then grow relatively tall. This contrasts with wood sorrel which is confined to raised areas like tree bases and hummocks where litter does not accumulate (Sydes and Grime, 1981a, b). *Anemone nemorosa* avoids the consequences of the waterlogged soils it grows on by having superficial roots, and it has its own anti-herbivore chemical, called proto-anemonin. It largely avoids shade by shoot growth in late winter and early spring. This growth is mainly by cell expansion aided by the presence of high levels of DNA in each cell. By mid-July the above-ground parts of the plant have died back.

Figure 2.50 Vernal herbs: *Anemone nemorosa* (wood anemone); *Ranunculus ficaria* (lesser celandine); *Primula elatior* (oxlip).

There are some famous upland dwarf oak woods, like the Keskadale oaks above the road between Buttermere and Keswick in the Lake District, which are mainly *Q. petraea*, and Wistman's Wood on Dartmoor, which are mainly *Q. robur*. Both woodlands lie between 300 and 450 m above sea level. The oaks of Wistman's Wood are twisted and covered with epiphytic ferns, mosses and lichens. At these altitudes birch can rival oak in height and exceed it in importance. The species of birch present in these oakwoods is usually *B. pubescens*.

2.8.6 LOWLAND 'MIXED' OAK WOODS

Quercus petraea woodlands may have a different history to *Q. robur* woodlands. *Quercus petraea* is more common in the north and west of Britain. Here it was the dominant tree north of the area of natural lime dominance. In the south-east, *Q. petraea* woodlands probably existed as islands in the sea of *Tilia* woodland, on areas of infertile sandy soils. Several of these ancient woodlands survive today around London, on the acid Blackheath Beds, Bagshot Sands and Claygate Beds, now islands in the sea of urban and agricultural development. Kenwood, on Hampstead Heath in London, is one, a melancholy tamed reminder of the wildwood, fenced and pruned, and overrun by rhododendrons. Nevertheless, even here, the antiquity of the wood is suggested by the survival of *Sorbus torminalis* (wild service-tree) and *Crataegus laevigata* (Midland hawthorn). Another indication of their antiquity is the genetic diversity of the oaks themselves. Each tree has a different physiognomy, recognizable by its leaf and acorn shape.

In lowland Britain, mainly in England, two different oak-wood communities are found with a different range of shrub and field layer species. On the poorest soils, the rankers and acid podzols, one is characterized by the presence of *Deschampsia flexuosa* (wavy hair grass) as well as birch, especially *Betula pendula*. *Deschampsia flexuosa* is relatively shade tolerant, though it relies on sun flecks to photosynthesize. In experiments its seedlings tolerated 3 months in the dark without dying. *Quercus–Betula pendula–Deschampsia flexuosa* woodland is not much managed, and although it is usually found in marginal situations it may not be a primary woodland type (Rodwell, 1991; W16). *Pteridium aquilinum* (bracken) and *Calluna vulgaris* (heather) are common in this kind of woodland, indicating its close relationship to two other types of acid-soil vegetation whose development is directly related to mankind's activities: heathland and beech woodland. This kind of oak wood may be secondary woodland developed on heath. Certainly this is supported by *Q. robur* (pedunculate oak) being more common than *Q. petraea* (sessile oak) in these woodlands. It has the greater ability to colonize open areas.

The other kind of lowland oak woodland found in England has a much greater claim to ancient/primary status, although it has been much modified from the wildwood. This is the *Quercus–Pteridium aquilinum–Rubus fruticosa* woodland (Rodwell, 1991; W10). In the recent past many of these oak woods were managed as coppice or wood pasture. They are found on weakly acidic brown earths. These woodlands were dominated by *Tilia cordata* (small-leaved lime) 7000–5000 years ago. *Tilia cordata* still forms a small component of mixed woods, especially in East Anglia and on the limestone in north Lancashire.

Ireland has one of the lowest areas of woodland in the whole of Europe, with less than 3% of the total area, even including plantations of conifers. Nothing much remains of the wildwood of *Ulmus glabra* and *Corylus avellana*, which once covered most of the land, except perhaps some of the scrubby *Corylus avellana* woodlands like those in the Burren, and strips of woodland on the steep slopes of eskers above the central plain.

There is a surviving primary oak wood in the Gereagh, near Macroom, in County Cork

(O'Reilly, 1955). Here in a wet valley flood-plain, dissected and intersected by many streams, is a wet oak-woodland on islands of alluvium. Below the canopy of *Q. robur* is *Corylus avellana* (hazel) and *Ilex aquifolium* (holly). There is a rich growth of epiphytes. The ground flora is dominated by *Osmunda regalis* (royal fern), *Allium ursinum* (ramsons) and *Euphorbia hyberna* (Irish spurge).

In Killarney are the best surviving semi-natural *Q. petraea* woodlands in Ireland. Here, on open ground not shaded by the oaks, *Arbutus unedo* (the strawberry tree) grows. The presence of *Arbutus* in Ireland, a species which is otherwise a component of the maquis vegetation of the Mediterranean region, is problematical (Sealy and Webb, 1950). There is no fossil evidence for a presence in warmer interglacials. For example, it was not part of the Gortian flora. It suffers badly where it has been grown in eastern England during hard winters. Nevertheless it was once more widespread in Ireland in the Holocene. Its distribution in Ireland indicates that it has survived only on thin soils where there was an absence of deep shade from other trees.

Killarney also has the only large yew woodland in Ireland. Along with *Taxus baccata* (yew) the main shrub layer plants are *Ilex aquifolium* (holly) and *Rhododendron ponticum* (rhododendron), the latter introduced only recently. The field layer is dominated by *Allium ursinum* (ramsons) where the soils are slightly base rich, and *Hyacinthoides non-scripta* (bluebells) where they are slightly acid. In the deep summer shade there is *Lysimachia vulgaris* (yellow loosestrife), *Orchis mascula* (early purple orchid), *Oxalis acetosella* (wood sorrel), *Geum urbanum* (wood avens), *Sanicula europaea* (sanicle) and *Mercurialis perennis* (dog's mercury). However, the glory of these wet woodlands is the rich moss, liverwort and fern flora; 186 species of mosses and liverworts have been identified. They include the beautiful filigree of *Thuidium tamariscinum*

covering rocks and tree boles alike. In the yew wood can be found the 'woolly mop' of *Thamnobryum alopecurum*. The mosses festoon the tree trunks and larger branches, creating places where ferns can establish epiphytically. The ferns include the rare *Hymenophyllum wilsonii* (Wilson's filmy-fern).

2.8.7 ASH, HAZEL AND ELM WOODLANDS

Fraxinus excelsior (ash) was probably only a minor component in much of the wildwood, existing as scattered solitary trees in the oak or lime woodland, an opportunist taking advantage of gaps in the canopy to establish. It can still be found like this today. However, ash woods generally favour relatively infertile calcareous soils where competing trees do not grow well. Ash can produce abundant wind-dispersed fruits, but seedlings are suppressed in shade, susceptible to damping-off in moist, shady sites, especially under the shade of a field layer such as *Mercurialis perennis* (dog's mercury). If it manages to grow above the field layer, the sapling can survive for a long time in the shade of the tree canopy, scarcely growing but waiting for a canopy gap to appear into which to grow. Upland ash woods are today remnants on the margins of upland pastoral land and are found on very irregular or steep slopes, and especially on Carboniferous limestone.

Ash woods are very diverse and include a wide range of subcommunities. This is probably the result of a long history of human disturbance. Many ash woods are secondary woodlands. A broad geographical distinction can be made between southern ash woods where *Acer campestre* (field maple) is commonly present (Rodwell, 1991; W8) and those of the wetter uplands of north and west Britain where *Sorbus aucuparia* (rowan) and birch are more common (Rodwell, 1991; W9). *Corylus avellana* (hazel) is a particularly important component of ash woods. There is a close floristic affinity between British ash woods and the hazel woodlands of the west of Ireland.

A common tree in the Early Postglacial woodlands was elm. In the Postglacial it became generally important shortly before oak. It remained a very important tree, especially in western Britain and Ireland, until it was decimated by several episodes of elm disease about 5000 years ago. This decline marks the beginning of another phase in the development of the British vegetation (section 3.1). Elm partially recovered from disease and was still an important tree in some ash woods before the latest episodes of elm disease. This indicates that in some areas ash woods may be the descendants of the elm–hazel wildwood which 7000–5000 years ago grew especially in south-west Wales, Cornwall and over much of Ireland. Elm can cast a deep shade and must have suppressed the underflora in a similar way to *Tilia*.

There are a number of native species of elm (*Ulmus* spp.) of two distinct kinds; *U. glabra* (wych elm) and the others, the lineage elms. The latter reproduce largely by suckering. Different numbers of species have been recognized, including the English, small-leaved, Cornish, Jersey and Plot's elms, a taxonomic situation made more complex by widespread and frequent hybridization. However, the most likely candidate for the important elm of the Early Postglacial, and the one which remained important in northern and western ashwoods, was *Ulmus glabra* (wych elm). *Ulmus glabra* may even have entered the region very early. It formed at least a third of the elm–hazel wildwood of Ireland and south-western Britain, and even up to a tenth of the wildwood elsewhere.

Ulmus glabra is today much more common in north-western Britain and Ireland than other species of native elm. In this region a segregate subspecies called *U. glabra* subsp. *montana* is found. It has narrower leaves than the typical southern *U. glabra* subsp. *glabra*. The latter was an important component of the *Tilia*-dominated wildwood of 7000–5000

years ago. It remained a constant associate of *T. cordata* even after most of the *Tilia* wild-wood had disappeared, and is an indicator of ancient woodland today.

It has been suggested that the lineage elms found in Britain, *U. procera* (English elm), *U. plotii* (Plot's elm) and various subspecies of *U. minor* (small-leaved, Cornish and Jersey elms), are not native because they set seed so infrequently. These so-called lineage elms all have the remarkable facility for a tree to reproduce mainly by suckering. They are not generally part of woodlands today, but are found mainly in hedgerows and around habitations, where they have been planted. However, it is precisely this ability to sucker which may have enabled them to have survived the Loch Lomond interstadial within or in the close vicinity of Britain. It is likely that in the distant past the lineage elms were a constituent of natural woodlands on gravel river terraces (Rackham, 1980). No trace of these woodlands survives today because these were some of the earliest woodlands to be cleared by humans. The ancient status of some elm clones is indicated by their large extent; these are weakly sucker-ing, slowly expanding hybrids between *U. glabra* and *U. minor* ssp. *minor* (small-leaved elm) which must be thousands of years old.

2.8.8 THE SHADIEST TREES

Tilia cordata (small-leaved lime) became abundant at about the same time as *Alnus glutinosa* (alder). Its importance in our natural climax wildwood is underestimated by absolute pollen frequencies because, unlike most other native tree species which are wind pollinated, it is insect pollinated and there-fore produces less pollen per tree. Corrected pollen frequencies for the wildwood make *T. cordata* a very important species in south-eastern England. It rose to dominance at different times in different regions; in East Anglia about 7200 years ago and around

Oxford 6800 ^{14}C years ago. In north-east Yorkshire it expanded in the lowlands at a similar age to the Oxford area but, like *Alnus*, it expanded into the uplands only several hundred years later, and there it did not rise to dominance.

The other native species of lime, *Tilia platyphyllos* (large-leaved lime), is rarer, especially in eastern England, and is usually confined to limestone. However, one place it can be seen in fair quantity is in the woodlands above the River Wye in Gloucestershire. The *Tilia* most commonly seen today, planted frequently as a street tree despite its habit of dripping sticky 'honey-dew' on parked cars, is *T.* × *europaea* (common lime), the hybrid between *T. cordata* and *T. platyphyllos*. The dominant species of the wildwood in south-eastern England, *T. cordata*, can be identified from the others by the way it holds its flowers and young fruits above the foliage. In the others they hang down. Today *Tilia* woodlands are rare, but *T. cordata* is an important component of groups of woods in various areas of the country from Suffolk and Essex up to Silverdale in Lancashire (Rackham, 1980).

The character of the *Tilia* woods at the climatic optimum was very different from the mixed oak woods which are found in the same geographical region today (Rackham, 1976). *Tilia cordata* is potentially a very tall tree and casts a deep shade. Its canopy forms a uniform leaf mosaic with few light gaps. Its own seedlings can survive in this shade but those of most other species die. This makes it a very good competitor. Under the dense stands of *Tilia cordata* that once existed over much of England, the field layer was probably fairly poor. Under existing large stands of *Tilia*, *Pteridium aquilinum* (bracken) is much less frequent than under oak. Either *Rubus fruticosa* (brambles) or *Mercurialis perennis* (dog's mercury) are the most important plants, with the only other significant members of the underflora being the vernal herbs, *Hyacinthoides non-scripta* (bluebell) or

Primula vulgaris (primrose). In the *Tilia* wildwood the suppression of the underflora may have blurred any distinction between any potential subcommunities developed on different soils.

Fagus sylvatica (beech) has a similar ecological position in today's woodlands to that once occupied by *Tilia cordata*, but *F. sylvatica* was a very insignificant component of the wildwood. It inherited south-eastern England from *Tilia* very late in the Holocene and in many cases second-hand after the wildwood had been first cleared for arable or pastoral use and after a succession to a secondary oak wood. Many beech woods, seemingly ancient, have been planted only in the past few centuries. Nevertheless, *F. sylvatica* is the dominant tree, or potential dominant, in much of England where *Tilia* was once dominant, and it provides a useful comparison because of the way it suppresses the underflora (Figure 2.51).

The shade may be thickened by a well-developed *Rubus fruticosa* (bramble) layer (Rodwell, 1991; W14). The shade is so dense that *Pteridium aquilinum* (bracken) performs poorly. A similar situation is seen under *Tilia*.

If anything, *Fagus* may be an even more formidable competitor than *Tilia*. As well as casting a deep shade, its leaf litter accumulates, suppressing the field- and ground-layer plants. A mor humus develops which can lead to a decline in soil fertility through leaching, even leading to podzolization. *Fagus sylvatica* grows on a very wide range of soil pH, from <3.5 to >7.5; however, it develops best on the deeper acid soils. On thin rankers where there is a *Fagus sylvatica*–*Deschampsia flexuosa* woodland (Rodwell, 1991; W15), or on calcareous soils where there is a *Fagus sylvatica*–*Mercurialis perennis* woodland (Rodwell, 1991; W12) the trees often grow crookedly and do not achieve the magnificent heights of 30 m or more found in some beech woodlands on brown earths. *Fagus sylvatica* is surface rooting, a feature seen especially when trees were blown over in the hurricane

Figure 2.51 Pollarded beech trees (*Fagus sylvatica*) at the iron age site, Loughton Camp, Epping Forest, Essex.

of 1986. The surface roots compete for moisture, restricting the development of a field layer (Watt and Fraser, 1933). The cathedral-like interior of some beech woods owes no little debt to their sterile emptiness.

In contrast, *Tilia cordata*, though it is today found predominantly on ill-drained acid soils of pH 3.6–4.9, favours fertile soils, often derived from loess. Mull and mor humus are equally common humus types. These fertile soils were cleared of woodland for farming. *Tilia* woods were, anyway, vulnerable because *T. cordata* was no longer regenerating effectively. A slight climatic cooling has reduced the potential for fruit to mature.

Once clear-felled, if a secondary woodland developed, *Tilia cordata* was replaced at first by birch and then by *Q. robur*. In some areas there may have been a gradual decline of *Tilia*

as it failed to regenerate, so that *Q. robur*, already present within the woodland as scattered individual trees, increased to dominance. This would have been accompanied by the development of a more extensive field layer. In the lowland oak woods of Britain today, *Q. robur* is still today the more common species of oak, especially on the wetter, gleyed soils.

The development of oak woods was, in some cases, only a temporary stage. *Quercus robur* finds it difficult to regenerate under an established tree canopy, so that in southern Britain, on the drier soils, *Fagus sylvatica* later became the new dominant tree. *Fagus sylvatica* seedlings are shade tolerant. It out-competed oak and changed the soils it was growing on. This sequence of succession has been recorded from Epping Forest (Baker, Moxey and Oxford, 1978). However, even after *Fagus* replaced *Tilia* as the climax tree in southern England, the natural spread of *Fagus* in northern Britain was hindered climatically. Nevertheless, once established as a plantation in the area of natural climax northern oak wood, it has proved itself successful well to the north of its natural range. It can out-compete *Quercus petraea* whose seedlings are more shade sensitive.

Similarly *Carpinus betulus* (hornbeam), a very minor component of the wildwood, has been slow to expand its range in Britain, despite being one of the most common trees in France and Germany. It favours wetter gley soils, and, like *Fagus*, out-competes oak. Its seedlings are very shade tolerant and a two-layered canopy of mature trees and saplings can create a very dense shade. In some woods its rise to dominance may have been delayed by centuries of coppicing. Since coppicing has declined, mature *Carpinus* woodland has begun to establish.

Taxus baccata (yew) is a native conifer, which is found quite often in beech woodlands. One particular subcommunity of beech woodlands is characterized by the presence of a well-developed undertier of yew. Very few other species grow in this subcommunity, which is deeply shady. One distinctive associate is *Buxus sepervirens* (box). In several places on steep, dry chalk slopes, especially at Kingley Vale in Sussex, and in one place on the Magnesian limestone at Castle Eden Dene in County Durham, yew woodlands have developed (Rodwell, 1991; W13). Once established, yew woodlands are the most species-poor of all our woodlands. They were probably not a feature of the wildwood but may have established by invading grassland in the dry Sub-Boreal period.

2.9 THE ECOLOGY OF THE WILDWOOD

The picture we can put together of the wildwood from the pollen record and by comparison with existing woodlands is that of a dark and gloomy place. With the exception of parts of northern and western Britain, which had either a *Pinus sylvestris* (Scots pine) or an oak-dominated woodland, and some upland areas, which were slightly more open, the wildwood was dominated by *Tilia cordata* or else by *Ulmus glabra* with *Corylus avellana*, each of which casts a dense shade. In the south, except in scattered localities on thin, peculiar or waterlogged soils, on deeper, well-drained soils, trees were tall and closely spaced. The underflora was suppressed so that, apart from a flush of vernal herbs, the field layer was a monotonous carpet of shade-tolerant species such as *Mercurialis perennis* (dog's mercury) or a tangle of *Rubus fruticosa* (brambles).

The amount of dead wood lying about in the wildwood would have been very striking to the modern observer. We can see something like this in parts of the New Forest or Windsor Great Park, where massive trees have been left lying where they have fallen. A fallen tree may have made an opening in the canopy, a temporary glade colonized by grasses, later by birch and oak, and attracting grazing animals. However, dead trees may not have always fallen to make an opening in the canopy. Even today in the New Forest dead trees can be seen standing upright, supported by the boughs of their neighbours, with the canopy closing around them.

The wildwood formed a green monochrome canopy, but on closer inspection there was a pattern concealed within the green. Some patterns had arisen from competition between species. Species diversity was maintained to a degree as the result of a partitioning of the environment, seasonally and spatially, so that plants could coexist. Plants of different species had different sizes, shapes and behaviours, so they fitted together in a diverse mosaic (Packham and Harding, 1982). Differences in the time of producing leaves and flowers spread the requirement for nutrients and light. There were temporal and seasonal fluctuations in microclimate, and in the activities of fungal pathogens or insect and mammalian herbivores. In summer 11% of the sunlight reaches the soil in an oak woodland, but in winter, 56%.

The size of the gap in the canopy is critical in regeneration. *Betula*, *Pinus sylvestris* and *Fraxinus excelsior* are shade intolerant and require a larger gap in the canopy to regenerate than other species. Trees of the same species of different ages exploit the environment in different ways. Some trees, such as *Carpinus betulus* (hornbeam), produce a seedling bank, a cohort of seedlings and saplings which survive for a long time hardly growing, waiting for an opening in the canopy. Some species, especially herbs of woodland glades, produce a long-lived seed bank in the soil.

There may be considerable variation in soil pH and moisture. Woodland calcicoles include *Sanicula europaea* (wood sanicle), *Allium ursinum* (ramsons), *Daphne laureola* (spurge laurel), and calcifuges include *Vaccinium myrtilus* (bilberry) and *Calluna vulgaris* (heather). Root development of most of tree species of the wildwood is inhibited in waterlogged soils, so that in the most waterlogged areas there would have been a different vegetation dominated by *Alnus glutinosa* (alder). The presence of tree roots influences the composition of the field layer (Watt and Fraser, 1933).

Different tree species lose their leaves at different times and there are large differences in the persistence of the litter. Tree litter collects unevenly on the woodland floor. A dense litter may suppress seed germination and seedling establishment of some species. Under *Acer pseudoplatanus* (sycamore), for example, there is a succession in the ground flora with a decrease in leaf litter from *Lamiastrum galeobdolon* (yellow archangel) to *Hyacinthoides non-scripta* (bluebell), to *Anemone nemorosa* (wood anemone), to the grasses *Milium effusum* (wood millet), *Holcus mollis* (creeping soft-grass) and *Poa trivialis*

(rough meadow-grass) and to *Mnium hornum* (a moss) (Sydes and Grime, 1981a, b). In deep litter, oak acorns are less likely to suffer predation.

2.9.1 EVADING OR TOLERATING THE SHADE

Oak woods provide a magnificent part of our native flora, most gloriously seen in spring when there is a carpet of *Hyacinthoides non-scripta* (bluebells) so blue that they almost seems to drift out of visibility (Figure 2.52).

Hyacinthoides non-scripta is an 'Atlantic' species *par excellence*, growing only in the moist conditions of western Europe, and reaching its peak of abundance in the British Isles. Like many woodland species, in the moist west it can be found growing out in the open (Blackman and Rutter, 1954). It is restricted to sites where, between April and mid-June, light levels do not fall below 10% of daylight. Like *Anemone nemorosa* (wood anemone) it is aided in the early expansion of its leaves by a high DNA content. It can share the same areas with *Pteridium aquilinum* (bracken) and *Holcus mollis* (creeping soft-grass), but rapidly dies back, removing nutrients from its leaves into its bulbs, as they grow up. It has a high content of the storage sugars fructan and fructose in its bulb in winter. According to Gerard's herbal of 1597, the fructan was used to stiffen Elizabethan ruffs.

Hyacinthoides non-scripta is vulnerable to predation because it cannot recover easily from damage. The number of leaves produced each year is predetermined the previous season. It contains toxic glycosides to discourage herbivores. The young shoot is spear-shaped and can push its way up through the leaf-litter layer. The bulb produces contractile roots which pull it into the soil so that the bulb may be very deeply buried, 25 cm or more down. This makes it somewhat vulnerable to waterlogging but allows it to grow with *Pteridium* by rooting below the *Pteridium* rhizomes. However, in

Figure 2.52 Bluebells (*Hyacinthoides non-scripta*) in full flower. (Copyright Jon B. Wilson.)

wetter areas it is replaced by *Allium ursinum* (ramsons). *Hyacinthoides non-scripta* has a weak root system and relies upon a strong mycorrhizal relationship for obtaining nutrients.

Bluebells are also frequently found in another major lowland woodland type, the *Fraxinus excelsior–Acer campestre–Mercurialis perennis* (Rodwell, 1991; W8) woodland, which is found on base-rich soils. These woodlands have often been called mixed oak woodlands, but although oak is commonly present, it is very variable in quantity and never the dominant species. These ash woods are today among the most rich floristically. The mainly calcicole herb layer includes *Geranium robertianum* (herb robert), *Potentilla sterilis* (barren strawberry), *Brachypodium sylvaticum* (false brome), *Teucrium scorodonia* (wood sage) and *Melica uniflora*

(wood melick). *Tilia cordata* is also a minor component. When *Tilia* was more dominant in the wildwood, the calcicole field-layer may have been suppressed. In the wildwood, perhaps only the shift from *Rubus fruticosa* agg. (brambles) to the calcicole species *Mercurialis perennis* (dog's mercury), which is relatively shade tolerant, marked the transition from acid to calcareous soils. A similar transition is seen between *Fagus sylvatica–Rubus fruticosa* agg. woodlands (Rodwell, 1991; W14) on acid soils to *Fagus sylvatica–Mercurialis perennis* woodlands (Rodwell, 1991; W12) on calcareous soil.

Mercurialis perennis is a dominant member of the field layer of many woods, not just because of its tolerance of shade but for the remarkable way it spreads vegetatively (Figure 2.53). An integrated network of rhizomes connects different parts of the clone.

(a)

(b)

Figure 2.53 *Mercurialis perennis* (dog's mercury): (a) clones in an oak wood; (b) in flower.

Large patches are occupied by individual plants (Hutchings and Barkham, 1976). There are few openings for other plants to establish within a *M. perennis* patch and it casts a deep shade. This makes it a formidable field-layer competitor. *Mercurialis perennis* is dioecious, having separate male and female plants. Male and female plants have slightly different ecological preferences, so that there are more females in the shady areas.

2.9.2 WET WOODLANDS

Gley soils developed in valley bottoms or in poorly drained areas are a very important category of soil, covering 40% of England and Wales and 30% of Scotland and Ireland. They are very diverse, depending upon the mineral nature or the source of soil material. Two major types can be distinguished; ground-water gleys and surface-water or stagnogleys. The former have seasonally fluctuating water tables. Unlike podzols which may also be waterlogged, gley soils are not stratified. The exclusion of air produces a grey or mottled grey/red–brown appearance due to the reduction of ferric ions into ferrous ions.

The degree of waterlogging is important in determining the distribution of field-layer species. *Mercurialis perennis* (dog's mercury) commonly roots at 10–15 cm or deeper on calcareous clays, and is more susceptible to waterlogging than *Hyacinthoides non-scripta* (bluebell). Lack of oxygen in waterlogged soils prevents root growth, and also the presence of ferrous ions (Fe^{2+}) is toxic to plants. *Mercurialis perennis* is particularly susceptible to ferrous ion toxicity. Arranging plants in order of increased tolerance to ferrous ions gives: *Mercurialis perennis*, *Hyacinthoides non-scripta*, *Brachypodium sylvaticum* (false brome), *Geum urbanum* (wood avens), *Circaea lutetiana* (enchanter's nightshade), *Primula vulgaris* (primrose), *P. elatior* (Oxlip), *Carex sylvatica* (wood sedge) and *Deschampsia cespitosa* (tufted hair grass) (Martin, 1968). In part this tolerance is related to the different degree of aerenchyma development in the roots. *Primula elatior* is one of those species with a well-developed aerenchyma. *Allium ursinum* (ramsons) is very susceptible to drought but will not grow on waterlogged soils (Tutin, 1957). It favours damp, well-drained areas such as stream banks.

On very waterlogged soils *Alnus glutinosa* (alder) and *Betula pubescens* (downy birch) become the most common tree species. *Alnus glutinosa* seedlings are more tolerant of waterlogging even than those of *B. pubescens* (McVean, 1953). The roots of adult *A. glutinosa* can allow oxygen to diffuse down into the waterlogged soils. *Alnus glutinosa* is often found with *Fraxinus excelsior* (ash) on wooded slopes where soils are waterlogged because of a change in the underlying bedrock. Here the tree canopy is often a little more open and the herb layer enriched. *Filipendula ulmaria* (meadowsweet) and *Lysimachia nemorum* (yellow pimpernel) are very common and there is a rich field layer containing a range of other species common in meadows and damp pastures or on woodland margins (McVean, 1956). The richest communities are found where flushes are least acid. Along the margin of streams, where there is fertile alluvium so that there can be the development of a tall-herb field layer, *Urtica dioica* (nettle) can inhibit the development of shorter herbs. A transition to the drier soils of the surrounding woodland is often marked by a decline of *Ranunculus repens* and *Lysimachia nemorum* and an increase of *Deschampsia cespitosa*, *Dryopteris dilatata* (broad buckler-fern) and *Oxalis acetosella* (wood sorrel).

The climatic optimum was a period of heavier rainfall, 120% of the present rainfall in the lowlands. The vegetation history of Cothill Fen near Oxford shows colonization of *Alnus glutinosa* (alder) (Day, 1991). Cothill Fen is sited in a valley where 10 000 to 9500 years ago there was a swampy vegetation whose margins were in the process of being colonized by woodland. Thereafter a marly

peat developed as the woodland was pushed back and ferns like *Thelypteris palustris* (marsh fern) became dominant. About 8800 years ago rushes became abundant and then aquatic bryophytes from 7700 years ago. Conditions continued to get wetter with a greater depth of water, indicated by the development of pure marl from 6850 years ago. Then *Alnus glutinosa* began to increase.

This kind of wetland within the woodlands provided a small but important local refuge for herbs, within the wildwood. Herbs recorded from the period 10 000–5000 years ago show a preponderance of wetland species such as *Bidens* (bur-marigold) (Godwin, 1975). An unknown factor in the ecology of the wildwood, but possibly very important in many areas, was the activity of beavers, damming valleys and creating waterlogged areas. A site inhabited by meso-lithic people at Thatcham in the Kennet Valley, Berkshire, for example, was on the edge of a lake even though there was no true lake basin present. Probably drainage had been impeded by dams built by beavers.

The wettest woodlands are found on ground that is flooded regularly. The national vegetation survey has identified seven different kinds of wet woodlands (Rodwell, 1991). Some kinds of swamp-woodland are characteristically developed on river terraces and valley bottoms subject to regular flooding. They are variously dominated by *Alnus*, *Betula pubescens* (downy birch) or *Salix* (willow), depending upon the degree of wetness, the base richness and fertility of the soil. In fertile conditions an *Alnus* and *Urtica dioica* (nettle) dominated woodland (Rodwell, 1991; W6) can develop from the alder carr or more directly from the reed fen. In some fens a stingless variant of *U. dioica* can be found.

Most woody plants cannot establish because the ground is flooded for more than a few weeks each year. Even *Alnus glutinosa* seeds require well-aerated conditions to germinate. Sedge tussocks provide sites for germination of *Alnus glutinosa* (alder) seeds as well as the seeds of herbs, so that a swamp woodland (carr) develops. Stability of *Alnus glutinosa* on the unstable soils is aided by the production of 'strut roots' which run down diagonally to a depth of 30–40 cm, where they end in a fan of stout rootlets (McVean, 1953). *Alnus* has nodules on its superficial roots containing a nitrogen-fixing microbe.

Salix and *Populus* (willows and poplar) are very drought sensitive and require continuously moist ground. Shrubs, especially *Rhamnus catharticus* (purging buckthorn) and *Frangula alnus* (alder buckthorn), are important in some woods. The latter is especially common at Wicken Fen. *Betula pubescens* (downy birch) and *Salix cinerea* (grey willow) are the first of the trees after *Alnus glutinosa* to invade the swamp woodland. They can also directly invade reed swamp. Succession can continue in a fen wood towards a shady oak-fen woodland. All the time, more leaf litter accumulates, removing the soil surface further and further away from the possibility of seasonal inundation. Different species of moss are especially important in this process. In time the accumulation of plant litter raises the soil surface so that it dries and allows dry-land species to colonize. Some species of tree, especially *Fagus sylvatica* (beech) and *Acer pseudoplatanus* (sycamore), are almost completely excluded from fen woods because they are particularly sensitive to water-logging. *Betula* (birch), *Quercus* (oak) and *Ulmus glabra* (wych elm) thrive as long as the water table is not too near the surface. For example, in parts of Wicken Fen birches, which colonized in a time of drought, have died because of renewed high water levels. The succession seems to lead inexorably to the development of a dry woodland and the elimination of the herb flora, but various factors can halt or reverse the succession.

2.10 'NATURAL' VEGETATION?

It takes a great leap to imagine what the natural vegetation of the British Isles might have been like 6000 years ago. Scarcely anything of the wildwood remains, and the surviving patches of woodland have been subject to thousands of years of use and modification. Even a seemingly timeless woodland like Wistman's Wood has suffered interference from man or his animals. There is a better case for describing many of the refuges of open-habitat species as natural. However, in most cases, they, too, have been shaped by human activity. Uplands have been grazed and limestone areas have been quarried. Coastal and wetland vegetation might seem to be the most significant remnant of our natural vegetation. The strong forces that mould them are the natural ones of climate and soils. Vegetation development seems to be a regular natural process, progressing in the absence of humans. There are clear successions and obvious zonations. However, the development of mire communities is by no means simple or one-way. Mire communities are very sensitive to human interference, so sensitive that natural or seminatural fen woods, once very widespread in our landscape, are very rare in today's landscape. Many wetland communities owe their existence to hundreds of years of grazing and cutting.

One example of how human activity has altered the composition of vegetation comes from a comparison of the Machair of north-west Ireland and Scotland. They are floristically distinct. *Carex arenaria* (sand sedge), *Luzula campestris* (field wood-rush), the winter-annuals *Saxifraga tridactylites* (rue-leaved saxifrage), *Geranium molle* (dove's-foot cranesbill) and *Erophila verna* (common whitlowgrass), and the mosses *Brachythecium albicans* and *Tortula ruralis* are more common in Irish than the Scottish Machair, and *Achillea millefolium* (yarrow) and *Euphrasia* spp. (eyebright) are much less common (Bassett and Curtis, 1985). The soil of Irish Machair has a lower calcium carbonate content, but differences between it and

Scottish Machair are mostly related to their use (Gimmingham, Gemmell and Greig-Smith, 1948). In Scotland the Machair is important for the cultivation of winter-feed, mostly rye and oats, and providing winter grazing. It is regularly mulched with sea-weed. Irish Machair was probably cultivated as intensively as that in Scotland up to the time of the potato famine, but now it is rarely mulched or used to cultivate hay but is used only for grazing. Some Irish Machair also suffers intense amenity use. Machair and other dune systems can also be intensively grazed by rabbits, introduced by people.

But what about saltmarshes? With their strong zonation edaphically controlled, they seem the most natural kind of vegetation. In fact, a number of different saltmarsh types, with different geographical distributions, have been recognized (Adam, 1978). Some variations can be related to 'natural' factors and others to human influence. Scottish marshes are different because they are north of the climatic limits of many salt-tolerant marsh species, such as *Limonium vulgare* and *L. humile* (common and lax-flowered sea-lavender), *Atriplex portulacoides* (sea purslane), *Seriphidium maritimum* (sea wormwood), *Spartina anglica* (common cord-grass), *Elytrigia atherica* (sea couch) and *Parapholis strigosa* (hard grass). They are subject to high rainfall and freshwater runoff so that they are not so markedly differentiated from grass-lands above the tidal range. They have a diverse middle and upper marsh flora. Further south the distribution of saltmarsh types is most closely related to patterns of grazing. One type is found mainly in south-east England and another type in western Britain, especially around the Irish Sea. The most heavily grazed marshes are found around the Irish Sea. Here grass-dominated communities with *Juncus gerardii* (saltmarsh rush) have arisen. *Festuca rubra* (red fescue) reaches further down the marsh and *Puccinellia maritima* (common saltmarsh grass) further up the marsh.

The British vegetation has developed from the interplay between natural forces and human influence. The story of mankind's increasing influence on our vegetation is told in the next chapter.

3.1 THE ELM DECLINE

The postglacial colonization of the British Isles occurred in two phases, at first from about 14 000 years ago and then, after a short cold stage, from 10 000 years ago. It has been traditional to regard the Postglacial, the Flandrian or Holocene, as starting 10 000 years ago, and to divide it into two halves, the first ending about 5200 years ago with the advent of neolithic man and agriculture, and a change of climate from a warm, wet Atlantic phase to the drier Sub-Boreal phase. The division of the Postglacial into two halves has been convenient because it is marked in the pollen record by a sharp decline in *Ulmus* (elm) pollen abundance, which seemed to have occurred relatively simultaneously over much of north-western Europe (Figure 3.1).

However, as more evidence accumulates, this too seems less certain, with radiocarbon dates for the elm decline ranging over nearly 1000 years: 6000 BP at Fallahogy in Co. Londonderry; 5800 BP at Rimsmoor in Dorset; 5600 BP at Hockham Mere in Norfolk; 5400BP at Gatcombe Withy in the Isle of Wight; and 5100 BP at Flanders Moss in Ayrshire (Smith and Willis, 1962; Scaife, 1988).

The decline of elm is an attractive marker for studying the development of our landscape because it has been associated variously with a climatic change, a pathogen and a selective use of *Ulmus* by the new neolithic human culture. *Ulmus* may have provided nutrient-rich foliage to feed herds of domesticated animals kept in a wooded landscape which had little grassland. Alternatively, because elms were growing on the more fertile soils, they were selectively felled or killed by barking to open the woodland for agriculture. In the elm decline, the abundance of *Ulmus* pollen fell to 50% or less of its former level. This would have required the felling of 47–80 million trees in a very short period. This would seem a tall order for the small population of that time unless one remembers what was achieved in less than 200 years in parts of the USA and New Zealand with scarcely better technology. Stone axes are said to be as effective as steel ones, if not as durable.

The use of woodland for browse, if horses and cattle also damaged the bark, which pulls off in strips, at the same time was potentially very damaging. However, other tree species, such as *Tilia cordata* (small-leaved lime), were also potential providers of browse and they do not show such a marked decline. Partial lopping would discourage flowering, but it would have had to be extremely extensive to have a marked effect. Elm bark is very stringy and could have been stripped to provide twine and rope, or even woven. The inner bark is a nutritious food but, again, harvesting of this could not have been happening to the extent necessary. An anthropogenic cause of the elm decline is unlikely because it occurred universally, even in areas marginal for human habitation. More importantly, it is increasingly obvious that the transition from

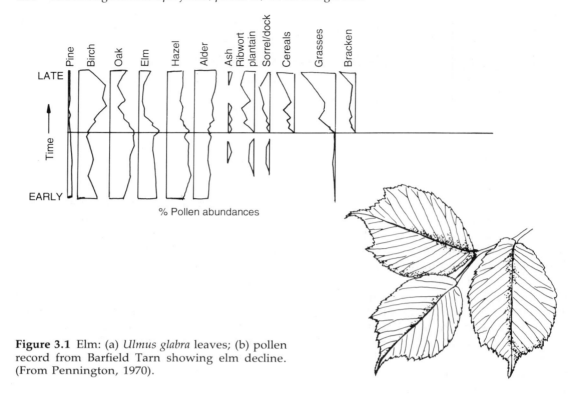

Figure 3.1 Elm: (a) *Ulmus glabra* leaves; (b) pollen record from Barfield Tarn showing elm decline. (From Pennington, 1970).

mesolithic to neolithic culture was more a gradual transition than a sharp change, and did not always coincide with the elm decline. In north-east Yorkshire, on the basis of a change in cultural artefacts, it may predate the elm decline by several hundred years (Figure 3.2).

In several places cereal pollen, indicative of the arrival of agriculture, is recorded before the elm decline (Edwards and Hirons, 1984; Scaife, 1988). The earliest record of cereal pollen dates to 500 years before the elm decline.

Perhaps a more likely explanation of the elm decline, given our experience of 'Dutch elm disease' from the early 1970s, is that the elm suffered an episode or a succession of episodes of elm disease (Perry and Moore, 1987). Elm disease has been recorded many times in history, marking phases of renewed virulence. For example, dead elms were an

obvious feature of the landscape of Shakespeare's age. Falstaff is called a 'dead elme' by Poins. The present epidemic may date back as far as the 1920s when the disease was noticed in the Low Countries, blamed on the use of poison gas. From here it was exported to Canada only to be re-imported in timber from there in the 1960s, and now it had enhanced virulence. Patches of dead elms may have opened the woodland canopy, providing places in which small patches of cereals could be cultivated.

The fungus *Ceratocystis ulmi* invades the wood tissue, blocking the conduction of water and interfering with the tree's hormones. It is transported from tree to tree by the elm bark beetle, *Scolytus*. The disease is more virulent in drier years, partly because the activity of the beetle is aided by lower levels of sap. In prehistoric times in Ireland and in south-east England, but not obviously

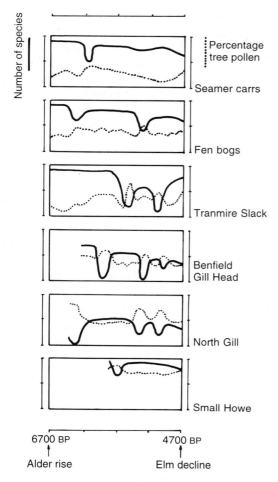

very rapid decline in elm pollen. For the remaining 85 years of the record there was extensive cultivation of cereals, indicated by cereal pollen and erosion of cultivated soils. On this occasion the elms did not recover, possibly because areas of dead elms were being cleared and any suckering regrowth was grazed back or coppiced.

3.2 CLEARINGS IN THE WILDWOOD

The elm decline is only a marker in the middle of the development of mankind's landscape. There is also a rich archaeological landscape. Field margins and woodland outlines record a history of mankind's use. Fragments of pottery or other relics buried in the soil provide clues about patterns of landscape use and the density of settlement. Styles of pottery or prehistoric architecture can date deposits and indicate particular cultures associated with different kinds of land use. Bones and seeds, the remnants of meals, found in middens indicate more directly the use of plants and animals. In the Holocene the stone age culture changed from what is called palaeolithic to mesolithic. This represents a change from a culture that exploited the large herds of deer and horses of the open plains of the Late Devensian to a woodland-margin culture. The culture is recognized by the use of microliths, small flints mounted on wooden hafts to make cutting blades. They also had large flint axes to clear the woodland.

There were a number of different mesolithic cultures (Laing and Laing, 1982b). The first Mesolithic culture was that of the Magleosians, 'Big Bog People'. Their culture reached as far east as Estonia. At Star Carr, in Yorkshire, they lived in a campsite dated to 9500 years ago on a gravel hillock beside a lake (Clark, 1954). They reached across the soft ground of reed swamp to the open water by building a platform of birch-tree trunks and branches. The campsite was home to about 25 people.

Figure 3.2 Pre-elm-decline woodland clearance phases from north-east Yorkshire. (From Innes and Simmons, 1988.)

elsewhere, there was a marked regeneration of elm, followed by a second decline. An analysis of the pollen preserved in annually laminated sediments of Diss Mere, in Norfolk, from the time of the elm decline suggests two stages of woodland clearance (Peglar, 1993). The initial disturbance lasted 50 years as small clearings were made for the cultivation of cereals. This may have been accompanied by an outbreak of disease but subsequently the elms recovered. For about 60 years there was less disturbance which ended with a

Mesolithic disturbance of the vegetation has been noted elsewhere, on northern Dartmoor before 8785 radiocarbon years BP (Caseldine and Maguire, 1986) and South Wales before 8000 radiocarbon years BP (Smith and Cloutman, 1988). The extent of mesolithic disturbance of woodland communities may have been underestimated. Disturbance in a forested landscape can in some circumstances increase pollen productivity of trees, which flower more profusely at the woodland edge, so that the percentage of pollen of herbaceous species actually declines (Edwards, 1993). Mesolithic man hunted red and roe deer, aurochs, wild boar and horses, as well as smaller game like hare and beaver (Simmons and Tooley, 1981). Elk became extinct by the Late Boreal. The occupation and utilization of different sites disturbed the natural plant communities, opening them up for weedy species. Glades were created within the woodland, allowing the survival of some herbs. There are occasional pollen grains of plants such as *Knautia arvensis* (field scabious), *Ajuga reptans* (bugle) and *Cardamine pratensis* (cuckoo flower) in woodland pollen floras throughout the Early Holocene. Sometimes there is a marked in-wash of silt to deposits, showing localized erosion of exposed soils. Exposed soils were colonized by *Persicaria*, *Fallopia* and *Polygonum* spp. (knotweeds), *Artemisia* spp. (mugworts), *Rumex* spp. (sorrels and docks) and *Plantago* spp. (plantains). Microscopic charcoal particles are associated with many deposits of mesolithic age. Mesolithic hunters were using fire to provide woodland openings which attracted game. However, in most places relatively soon the spread of *Corylus avellana* (hazel) out-shaded the glade herbs and grasses. The woodland canopy closed.

The decline of tree pollen at mesolithic sites is sometimes accompanied by an increase in heather and bracken. At Iping in Sussex there is more evidence of a lasting change in the vegetation about 8000 years ago (Figure 3.3) (Keef, Wymer and Dimbleby, 1965).

Here *Corylus* was growing on a rich brown earth, with strong earthworm activity maintaining soil fertility. Clearing of the *Corylus* by fire and/or by overgrazing, led to a catastrophic change in the soil. Earthworm activity ceased, erosion and soil leaching occurred and a *Calluna vulgaris* (heather) dominated vegetation arose on an impoverished acid soil inimical to many herb species.

Different kinds of disturbance sites have

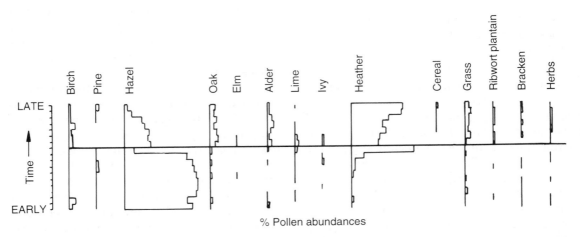

Figure 3.3 Pollen distribution in the soil profile from a mesolithic site at Iping, Sussex. (From Keef, Wymer and Dimbleby, 1965.)

been described in the Thames Valley (Devoy, 1980; Holgate, 1988). Some, on the edge of the gravel terrace overlooking the rivers and at the margin of the woodland, were either occupied on a temporary or a more continuous basis. Other disturbance sites on the flood plain and on the upper valley slopes of the Cotswolds, Marlborough Downs, Chilterns and North Downs were activity sites, used for hunting or flint collecting. In many places there were repeated episodes of clearing. This has been shown particularly effectively in sites in north-east Yorkshire (Figure 3.2). By about 6500 years ago the population was taking advantage of the uplands where, perhaps as in north-east Yorkshire, the vegetation was more open and the game more concentrated (Innes and Simmons, 1988). Summer hunting grounds were occupied in the hills. In the autumn, when hazel nuts were abundant, the small bands of hunters gathered together in larger groups. As winter progressed they dispersed to the lowlands and sea-shore. In most places, woodland eventually regenerated following disturbance, but meanwhile adjacent areas were being cleared. Microscopic fragments of charcoal dispersed by the wind from new clearings are found throughout woodland pollen assemblages.

A shift from valley bottoms in the Late Mesolithic has been noticed in several places. The abandonment of the valley bottoms may have been encouraged by their increasing wetness due to the down-warping of the land in the south-east but a similar gradual shift on to more mid-altitude and upland sites in the Late Mesolithic has been noted in Yorkshire. In the south a shift on to the loessic soils present on the chalk and limestone placed the population in an excellent situation for the establishment of agriculture.

3.2.1 THE BEGINNINGS OF AGRICULTURE

The creation of woodland clearings for game leads inexorably to their use for domesticated animals. Small Dexter-like cattle were introduced, as well as small pigs, sheep and goats. The harvesting of wild plants leads inexorably to the cultivation and harvesting of domesticated plants (Evans, 1975; Edwards and Hirons, 1984). The shift from the mesolithic woodland hunting-gathering culture to neolithic agriculture is marked by the presence of cereal pollens, especially of einkorn (*Triticum monococcum*) and emmer wheats (*T. dicoccon*) and barley in pollen floras. Agriculture and its crops spread across mainland Europe from the Near East 9000–8000 years ago. The crops were accompanied by weeds of diverse origin (Figure 3.4).

Some, such as *Fallopia convolvulus* (black bindweed), may have accompanied the cereals from their site of origin. Another group, including *Lapsana communis* (nipplewort) and the meadow grasses colonized the fields from woodland margins. *Chenopodium* spp. (goosefoot and fat hen), and *Persicaria maculosa* (redshank) or *Polygonum aviculare* (knotweed) probably colonized the fields from river banks or lake shores.

The introduction of agriculture to the British Isles in the sixth millennium BP did not cause a great upheaval. Cereal pollen is recorded from Arran in Scotland at 5880 radiocarbon years BP (Edwards and McIntosh, 1988) and at Cashelkeelty in south-western Ireland at 5845 radiocarbon years BP (Lynch, 1981). Farmers settled especially in southern England, the Lake District and northern and eastern Ireland. In Northern Ireland and the Lake District forest clearance and cultivation is recorded in the pollen record. Sometimes they are associated with archaeological remains. The earliest, perhaps, are the remains of a rectangular house dating from about 6000 years ago with stout, upright timbers and split-oak plank walls, excavated at Ballynagilly in Co. Tyrone. The signs of habitation, earlier pits and hearths, date back as far as 6400 years ago. In southern England it is mainly archaeological remains, cairns and long barrows, which record the presence

Figure 3.4 Characteristic plants of a woodland clearance phase in the Neolithic: (a) *Plantago lanceolata* (ribwort plantain); (b) *Chenopodium rubrum* (red goosefoot); (c) *Polygonum aviculare* (knotweed). (Copyright Jon B. Wilson.)

of the early neolithic population. Here settlement was concentrated on relatively discrete areas of the Cotswolds, Berkshire Downs, Marlborough Downs and North Downs, and perhaps the Chilterns. Barrows, designed to be seen, were constructed in an open landscape. On the margins of these areas, on the woodland edge or in clearings, causewayed

enclosures were created, perhaps as meeting places.

Many other monuments scattered all over the country are the tangible evidence of the presence of agricultural man living in established, well-organized communities. The concentration of archaeological sites in particular areas is derived in two ways. It indicates those areas of dense settlement in the past but it also indicates areas where a particular kind of land usage was permanently established, favourable to the survival of the monuments. This usually means that a pastoral use of the land had been established at the time of construction, because, although large monuments like Stonehenge or Silbury Hill are, of course, very resistant to change, lesser tumuli can be, and have been, destroyed by the plough. Hence the smaller number of tumuli and other prehistoric archaeological sites in the lowlands may not be an expression of the avoidance of these areas by early agricultural communities, but the result of a more intense agricultural usage at some later stage. That this is sometimes the case is shown by the rare survivals like those at the Port Meadow and Wolvercote Common on the eastern bank of the Thames just above Oxford (Holgate, 1988). The flood plain is scattered with bronze age and iron age ring ditches and enclosures of different farmsteads, which were only revealed as crop marks in aerial photographs taken in the 1930s and also especially in 1976 after the drought.

Aerial photographs show the presence of prehistoric sites only because of the different behaviour of plants. It is fascinating that plants themselves have been used to detect the presence of prehistoric sites in fields that otherwise look completely uniform. Crop marks are produced in spring or late summer because of the stunted growth of the crop over walls and other prehistoric masonry, while over ditches, since filled, the crop grows more lushly. In the drought summer of 1976 the contrasts between lush and stunted growth were exaggerated and many new crop marks were revealed in aerial photographs. In established grassland where there is a range of species present, the distribution and performance of different species can be related to the archaeological remains below.

The transition between the mesolithic and neolithic culture was a gentle one. In the Early Neolithic, wild foods continued to be a significant part of the diet, not just hazel nuts and the obvious fruits like crab apples and elder berries. Goransson has suggested that the beginning of the Neolithic is marked by extensive forest utilization (Edwards, 1993). The woodlands provided many utilizable products, and there was extensive encouragement of leafy growth by coppicing and girdling of trees to provide fodder. Gradually, through use, the uniform woodland was transformed into a mosaic of coppice wood and arable lands. Leaf-foddering of cattle kept stalled in winter may have been a significant use of the woodland. *Hedera helix* (ivy) and *Viscum album* (mistletoe) were used in medieval times and probably before. *Ilex aquifolium* (holly) was used as fodder for sheep, and even today *Fraxinus excelsior* (ash) twigs are fed as a food supplement in some farms on the North Yorkshire moors. The necessity of finding fodder for cattle may have driven much agricultural development. Even cereals may have been cultivated to provide winter food for them.

The fields must have soon become infested with weeds but even they may have provided an important source of food. The variant of *Arrhenatherum elatius* (false oat-grass) called onion couch (*A. elatius* var. *bulbosum*) was harvested for its swollen, edible base. Many weeds, such as *Chenopodium album* (fat hen) and *Persicaria maculosa* (redshank), have edible seeds. The seeds of *Chenopodium album* contain albumin, but the green parts of the plant were also eaten as a vegetable (J.T. Williams, 1963). Young plants of *Sinapis arvensis* (charlock) could be eaten when boiled (Fogg, 1950).

A characteristic pattern of cultivation has been recognized widely. It was first described by the Swedish botanist Iversen and called a 'landnam'. It is marked by a clearance phase where there is a marked decline of tree pollen. The first 'landnam' type clearance in the British Isles dates to about 6000 years ago at Ballynagilly, the same date as the construction of the plank house. There followed an arable phase of perhaps 300 years. Cereal pollen are present and also those of weeds such as *Plantago* spp. (plantains). *Plantago lanceolata* (ribwort plantain) showed a tremendous increase in the second half of the Postglacial, and has been used as an indicator of forest clearance. It reproduces rapidly when without competitors (Sagar and Harper, 1964). It had a similar rapid increase when introduced to North America by the colonials and was recognized as the 'footprint of the English man' by the native people.

There is variation in the timing and length of landnam phases between sites. A similar but shorter arable phase to that at Ballynagilly occurred at Beaghmore, Co. Tyrone at roughly the same time. Forest clearance of approximately similar date has been described at Barfield Tarn in the Lake District and at Shippea Hill in Cambridgeshire (Clark and Godwin, 1962). At Fallahogy in Co. Londonderry there were three phases of forest clearance (Figure 3.5) (Pilcher, 1971).

The first phase, before 6000 years ago, was a period of minor disturbance marked by a slight decline of *Ulmus* and the presence of *Pteridium aquilinum* (bracken) spores (Smith and Willis, 1962). The second is from a similar age to that at Ballynagilly and was much more severe. The short clearance phase was marked by a steep decline of *Ulmus*, and the rise of grass and *Urtica dioica* (common nettle). It was followed by a short cultivation phase when weeds like *Plantago lanceolata* (ribwort plantain) and *Rumex* spp. became important (Figure 3.4). This was followed by a longer phase, lasting 300 years or more, when the woodland, including *Ulmus* regenerated. A third and more lasting clearance happened about 5000 years ago.

At Ballynagilly and some other places forest clearance is marked by a charcoal deposit showing that fire was used to help clear the woodland (Pilcher, 1971). Weeds may have been a severe problem and if fire continued to be used to control them this may explain the spread of *Pteridium aquilinum* (bracken) and then *Corylus avellana* (hazel) scrub. The spread of *Pteridium* may have been a particular problem which led to the

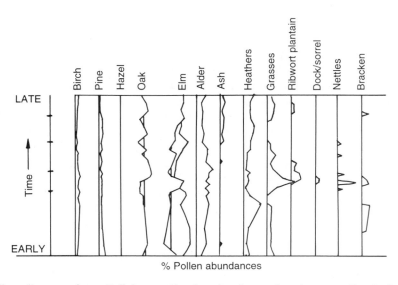

Figure 3.5 Pollen diagram from Fallahogy, Co. Londonderry showing woodland clearance phases. (After Smith and Willis, 1962.)

abandonment of the land for arable use. The rise of *Corylus* may have occurred because hazel foliage is relatively unpalatable to cattle so that it was the first woody species to re-establish.

At Ballyscullion, in Co. Antrim, forest clearance occurred a little later, at about 5500 years ago, than the other sites in Northern Ireland, and it occurred more gradually, without much of an arable phase. It is as if the woodland was being cleared piecemeal, creating open areas for grazing. Domestic-ated cattle and pigs could utilize the wood-land mosaic in much the same way as aurochs and wild boar, but the intensity of grazing encouraged by man may have halted regeneration of woodland for long periods. Cattle were by far the most important domest-icated animals throughout the Neolithic.

Whatever the period of clearance and fre-quency of disturbance, a new diversity in the flora of the British Isles was encouraged by the activity of mankind. The length of the open phase varies a good deal, depending probably on how intensively the area was used for pasture after cultivation ceased. It could last 20 years or 500 years or more. With each phase of clearance there was the potential for a marked increase in floristic diversity as areas of herb-rich grassland were established.

In some places there was no regeneration of woodland. This was seen both in upland areas and in the lowlands. At Blea Tarn in Cumbria, for example, there was a minor forest clearance at 6500 years ago and then a more significant one 5800 years ago (Pennington, 1970, 1975). This time open conditions lasted as long as 5000 years ago and may have become permanent with a decline of soil fertility and the development of blanket peat.

3.2.2 THE GHOSTS OF FIELDS

One idea of the landnam is that it was part of a pattern of shifting cultivation where small fields suffered a short period of intensive cultivation but were soon abandoned because of a decline in soil fertility. Neolithic cultiva-tion of the soil was limited to hoeing the surface layer with mattocks, thereby encouraging leaching, soil erosion and a decline in fertility because the soil was not turned deeply enough to bring nutrients up from deeper layers. Turning the superficial soil released iron from upper layers, and in some cases this was carried down to precipit-ate as an iron pan lower in the soil, creating a podzol. In addition, trees, with their deep roots, which might have brought nutrients to the surface, were now absent. Thin upland soils were cultivated because here the wood-land was more easily cleared, but these were more vulnerable to a decline in fertility.

Studies of pollen floras and snail and animal faunas, at the many sites in the Avebury area of the Kennet Valley in Wilt-shire, show a contrasting pattern, which may have been more widespread in the lowlands and south (Ashbee, Smith and Evans, 1979; Smith 1984). They show how a mosaic of woodlands, fields and pasture changed over time (Figure 3.6).

The sites date from the earliest, the Horslip, an earthen long barrow of about 5800 years ago, through to the latest stage of the Sanctuary 4200 years ago, and include the magnificent monuments of Windmill Hill, Silbury Hill and Avebury Ring.

Of the many monuments and earthworks in the region of Avebury, *in situ* cereal growing is proved from the two earliest sites, the Horslip long barrow and the South Street Barrow, dating from about 5300 years ago. At the South Street long barrow, a prehistoric soil has been preserved beneath the barrow mound. The whole history of the soil from the glacial period onwards is preserved. There are involutions formed in the freezing conditions of the tundra of 14 000 years ago and subsoil hollows, the casts of tree roots of the wildwood of 6000 years ago in which woodland snail shells are preserved. Above

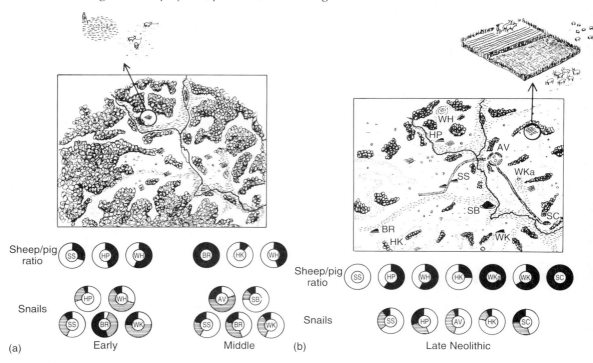

Figure 3.6 The changing neolithic landscape of the region around Avebury (adapted from Smith, 1984): (a) early and middle neolithic; (b) late neolithic. Pie diagrams show relative importance of sheep (unshaded) to pigs (shaded), and of snails of different kinds of habitat (dark shading, woodland; light shading, open country; lined shading, intermediate country) at different ages. AV, Avebury; BR, Beckhampton Road; SS, South Street; HP, Horslip; WH, Windmill Hill; WK, West Kennet; SB, Silbury; HK, Hemp Knoll.

this there are layers showing some of the different stages in the preparation and tillage of the soil. The first stage of the conversion of woodland to tilled fields was when trees were killed by ring-barking or felling. Woodland regeneration was then prevented by a phase of cattle grazing. Conversion to rough pasture was aided by the use of pigs with their propensity for rooting behaviour which would eliminate troublesome weeds such as bracken. Below the South Street long barrow a fence line is preserved, suggesting the way animals were concentrated in small areas to prepare the soils for cultivation.

At this stage, as well as later, sheep would be folded on the land to control weeds and tread the soil. Both South Street and Horslip have a higher proportion of sheep remains than other sites of the Avebury region, illustrating the close connection between sheep and cereal cultivation in the Neolithic. While cattle and pigs can survive in a wooded landscape, sheep require open, well-drained conditions and so they were concentrated on the cleared land either after harvest or during fallow periods. The sheep were a variety resembling the Soay sheep which are feral on St. Kilda. They moult their fleece in spring and had to be plucked and combed to get the wool. Bone combs for doing this are found quite commonly in archaeological sites.

The criss-cross scratchmarks of the plough

are preserved in the South Street soil, showing the final stage of field preparation, in which the rooting systems of the grassland plants were destroyed by ploughing (Dimbleby and Evans, 1974). Turfs may have been removed (Figure 3.7).

They could be used to construct tumuli and barrows, burial mounds, or added to the field boundary. At this stage, also, sarsens, blocks of sandstone found scattered over the surface of the chalk, and other large stones were removed to cairns or the field margin. Large ones may have been shattered first by fire. Fire may also have been used to burn off weeds. Above the ploughed soil there is a cultivation soil in which stones have been sorted into layers by intensive cultivation with spades and hoes on at least two occasions with an intervening fallow period. The field was evidently cultivated intensively

as an 'infield'. After harvest, sheep and pigs grazed it, controlling weed development and providing manure. At the top of the soil profile at the South Street barrow, before the soil was capped and preserved by the construction of the barrow, there is the sign of another period of fallow, with the development of a sheep-grazed turf.

What was the shape and size of these fields? Possibly early fields were protected by a dead hedge of thorn or, where longer lasting, by a living hedgerow or by piles of stones removed from the field. In the Late Neolithic larger, more permanent field systems, marked by permanent boundaries, became more common. A famous area for preserved neolithic fields is Co. Mayo. They are preserved entombed in peat at Belberg, west of Behy. The fields are walled and are roughly rectangular in shape and quite large,

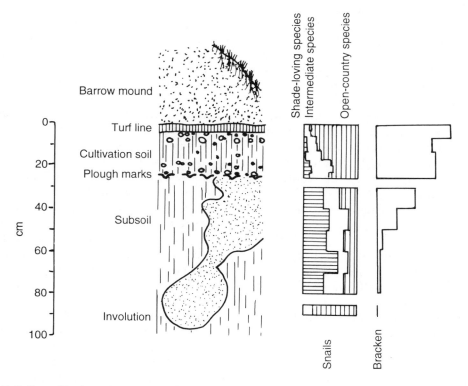

Figure 3.7 Soil profile from beneath South Street Long Barrow in Wiltshire. (From Evans, 1975.)

between 1.5 and 2.5 ha. About 0.4 ha could be ploughed in a day. Ploughs were of several different sorts of scratch plough, called ards, pulled by long-legged Dexter type cattle. The rip ard broke the soil but was probably used only initially. This kind of plough left scratch marks in the subsoil. Subsequently the soil was stirred by a bow ard to create a till. A third kind of ard drilled a row for seed. Alternatively, after the use of a rip ard the field was tilled with mattocks and hoes. There is evidence in the Irish fields that raised lazy beds were constructed.

In the Middle Neolithic some fields were abandoned while new areas were taken into cultivation. Changes in patterns of cultivation in the Mid-Neolithic are unlikely to have been due to climatic change, although there may have been a period of greater rainfall. A more important factor was the failure of techniques of cultivation to deal effectively with the surface vegetation. It had to be weeded by hand or burned off. Bracken may have become a very significant problem. In time, through the loss of crop productivity and the greater infestation by weeds, some fields were abandoned and new areas taken into cultivation. It is unlikely that loss of fertility was the main reason for abandonment of fields, although there are some signs of soil erosion with the creation of stepped lynchetts. Burning would create a flush of nutrients, but done repeatedly would result in nutrient loss from the system. Clearing areas for cereal cultivation was an arduous process, which made existing fields too valuable to abandon easily. The availability of stock to manure the field and control the weeds was very important.

The presence of small amounts of cereal pollen at sites such as Windmill Hill causewayed enclosure and Beckhampton Road long barrow, without any evidence of intensive soil preparation, indicates that cultivation was being maintained in distant lower-valley land. Perhaps these areas can be likened to the 'outfield' of later agricultural systems. 'Outfield' areas were actively cleared of trees by felling and by the use of fire. Once cleared, the outfield may have been cultivated intermittently for cereals with long periods of fallow when grazing, especially by cattle and pigs, helped to prevent the regeneration of scrub.

3.2.3 A SHIFT TO PASTORALISM

Meanwhile, grasslands were developing. Some fields may have been cleared of woodland and stones not for cultivation but for pasture. In some upland areas, like Exmoor, the decline of woodland was rather gradual, and there is little evidence of there having been an arable phase. While sheep required clear-felled land, pigs and cattle could use the woodland directly and here they gradually suppressed the growth of shrubs and trees.

Grasslands are such a characteristic feature of our living landscape that it is strange to realize that they were such a late development. There are widely contrasting types (Rodwell, 1992b) (Figure 3.8).

Most grasslands have been created by a long history of grazing. At the time of the maximum development of the wildwood, grasslands were rare outside wetland or limestone areas or at high altitude. Most grasses of well-established pastures and hay-meadows can also be found in natural or seminatural habitats, from where they colonized developing grasslands. Grasses colonized grasslands from refuges along woodland margins and other temporary openings in the woodland, or from cliffs, fens and sand dunes. For example, *Arrhenatherum elatius* is an important colonist of unstable limestone scree (Pfitzenmeyer, 1962). Several other species can be found to some extent in woodlands, near the woodland margin or where the tree canopy is thinned at stream sides and in glades. *Deschampsia flexuosa* is relatively exceptional, abundant both in the shade of woodlands and in open situations

Figure 3.8 Grasslands: (a) see plate section; (b) hay meadow, western Ireland; (c) wet pasture, Somerset Levels; (d) see plate section.

such as heaths and upland pastures (Scurfield, 1954). It thrives in full light but can withstand deep shade. In moderate shade it produces long, arching leaves. In deep shade the leaves are slender and limp.

The fossil evidence for the development of grasslands is thin and made difficult to study because of our inability to identify individual species of grass from their pollen. Some very common species of broad-leaved herbs, such as *Vicia cracca* (tufted vetch), are not recorded at all. Others are recorded in Europe but not the British Isles. There is a short list of those that are preserved as macrofossils (Greig, 1988). These can be used as an index of growing grassland diversity. In the Neolithic period, lowland dry grasslands may have had a low diversity, with only 21 different species recorded from Europe and only four from Britain. These were *Ranunculus acris* (meadow buttercup), *Lychnis flos-cuculi* (ragged robin), *Pastinaca sativa* var. *sylvestris* (wild parsnip) and *Plantago lanceolata* (ribwort plantain). In the Bronze Age, grasslands had greater diversity, with 25 (22%) species recorded from the British Isles. Some species of grassland herbs common now, such as *Leucanthemum vulgare* (ox-eye daisy) and *Hypochaeris radicata* (cat's ear) are not recorded. Others, such as *Centaurea nigra* (common knapweed), were rare. Nevertheless, at Runnymede on the Thames a calcicolous grassland had *Polygala* spp. (milkwort), *Sanguisorba minor* (salad burnet) and *Scabiosa columbaria* (small scabious) (Lambrick and Robinson, 1988). In the Roman period the British grasslands had 48 species (42%) including a range of grasses such as *Festuca ovina* (sheep's fescue), *Poa pratensis* and *P. annua* (smooth and annual meadow-grasses) and *Lolium perenne* (perennial rye-grass).

Of course, this analysis grossly underestimates the diversity of the grassland at different ages, but it does give an idea about the gradually increasing diversity of grassland. The development of distinct grassland communities can be dated from the Bronze and Iron Ages. A broad division can be made between wet grasslands, with *Molinia caerulea* (purple moor-grass), mesotrophic grasslands and chalk grasslands.

The English chalklands were probably covered by woodland in the Early Holocene but there is a lack of evidence about the high chalklands (Evans, 1993). After clearance in the Neolithic, natural communities could develop either as secondary woodland or as grassland. Impoverished grassland could develop as soils became decalcified. The development of species-rich grasslands on calcareous soils was encouraged by loss of organic content and erosion, especially following episodes of cultivation. In the thin soils that resulted, the influence of the base-rich subsoil was more marked. Rendzinas became more widespread.

It is unlikely that much grassland was harvested for hay in the Neolithic or Bronze Age. Evidence from Switzerland indicates that at least in the Early Neolithic, stock relied as much on the foliage of trees and shrubs as herbs. Nevertheless, herbage allowed to grow long and rank, perhaps even protected from grazing, during the growing season, and allowed to stand as 'fog', may have provided winter grazing. This kind of use is recorded in the vernacular name of *Holcus lanatus*, Yorkshire fog. *Holcus lanatus* is rather lax-growing, and close grazing removes many shoot buds for regrowth. Late mowing allows the seed to ripen (Beddows, 1961). *Holcus mollis* (creeping soft-grass) may also have been important. A dense thatch of grass would have restricted the growth of herbaceous dicots. These grasslands may have been relatively sensitive to grazing and more like the ungrazed grasslands found on road margins today than close-grazed pastures. These are often dominated by *Arrhenatherum elatius* (false oat-grass), *Bromopsis erecta* (upright brome) or *Brachypodium pinnatum* (tor grass).

In today's landscape some species are

particularly characteristic of ancient grassland but it is dangerous to use such species as indicators of antiquity. For example, *Carex humilis* (dwarf sedge) is particularly abundant on ancient grassland in Wiltshire and Dorset. However, this rare species can also be found in habitats that are very new. The composition of grassland is very susceptible to different grazing regimes, and for this reason the grasslands that clothe many ancient monuments are not necessarily relics of the ancient grassland that was present at the time of the construction of monuments. When they came into being, shortly after the monuments were first constructed, there were no rabbits in Britain to graze them. Nevertheless, ancient monuments often preserve a very interesting flora in today's landscape. The slopes of tumuli and the banks of hill forts on the chalklands of the south can be islands of seminatural grassland in an intensively farmed landscape. Eight of the present 27 recorded sites of *Pulsatilla vulgaris* (pasque flower) are on ancient earthworks. Extinction from as many as 25 other sites has been due to ploughing (Wells, 1968). The Devil's Dyke in Cambridgeshire provides a refuge for the chalk grassland flora since the destruction of the Cambridgeshire Downs. These grasslands are very ancient but they have been much modified over the millennia.

A vegetation succession does occur in grassland. On the Porton Ranges, grasslands less than 50 years old are characterized by *Arrhenatherum elatius* (false oat-grass), *Anthyllis vulneraria* (kidney vetch), *Vicia cracca*, *V. sativa* and *V. hirsuta* (tufted and common vetch, and hairy tare), *Silene vulgaris* (bladder campion), *Potentilla reptans* (creeping cinquefoil), *Pastinaca sativa* (wild parsnip), *Linaria vulgaris* (common toadflax), *Cerastium arvense* (field mouse-ear) and *Agrimonia eupatoria* (agrimony). In grassland 50–100 years old there is *Carex caryophyllea* (spring sedge), *Pimpinella saxifraga* (burnet saxifrage), *Polygala vulgaris* (common milkwort), *Helictotrichon pratense* (meadow oat-grass), *Helianthemum nummularium* (common rock-rose), *Filipendula vulgaris* (dropwort), and *Asperula cynanchica* (squinancywort). This represents a shift from fast-growing and fast-reproducing competing species to slower-growing stress-tolerators.

3.2.4 GRASSLAND ECOLOGY

Grassland can be a particularly rich vegetation because of the range of different strategies grassland species adopt. Species compete for light, water and nutrients, but no one species completely excludes others because the most common species are most heavily grazed. There is competition for limited resources, but each species has a different niche. Differences between species and between individuals allow them to coexist. *Anthoxanthum odoratum* (sweet vernal grass) and *Phleum pratense* (timothy) produce more tillers when growing together than when apart; maximum yield is achieved with a ratio of 34:1 *Anthoxanthum:Phleum*. They utilize space differently, minimizing interference. Differences between, and the interactions between, species are obvious, but many of the interactions that take place in a grassland occur between genetically different clones of the same species. There is some evidence to suggest that plants may be specially adapted to their own local area and to the individual plants of the same or other species with which they grow. Certainly, they seem to do best in competition experiments when grown with individual neighbours rather than plants from elsewhere in the field. This has been shown to happen with different clones of *Trifolium repens* (white clover) grown with *Cynosurus cristatus* (crested dog's-tail), *Lolium perenne* (perennial rye-grass) and *Agrostis capillaris* (common bent). They grow best along with their original partner (Harper, 1977).

Species differ according to whether they are annual, biennial or perennial. Annual species have potentially a very high rate of reproduction but suffer a high risk of not finding a safe site for seed germination and seedling establishment. Perennial species commit most of their resources to vegetative reproduction by tillers, bulbs, runners or other means. Only a small proportion of resources are committed to seed. In an established grassland there is very little recruitment of new plants as seedlings. Only in scrapes, hoof prints, worm casts and dung are open sites for colonization created.

The three meadowland buttercups provide an interesting example of contrasting strategies (Harper, 1957). *Ranunculus repens* (creeping buttercup) reproduces mostly vegetatively by producing daughter rosettes at the end of runners. Each year about one vegetative propagule, a ramet, is produced and, on average, only one seed per plant. In contrast *Ranunculus bulbosus* (bulbous butter-cup) reproduces only by seed. Each year it produces about 15 seeds per plant. This is a very low number for a plant reproducing only by seed but it is a perennial and survives the winter because of its basal corm. *Ranunculus acris* (meadow buttercup) is inter-mediate between the others. It produces about 10 seeds per plant and one ramet, a rosette on a very short runner, each year. Each buttercup differs in its pattern of establishment and mortality. Seeds of *R. bulbosus* germinate in autumn, of *R. acris* in spring and its ramets are produced in winter. *Ranunculus repens* produces ramets in autumn and seeds in spring. A large proportion of its seed, about 30%, do not germinate immediately but enter the soil seed bank where they remain dormant until a gap in the canopy arises.

A few species, like *Helianthemum nummularium* (common rock-rose) and *Thymus polytrichus* (wild thyme) are small procumbent or prostrate shrubs and survive by being laterally extensive, with a free-rooting diffuse growth form. Some grassland species have deep roots, but many root mainly in the top few centimetres. Most of them are relatively sensitive to waterlogging, especially as seedlings. A few, including *Plantago media* (hoary plantain) and *Sanguisorba minor* (salad burnet) can tolerate some waterlogging. The species that occupy the friable soil of ant hills, such as *Erophila verna* (common whitlowgrass), avoid the worst of summer drought by growing rapidly in the spring.

Species like *Tephroseris integrifolia* (field fleawort), which prefer slightly acidic conditions, can be found on slightly deeper soils where there has been some leaching. Some chalk grassland species are fairly tolerant of slightly acidic soils, pH 5–7, but seedlings are generally intolerant of pH < 5. Most need a fairly high concentration of calcium ions in the soil solution. A smaller number, including *Anthyllis vulneraria* (kidney vetch), *Filipendula vulgaris* (dropwort), and *Pimpinella saxifraga* (burnet saxifrage) can grow on calcium-deficient soils if they are high in magnesium.

The buttercups also differ in their palat-ability. *Ranunculus acris* and *R. bulbosus* are avoided by cattle, sheep, horses and rabbits, probably because of the presence of the acrid poisonous compound, protoanemonin, released in the sap. *Ranunculus repens* has a low concentration of the precursor com-pound to protoanemonin, called ranunculin, and is more palatable. Many common species found in pasture, including *Trifolium repens* (white clover) and *Lotus corniculatus* (bird's-foot trefoil), have variants that produce cyanide when they are chewed, probably providing a protection against predation by slugs and snails. *Senecio jacobaea* (common ragwort) is one of those plant species which flourishes in grazed areas because it is un-palatable. Cattle and horses have frequently been poisoned, dying 1–5 months after eating it. Given a choice cattle will avoid it but the toxic alkaloids it contains are not destroyed on drying and stock can be poisoned by eating contaminated hay. Sheep are rarely poisoned but make poor growth if they eat it. Rabbits avoid it altogether so that its tall, yellow inflorescence can be seen conspicuously flourishing in rabbit-ridden grassland (Harper and Wood, 1957). Since 1921 laws have required farmers to cut and clear it from their land. It is one of five injurious weeds listed under the Weeds Act of 1959 which allows the Ministry of Agriculture to require the occupier to cut it down and destroy it. The other species listed are thistles (*Cirsium arvense* and *C. vulgare*) and docks (*Rumex crispus* and *R. obtusifolius*). Clearing of *Senecio jacobaea* with plant hormone herbicides has the danger of leaving distorted-looking plants which cattle fail to recognize. *Senecio jacobaea* probably colonized grasslands from the dune systems where it is widespread.

The grassland of chalk downlands has a particularly diverse range of species. One of the most important features of chalk grassland herbs is their relative intolerance of competition by large grasses such as *Dactylis glomerata* (cocksfoot). These grasses have restricted growth because of the deficiency of major nutrients, like nitrogen, phosphorous and potassium, and grazing. Where grazing is restricted, a tussocky grassland results. Close grazing by sheep, and much later by rabbits, was very important in the establishment of species-rich chalk grassland. It held at bay the scrub and woodland, and small gaps in a short-grazed turf provide sites for seedling establishment. The activities of animals – ant hills, mole hills, sheep and rabbit scratchings – and trackways provide other sites (Duffey, 1974).

Many grassland species have a life form that allows them to survive grazing and regrow. Nearly 70% are hemicryptophytes, perennials with buds at the soil surface. Some, like *Cirsium acaule* (dwarf thistle), are rosette plants. *Cirsium acaule* has the ability to grow and flower dwarfed, less than 5 cm tall, although it can grow more than 30 cm tall in an ungrazed sward. These plants are favoured in grazed pastures where the sward is less than 10–15 cm tall (Pigott, 1968). The rosette is pressed to the ground and the spiny

lobes of the leaves are directed upwards. It is avoided by cattle and, although it is eaten by rabbits and sheep, the rosette is rarely destroyed. Growing as a rosette also protects *Plantago lanceolata* against being grazed by cattle, but sheep favour it particularly and will chisel out the crown (Sagar and Harper, 1964).

The size of mouth-parts and the favoured food plants of different grazers varies. Horses are choosy eaters. Cattle were by far the most important domesticated animal in the Middle Neolithic, outnumbering sheep and goats by 4 to 1 in most places and pigs by even more. Cattle create a rough pasture by wrapping their tongue rather indiscriminately around a tussock of vegetation. As a result they favour the tussock grasses such as *Festuca pratensis* (meadow fescue), but in experiments they were less prepared to eat the tussocky *Dactylis glomerata* (cocksfoot) than sheep (Cowlishaw and Alder, 1960). Sheep are more delicate eaters and discriminatory grazers. They particularly favour *Lolium perenne* (perennial rye-grass) and *Trifolium repens* (white clover). Sheep even discriminate between *Trifolium repens* variants with different leaf marks (Cahn and Harper, 1976). The timing of grazing is also important because sheep generally favour young, green, leafy tissues over old, dry, stem tissue. Intense grazing in spring favours *Trifolium repens* which would otherwise be shaded by the grasses (Jones, 1933). Once tussock grasses have achieved a certain size they are preferentially ignored by sheep because they contain a large proportion of older plant tissue. However, sheep can produce a closely cropped vegetation by preventing large tussocks from ever developing. They will graze *Cynosurus cristatus* (crested dog's-tail) down so close that it would be eliminated except that they ignore the tough seeding stems (Lodge, 1959). Each kind of livestock helps to maintain a particular kind of diversity.

On chalk and limestone, growth of the

coarser grasses, which are mostly calcifuge, is restricted anyway so that sheep grazing can create a very diverse fine turf with tightly packed herbs and fine grasses, growing as tiny plants. However, on acid soils sheep are likely to create a less diverse grassland than cattle because they selectively graze the finer grasses and more palatable herbs back and coarser, competitive grasses flourish. Deer in the uplands can reduce floristic diversity. Goats are damaging because of their agility, they reach ledges which might provide a refuge for herbs. Horses are very selective grazers and so they can markedly change the floristic composition of a grass sward. Harper (1977) has observed that the spread of *Rubus fruticosa* (brambles) in the New Forest was effectively prevented by the grazing of tough New Forest ponies until they became effete through cross-breeding with introduced Arab stock with more finicky habits.

The effect of grazing depends very largely on the stocking rate and the season the grazing takes place. Trampling also affects the species composition. It encourages grasses like *Poa pratensis* (smooth meadow-grass) and *Lolium perenne* (perennial rye-grass), with their trampling-resistant tillers, over other species. The grasses have a flat, linear leaf folded around the sheath, and at the nodes there are meristems which can grow to resurrect the plant after trampling. *Trifolium repens* (white clover) is unusual amongst broad-leaved herbs in being somewhat trampling resistant because of the flattened, prostrate runners it produces. It also prefers the higher light levels of trampled sites. In wet weather, heavy, hoofed cattle can cause poaching, the crushing of soil structure. On steep slopes they can cause erosion. In a mixed sward, grazing in late spring and summer favours *Dactylis glomerata* (cocksfoot) but heavy grazing from autumn to early spring eliminates it (Beddows, 1959).

In a pasture the distribution of some species is related to latrine areas. *Cirsium*

vulgare and *C. arvense* (spear and creeping thistles) and *Senecio jacobaea* (common ragwort) are largely confined to latrine areas, and *Lolium perenne* (perennial rye-grass), *Trifolium repens* (white clover) and *Hypochaeris radicata* (cat's ear) are more common there. *Poa compressa* (flattened meadow-grass), *Leontodon autumnalis* (autumn hawkbit) and *Sagina procumbens* (procumbent pearlwort) avoid the latrine areas, preferring the shorter turf elsewhere.

A time-lapse film of a meadow would show a fascinating story. The vegetation would appear to be in continuous flux as established plants grew up and spread, jockeying for position. Battles for space would be fought out over the season. Very occasionally a local catastrophe, like a dollop of dung, would create a space for colonization by a new individual. A seedling would take hold and begin to spread. The grassland would be in constant movement, like waves on a pond, but fundamentally nothing would change.

The composition of grasslands may change as a result of changes in climate and changes in grazing. Grasses tend to have been favoured over herbs by sequences of dry summers on thin soils. In wet, upland conditions, leaching of soils and the development of podzolic conditions has encouraged the development of an acid montane grassland. The hypothetical dominance relationships of grasses in Upper Teesdale is illustrated in Figure 3.9.

In the wettest areas rushes become co-dominant. In drier areas bracken is co-dominant. Throughout, human influence has been pervasive, because dominance relationships are modified by grazing, controlled by people. Changes in species composition can occur very rapidly in grassland with a change in the intensity of grazing, even over a couple of years (Jones, 1967). There have been many shifts in the nature and intensity of grazing. By allowing overgrazing, relatively unpalatable grass species such as *Nardus stricta* (mat grass) have been encouraged. Too little

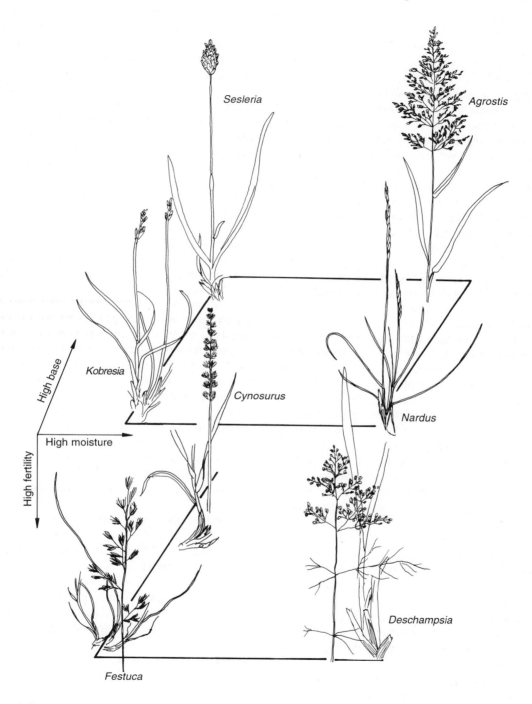

Figure 3.9 Relationships between grassland dominants in upland grassland in Upper Teesdale. (After Jeffrey, 1987.)

grazing allows colonization by *Pteridium aquilinum* (bracken).

3.2.5 SCRUBLAND AND FIELD MARGINS

Imagine a middle Neolithic landscape. It is mostly wooded but as time passes, more and more of the primeval woodland is cleared. Here and there, in clearings in the woodland, wisps of smoke rise up from clusters of circular huts. Near each settlement there is a small cultivated patch of ground, a garden in which wheat and a few vegetables are cultivated. The crop is being weeded by hoe. Nearby there are a few tethered cattle for milk. A pig, tethered by its foot, makes attempts to get amongst the crop but there is a low thorn fence around the plot. In early spring it will be taken into the wood to root for bulbs. The young foliage of trees will be cut to feed the cattle.

There is a patchwork of garden, woodland, scrub and grass. The garden settlements are connected by paths trailing through the wood. Along some of these paths there are clearings in the wood which were once occupied. Here and along the paths the grass is well developed. Cattle are tethered in the grassy clearings, enclosed by the spiny scrub around or shut in by a dead hedge of thorn. In the autumn these animals will be brought to manure the garden. In the wetter parts of the clearing the grass is allowed to grow for hay.

From this patchwork of woodland, scrub, grass and cultivated garden, by the end of the Neolithic period, the landscape had changed. There was less primeval woodland and greater areas of grass. There were larger numbers of grazers and it had become more important to protect the areas of arable crop. Fields were slightly larger and more permanent, some with a wall, or a hedge formed from the scrub as a boundary. In some areas clearance cairns, piles of stones cleared from the field were common. It was no longer feasible to tether animals individually. They were allowed to wander freely outside the cultivated fields. By their activity the grassland had spread and coalesced and scrub was kept back. After harvest the animals were enclosed in the fields to manure the soil and to control the development of weeds. Their manure made it no longer necessary to shift the area cultivated as often as before. Some fields were left to fallow. Others were kept as hay meadows. By the end of the Neolithic in some areas the landscape had been transformed, with large open areas of grassland and fields marked by permanent boundaries (Muir and Muir, 1987).

Probably very significant in the establishment of a settled landscape, but difficult to be precise about, were the social changes which occurred in the Late Neolithic. They allowed the building of the huge monuments of Silbury Hill, the Avebury Ring and the earlier stages of Stonehenge. The ability to organize and construct these monuments suggests a greater stratification of society with a few chiefs or priest-lords and many subservient underlings. Greater status required the greater display of wealth. In many cultures this has been primarily through the display of wealth on the hoof, in the number of cattle or pigs possessed by the chief. Pigs outnumbered cattle in Late Neolithic sites by 2 to 1 and sheep and goats were still relatively unimportant. The greater proportion of pigs may be a measure of their importance in controlling the spread of *Pteridium aquilinum* (bracken).

A group of plants, including *Agrimonia eupatoria* (agrimony), *Ballota nigra* (black horehound), *Stachys officinalis* (betony) and *Anthriscus caucalis* (bur parsley), became prominent with the advent of agriculture. They include many that are not necessarily strongly weedy but are characteristic of field margins, hedge banks and ditches. *Alliaria petiolata*, called Jack-by-the-hedge or hedge garlic, is one example. *Alliaria petiolata* favours the moderately shaded, moist conditions of the hedgebank (Grime, Hodgson and

Hunt, 1988). It can respond to some disturbance by producing lateral inflorescences. Leaves are produced both in the spring and autumn, before and after the period of maximum shade.

The amount of scrub present varied considerably with the intensity of grazing. The national vegetation survey identifies five different kinds of scrub in today's vegetation. On base-rich soils a *Crataegus monogyna–Hedera helix* (hawthorn/ivy) scrub (Rodwell, 1991; W21) is most common. It is a very rich and diverse vegetation type and includes many species of the woodland underflora and hedgerow species. Forming a fringe below it there can be a community dominated by *Rubus fruticosa* (brambles) and *Holcus lanatus* (Yorkshire fog) (Rodwell, 1991; W24). On deep and moist soils there is another community, dominated by *Prunus spinosa–Rubus fruticosa* (blackthorn and brambles) (Rodwell, 1991; W22). *Prunus spinosa* (blackthorn) casts a heavy shade and is highly competitive, because of its ability to sucker, so that this community is much less rich floristically. It is also resistant to exposure and tends to dominate scrub at the coast. On acid brown soils *Ulex europaeus* (gorse) and *Rubus fruticosa* (brambles) (Rodwell, 1991; W23) form a scrub vegetation. Ghosts of past woodlands may be occupied by a *Pteridium aquilinum* (bracken) and *Rubus fruticosa* (brambles) scrub (Rodwell, 1991; W25).

If grazing pressure decreased, scrubland gave way to a secondary woodland. On the downlands, on exposed slopes where the soil is highly calcareous and slightly compacted, juniper scrub arose. A richer hawthorn scrub, sometimes including *Juniperus communis* (juniper) developed on deeper, less compacted soils. This might have included a range of other shrubs and trees like *Cornus sanguinea* (dogwood), *Ligustrum vulgare* (privet), *Prunus spinosa* (blackthorn), *Rhamnus cathartica* (buckthorn), *Clematis vitalba* (traveller's joy), *Rosa canina* (dog rose), *Euonymus europaeus* (spindle), *Sorbus aria*

(whitebeam), *Sambucus nigra* (elder), *Corylus avellana* (hazel) and *Acer campestre* (field maple). *Fagus sylvatica* colonized this scrub vegetation, giving rise to many of the beech woods we see today. A few shrubs and other trees survived, at least at the margins of the wood. The field layer is poor: *Sanicula europaea* (sanicle) and rhizomatous or rosette herbs, such as *Fragaria vesca* (wild strawberry), *Viola hirta* (hairy violet), *Circaea lutetiana* (enchanter's nightshade) and *Mercurialis perennis* (dog's mercury), surviving in the wood. Alternatively, in a few cases, *Taxus baccata* (yew) dominated the secondary woodland, excluding almost everything else. At the other extreme in some areas on the chalk, on the South Downs especially, and on upland grasslands on limestone a diverse *Fraxinus excelsior* (ash) dominated woodland developed. The thin, flexible canopy allowed light to penetrate.

As permanent fields were established a vegetated field margin was established from scrub or woodland. The hedgerow can be regarded as a narrow strip of scrub or woodland. Related to this is the fact that the majority of broad-leaved hedgerow species are not important arable weeds. Some species, such as *Artemisia vulgaris* (mugwort), *Heracleum sphondylium* (hogweed) and *Anthriscus sylvestris* (cow parsley), habitually occupy the headland part of the field but cannot survive intensive cultivation. However, the uncultivated field margins, hedgerows, ditches and path sides provide an important reservoir for contaminating grass weeds.

3.2.6 USING THE WOODLAND

Meanwhile, surviving woodland was being utilized more intensively and in a more sophisticated way. Some of the best evidence for this comes from the wooden trackways laid down across the marshes of the Somerset Levels, connecting surrounding higher ground with the low islands of Edington

Burtle and Westhay Island (Coles and Coles, 1986). The earliest, the Sweet Track, is the most sophisticated. It dates from about 6000 years ago. It was a raised walkway of *Quercus* (oak), *Tilia* (lime) and *Fraxinus* (ash) planks, held up on a trestle of *Fraxinus* (ash) and *Corylus* (hazel) pins, all held in place by a foundation rail of elm. The ash and hazel pins are 1–1.5 m long and straight, with a diameter of about 40–50 mm. They were evidently produced from coppice stools which were harvested on a 7 year cycle. The entire trackway, more than 2 km long, was probably constructed in a single year, with all the trees felled at the same time. Some oak trees were 400 years old, others only 100 years old.

The main track was preceded by the Post Track, large planks of *Tilia* (lime) and *Fraxinus* (ash) laid directly on the surface. The Post Track was used by the work-gang to carry materials for the Sweet Track. The long, straight lime planks may indicate that *Tilia* was growing in a dense and shady wood; their trunks were tall and unbranched. Some of the oak trunks were 5 m long and 1 m in diameter. Other tracks were less robust. The Walton Track, which was laid about 5000 years ago, was a corrugated roadway of hurdles constructed from hazel poles, from a coppice, and willow withies. In the Avebury area there is evidence for the use of coppice poles and withies cut from pollarded willows. Even by the Middle Neolithic large timbers were rarely used. Perhaps they were not readily available in the area.

It is impossible to overestimate the importance of wood and timber as a constructional and craft material. A sample of small items, including bowls and bows, was lost and preserved beside the Somerset trackways. Larger items, such as the habitations constructed of wood, have not survived, but paradoxically we can get a hint of their sophisticated construction from surviving stone huts in Shetland. The lack of wood here forced construction with stone, but the forms may reflect those common in wooden habitations. The stone houses at Skara Brae even include stone furniture, such as a 'dresser' and 'box-beds' (Laing and Laing, 1982a).

3.3 MARKING OUT THE LAND IN THE BRONZE AGE

There is no sharp transition between Stone Age and the following metal-using cultures (Laing, 1979). Stone tools continued to be used along with the new metal ones. The Beaker People, named for their distinctive cord-marked pottery, brought copper smelting technology with them. Perhaps they entered Britain in a search for ore. The first seem to have arrived about 5000 years ago from the Rhineland and Low Countries, and still used stone tools. Another wave arrived about 4500 years ago from Spain and France, and around this time the first gold and copper appeared. In contrast to the resident neolithic population, the newcomers were short and robust with broad, round skulls. There were lots of them. There was a new phase of woodland clearance. They showed a preference for barley, perhaps to brew a kind of ale. No wonder some of the older residents seem to have adopted their life style.

In a few hundred years the Copper Age gave way to the Bronze Age. About this time the last and grandest stage of the construction of Stonehenge took place. As time passed, more and more woodland was cleared. In the Thames Basin there had been a spread of settlement into the valleys in the Late Neolithic (Holgate, 1988). The spread of settlement inevitably led to conflict over ownership. The marking out of fields with permanent boundaries became more important (Muir and Muir, 1987). In several regions great areas of land were marked out. A classic pattern is shown on Dartmoor. Here at one time, about 3700–3600 BC, an extensive system of linear fields was laid out, marked by stone walls, 'reaves'. Some reaves ran across country following ridges or marking the upper edge of cultivation at 400 m. Running down from these at right angles were other reaves marking out ribbons of land. These ribbons were subdivided by short cross-walls to make small, rectangular fields.

This pattern of coaxial fields was not a new development but had been present in the Neolithic of Co. Mayo, and coaxial field patterns are found elsewhere. The landscape was divided up, territories marked and fields enclosed. At Fengate near Peterborough, ribbons of land 50–150 m wide, running down into the fen, were marked out by ditches and hedges or droveways (Williamson, 1987). Again, fields were made by subdividing the ribbon of land with cross-walls. The marking out of the land indicates a well-organized society with a strong social hierarchy, with leaders able to impose their will on lower ranks.

This pattern of land division may indicate that some regions were largely unwooded. Open areas were no longer islands in a sea of wildwood but woodlands were islands in a sea of cultivated fields and grassland or scrub. The construction of ceremonial monuments in the Late Neolithic and Bronze Age also indicates an open landscape. Large open areas of arable fields and grassland had been created so that monuments could be seen clearly from a distance. In this open landscape there was an intensification of agriculture with, from the Middle Bronze Age, a greater range of cereals and legumes cultivated. Large granary stores were created. A storage pit could hold about a tonne of grain, the produce of 3–6 fields. Plugged with clay, elevated carbon dioxide levels kept the grain dormant. The agricultural system could be relied upon to feed the population. The wild flora became less important for survival. The proportion of wild fruits and nuts in the diet declined, and the bones of hunted animals are found less frequently in archaeological deposits. The population rose to a peak by about 3300 years ago (Burgess, 1985).

In some areas the pattern of 'Celtic' fields was more varied, mostly reflecting the more varied topography, and perhaps also a less rigid social hierarchy. Here ploughing and tillage encouraged soil erosion. Soil was shifted from the top of the field and trapped by the wall or hedge at the bottom of the field, creating a terrace or lynchet. Systems of stepped terraces were created. Squarish 'Celtic' fields of between 0.2 and 0.4 ha would have allowed a plough team of two oxen travelling at about 3 km/h to criss-cross plough them with an ard in about 8 hours.

By the Middle Bronze Age the landscape had become a diverse mosaic of cultivated, seminatural or natural plant communities which were exploited in different ways at different times. Around the Somerset Levels, for example, the settlements were on the lower slopes with arable fields above. Further up the slopes there were areas of coppice, grading into woodland with open glades in which pigs were herded (Coles and Coles, 1986). Below the settlements there was grassland, some of it seasonally flooded, on which cattle were grazed.

3.3.1 CHANGING CLIMATES AND RISING SEA LEVELS

Up to about 3000 years ago the climate was generally warmer than at present (Lamb, 1981) though it was rather unsteady, with periods of greater cold and wet. Oscillation in the development of bogs in Denmark occurred with a period of about 260 years (Aaby, 1976). During one cooler period about 4000 years ago the tree line retreated from an altitude of 800 m in Scotland (Moar, 1969). Further south it remained above its present limit for the time being. There was a warm period around 3000 years ago followed by about a 2°C decline. Glaciers in the Alps were advancing during this neo-glaciation. The growing season was shortened by about 5 weeks. Between 2750 and 2400 BP there was also a period of much greater wetness in the west (Moore, 1988). The peat at Tregaron Bog was growing at a rate of 4 cm/100 years, compared to its rate of 1 cm/100 years at other times.

Great tracts of moorland started to develop in the Atlantic period but they became widespread especially in the Bronze Age as the woodland was cleared. There were phases of rapid growth of the peat interspersed with drier phases. This was first noticed in the stratigraphy of Scottish bogs by James Geike in the late nineteenth century. He described the remains of trees, *Betula* (birch) and *Pinus sylvestris* (Scots pine), preserved in two horizons in the peat (Figure 3.10).

He named these the Upper and Lower Forestian zones. Each was overlain by a peat composed of *Sphagnum* (bog moss) and *Eriophorum* spp. (cotton grass), which were called the Lower and Upper Turbarian zones. The trees of the Upper Forestian zone were growing on a 'recurrence surface'. The peat surface had been colonized by trees when it had dried, and they were later overcome by the renewed spread of blanket peat. Geike's zones have been correlated to the stages of the Scandinavians, Blytt and Sernander, which have become the standard terminology: Boreal = Lower Forestian; Atlantic = Lower Turbarian; Sub-Boreal = Upper Forestian; and Sub-Atlantic = Upper Turbarian (Evans, 1975). However, radiocarbon dating of forest horizons has demonstrated a great range of dates which are not consistent with the Boreal or Sub-Boreal stages (Birks, 1970, 1972a, b; Blackford, 1993).

Oscillations in climate made cereal cultivation less certain and towards the end of the second millennium BC encouraged the abandonment of upland fields on Dartmoor and elsewhere (Askew, Payton and Shiel, 1985). Only hundreds of years later in the middle centuries of the first millennium BC is there some sign that the uplands of Devon and Cornwall were being recolonized to a limited extent. The abandonment of upland arable fields was only part of a general collapse of population and abandonment of settlements between 1250 and 900 BC, which occurred even in the lowlands. From the Late Bronze Age there was a switch from naked barley and emmer wheat to the more robust hulled barley and spelt wheat. These hardier varieties may have been sown in autumn to overcome a shorter growing season, but autumn cultivation would also have increased soil erosion. Legumes, such as beans (*Vicia faba*) and even peas (*Pisum sativum*), with their root nodules that fix atmospheric nitrogen, were cultivated for the first time. Oats and rye, tolerant of acid soils and colder conditions, were also being cultivated for the first time. The new crops were present sporadically at first but grew in importance.

It has been traditional to identify at this time a change from the dry Sub-Boreal to a wetter and cooler Sub-Atlantic stage, but changes in land use may not have been due entirely to climatic change. Associated with these changes was a widespread cultural

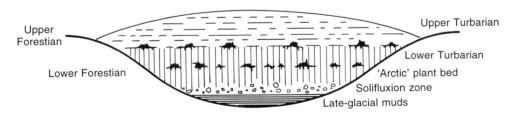

Figure 3.10 Profile through a Scottish peat bog. (Geike, from Evans, 1975.)

change, a profound shift towards pastoralism, which had been first expressed many centuries earlier in some areas. For example, about 3500 years ago in the southern 'Wessex' area, ranch boundaries were laid down as stockproof hedgebanks. Some boundaries survive as linear earthworks. In some places, as at Martin Down in Hampshire, the new boundaries ignore the boundaries of earlier fields, perhaps indicating a trend towards greater stratification of society. Chiefs could organize the construction of ranch boundaries on a broad scale. Round barrows were constructed for prestige burials. Increased social stratification and tribalism led to a period of conflict, so that about 3000 years ago many protected centralized hill-top settlements were constructed. In times of social upheaval, and perhaps even war, herds were easier to protect and were a mobile kind of wealth and sustenance. Grasslands became more widespread. Where ranch boundaries were established they indicate a fundamental shift from arable or mixed farming to livestock rearing. However, the construction of new boundaries was halted by a population collapse in the early centuries of the first millennium BC, because the population became too small for any large-scale construction activity.

The collapse in population at this date, if analogous to later ones in the late Roman period and in the thirteenth century, may have been due to a plague of some sort (Burgess, 1985). The decline in population and social upheaval had consequences for vegetation development. Soil fertility in the uplands relied upon continuous patterns of cultivation and manuring. Uplands without people to cultivate them became more marginal for cultivation. In the lowlands, fields were abandoned to scrub, and then secondary woodland with different species composition arose. *Fagus sylvatica* (beech) became more important in the south. *Tilia cordata* (small-leaved lime) continued its relative decline because it was now more limited by the shorter summers, too short for its fruits to mature. In the limestone uplands ash woodlands arose.

Around the wetlands of the Fenland and Somerset Levels different changes in the vegetation occurred. At first there were a large number of Bronze Age settlements around the south-eastern fenlands in East Anglia, established at a time when there may have been a drying of the area. People were taking advantage of the rich soil and open plant communities. Wetland grass and sedge communities are potentially very productive. Trackways and causeways were built between islands, above the fen, at the end of the Sub-Boreal stage. This activity may simply indicate the greater use of wetter areas, but later causeways may also have been constructed because of the onset of the wetter climatic conditions of the Sub-Atlantic stage.

Eventually, in the middle of the first millennium BC, the Fenland, including the fields of Fengate, were abandoned. There was a great marine transgression which started between 600 and 300 BC. Large coastal saltmarshes formed and there was an estuary all the way to Wisbech. Apart from a few gravel islands, the Fenland became too wet to occupy. During this period the rivers were tidal and flooded frequently. Silt was deposited on the river banks so that these became firmer and higher than the surrounding countryside where marshy peat was forming. In later ages, as the rivers dried and the peat shrank, the river banks, called roddons, became a prominent feature of countryside. Apart from the Fenland and Somerset Levels, there were extensive marshes in Kent, Essex and Yorkshire, and many other smaller areas scattered elsewhere. There were large areas of shallow water utilized by cormorant, pelican and crane. On the margins of the Glastonbury Lake a village was built on a massive platform of logs. In Ireland and northern Britain other, smaller, crannogs were established.

3.3.2 THE DEVELOPMENT OF HEATHLAND

At the root of many land-use changes that occurred in the Bronze Age were changes in the soils (Dimbleby, 1962, 1965). Some areas had experienced millennia of cultivation and grazing. While cultivation continued, soil fertility could be maintained by the plough or hoe, and by manuring, but once this ceased soils would mature very rapidly. Leaching led to podzolization. Change was most severe over non-calcareous subsoil. Agricultural use led to a loss of organic matter, which speeded the subsequent rate of podzolization.

Podzols are particularly influenced by the nature of the vegetation they support. They have developed especially under heathland, and under other vegetation. They cover about 5% of England and Wales, 25% of Scotland and 8% of Ireland. The development of a podzol is often marked by the invasion of acidophile (calcifuge) shrubs, especially *Calluna vulgaris* (heather), but including other members of the heather family, the Ericaceae, such as the heaths (*Erica* spp.) and bilberries (*Vaccinium* spp.). Podzols are also found under *Pinus sylvestris* (Scots pine) and even under broad-leaved woodland. The soil profile is of a leached, acidic soil, which is well stratified. Podzols have a mor humus; organic matter is slowly or only partially decomposed. There is a light (ash)-coloured eluvial horizon where aluminium and/or iron (sesquioxides) have been leached down, to precipitate in a lower black or reddish-brown illuvial Bs horizon. In conditions of poor aeration, an iron pan (Bf) may form just below the eluvial horizon. Humus may also move down the profile with the metals, as metal-organic complexes. When well developed, the iron pan is a barrier to root growth and drainage. Clay leaches down separately.

The timing of the development of heathland varies between sites. On the Breckland sands the activities of a dense Neolithic population resident around Grime's Graves began the process of podzolization early. In many other places the Bronze Age was the most important period in which lowland heaths developed. In other areas at the time of the Bronze Age and even later, in the Iron Age, the process of leaching of bases and podzolization was only in its early stages.

Barrows are particularly common on heathlands. Prehistoric soils buried and preserved beneath Bronze Age barrows and Iron Age embankments in widely separate areas, from North Yorkshire, Dorset, the New Forest and Surrey, have only slightly degraded brown soils (Dimbleby, 1962). When constructed these monuments were surrounded by woodland, scrub, grassland or agricultural land. Today these prehistoric monuments are often surrounded by an acid heath or moorland. In parts of Dorset the woodlands maintained an open character with an under-scrub of heather throughout the Early Holocene, surviving the expansion of *Corylus*, *Ulmus* and *Quercus*. The process of heathland formation was relatively quick, so that by 1000 BC they had been established (Moore, 1962, 1988; Waton, 1983). Here it seems that the heathscape survived relatively unchanged for nearly 3000 years, so that in 1706 part of Dorset was described (Defoe, 1971):

> . . . from Piddletown, Bere Regis, and Wimbourne Minster, to the Purbeck Hills, is a most dreary tract of heath land, and is scarcely capable of any improvement in the hands of the agriculturalist . . . A few cattle are kept on various parts of the heaths, some poor half-starved sheep are occasionally seen wandering about.

This is the landscape of Thomas Hardy's Egdon Heath, which had 'a lonely face, suggesting tragical possibilities'.

A range of lowland heath types have been recognized, distributed largely according to climate (Rodwell, 1992a). They are species-poor communities dominated variously by *Calluna vulgaris* (heather), *Erica cinerea* (bell heather) (Figure 3.11) and one of the species of *Ulex* (gorse) (Figure 3.12) (Webb, 1986).

In the driest part of the country there is a very poor community dominated by heather with the mosses *Hypnum cupressiforme* and *Dicranum scoparium* and the lichens *Cladonia* species. *Calluna* goes through a series of growth forms (Mohamed and Gimingham, 1970). The plant reaches maturity after about 15 years from seedling establishment, forming a dense bush under which little else but bryophytes and lichens can grow. Subsequently growth slows as the bush gets very leggy and collapses outwards. Eventually it degenerates so that branches lie along the ground, exposing the central area to colonization by seedlings.

This cycle is modified by burning, which halts the process in the pioneer and building stages when growth is most rapid. One method of management of heathland continued to this day has been through burning to maintain a youthful community which is preferred for grazing. On grouse moors small patches of heathland are burnt at a time so that there is a mosaic of patches of *Calluna* at different ages, providing grazing but also places for brooding chicks. It is possible that burning accelerated the development of heathland in the Bronze Age. Burning produces a flush of available nutrients but there is a net loss of nutrients in smoke and from runoff. The loss of nutrients is greater after a severe burn, so burning is better done in spring or autumn when wetter conditions prevent very high temperatures

(a)

(c)

(b)

Figure 3.11 Heathers: (a) *Calluna vulgaris* (heather or ling); (b) *Erica cinerea* (bell heather); (c) *E. tetralix* (cross-leaved heath). (Copyright Jon B. Wilson.)

Figure 3.12 *Ulex europaea* (gorse) (copyright Jon B. Wilson.)

being reached. Some heathland plants, such as *Ulex minor* (dwarf gorse), can show prolific re-sprouting after burning. *Deschampsia flexuosa* (wavy hair grass) is common on most moorlands and heaths. It is frequent as a colonizer of burnt heather areas. On the heaths of the south and west *Agrostis curtisii* (bristle bent) has been particularly favoured by regular burning (Ivimey-Cook, 1959). It has an interesting distribution, almost like a Lusitanian plant down the Atlantic coast of France, Spain and Portugal, but it is not known in Ireland. In the autumn following the fire there is rapid colonization of open ground by its seed. A pure sward is produced the following season and in some places it remains dominant.

However, burning was probably not a very important kind of management in many lowland heaths until the nineteenth century (Webb, 1986). Grazing and cutting were more important. *Deschampsia flexuosa* is eaten by sheep and rabbits but is capable of withstanding grazing pressure for a long time (Scurfield, 1954). *Calluna vulgaris* is grazed preferentially. Paths through heather are often marked by bands of *Deschampsia flexuosa* either side of the trampled middle. Grazing by sheep alone is said not to be sufficient to maintain a heathland. In the twentieth century, with the cessation of organized grazing, heaths like those in Surrey were colonized by woodland, especially by *Betula pendula* (silver birch) and then by *Quercus* (oak).

The cutting of turfs or peat has been very important in the maintenance of heathland, especially perhaps in the absence of fire. Turf cutting dates back to the Neolithic when many of the barrows were constructed out of turfs. From the Middle Ages turbary was a well-established right, providing an important source of fuel. In the New Forest a single 'turbary right' averaged about 4000 turfs annually, and up to half a million turves were cut annually. The consequence of turf cutting was to prevent the accumulation of nutrients and it thereby maintained heathland. In addition to turf cutting, the cutting of 'furze' (gorse) and 'fern' (bracken) for fuel was also carried out. The use of these alternative fuels is a powerful indication of the rarity of trees to provide wood for fuel in heathland areas.

In a well-developed heathland of the East Anglian sort (Rodwell, 1992a; H1) there is little opportunity for other vascular plants to establish, but there are tussocks of *Festuca ovina* (sheep's fescue) and occasionally *Ulex europaeus* (gorse), *Rumex acetosella* (sheep's sorrel), *Deschampsia flexuosa* (wavy hair grass), *Campanula rotundifolia* (harebell) and *Pteridium aquilinum* (bracken), among others, may establish. On slightly wetter soils *Erica cinerea* (bell heather) may be present. *Pteridium aquilinum* establishes on the slightly deeper, richer soils and may alternate dominance with *Calluna vulgaris*. On shifting sands *Carex arenaria* (sand sedge) may become dominant. On the soils of the Breckland, which have been patterned by periglacial action, heather grows in the sand-filled cracks around patches or stripes of grassland. The distribution and abundance of Breckland rarities has depended upon changes in the use of this land. Throughout the Middle Ages, and before, much of the area was acid heath dominated by *Calluna vulgaris* (heather). The traditional use of this poor land involved a long fallow of unenclosed fields or 'brecks'. About 80% of the land was fallow at any one time and used only for rough grazing by sheep. Surface soil disturbance by the sheep hooves created many open microsites where seedlings could establish.

Bracken can be a very strong competitor on the drier soils (Watt 1955; Grime, Hodgson and Hunt, 1988). Clones grow forward vegetatively, invading with deep rhizomes ahead of the fronds in a continuous 'phalanx'. Individual clones can live for hundreds of years. They cast a deep shade and produce a deep leaf litter. *Pteridium* has high disease resistance, low palatability and produces toxic/carcinogenic compounds. In particular, it produces in its leaves many phenolic compounds which have an allelopathic effect, preventing the growth of other species. Leachates from the living and dead fronds inhibit the root development of grasses such as *Deschampsia flexuosa* (wavy hair grass) or *Agrostis* spp. (bent grasses). In many areas *Pteridium* is rapidly becoming the dominant species of drier heaths. It is said to be increasing in extent 1–4% each year. This is a result of the decreasing use of lowland heaths for grazing or cutting. The cutting of bracken was regularly carried out in many areas in the past to provide farmyard litter and maintain grazing. The ability of *Pteridium* to recover rapidly after fire by growing from the rhizomes, which were protected from heat in the soil, is an important advantage. The

extensive rhizome system allows it to spread rapidly and dominate large areas of hillside wherever the ground is well drained. As well as waterlogging, *Pteridium* is also sensitive to spring frosts. Genetically and cytologically it is very variable. High propagule mobility ensures the exchange and recombination of genetic variation from different areas, so that newly adapted variants can arise, adapted to local conditions.

Heathland soils are notably deficient in nitrogen but the gorses have nodules containing the symbiotic bacterium *Rhizobium* which fixes atmospheric nitrogen making it available for the plant. The heaths on the acid sands south of London (Figure 3.13a) are richer than those in East Anglia.

In the slightly less continental conditions here, wetter with milder winters, *Calluna vulgaris* shares dominance with *Ulex minor* (dwarf gorse), and *Erica cinerea* (bell heather), *Deschampsia flexuosa* (wavy hair grass) and *Pteridium aquilinum* are commonly present (Rodwell, 1992a; H2). *Ulex minor* flowers in the summer and autumn, unlike *Ulex europaeus*, which flowers mainly in winter and spring. This kind of southern heath is common in the New Forest, but there is also an alternative community present where drainage is impeded. In this community *Agrostis curtisii* (bristle bent), *Molinia caerulea* (purple moor-grass), *Erica tetralix* (cross-leaved heath) and *E. cinerea* (bell heather) are co-dominants along with *Ulex minor* and *Calluna vulgaris* (Rodwell, 1992a; H3). This kind of community is common in the more oceanic conditions of the Dorset heathlands. Occasionally the rare *Erica ciliaris* (Dorset heath) is also present.

Further west, in heathland at a more oceanic extreme, *Ulex minor* is replaced by another kind of dwarf gorse, *Ulex gallii* (western gorse). Cutting and burning may have favoured the more prostrate *Ulex minor* and *U. gallii* (dwarf and western gorse) over *Ulex europaeus* (gorse). The distributions of the two short species scarcely overlaps except

Figure 3.13 Heathland and moorland: (a) see plate section; (b) Exmoor, Devon. (Copyright Jon B. Wilson.)

in the region of Poole Harbour (Proctor, 1965). They are hard to distinguish from each other, with variable growth forms, but the length of the calyx can be used to identify them: *Ulex minor* (dwarf gorse) <9.5 mm; *Ulex gallii* (western gorse) >9.5 mm. Apart from the difference in gorse species, this western heathland is very similar to the Dorset heathlands (Rodwell, 1992a; H4). On more freely draining soils, rankers and podzolized sands, throughout western England and Wales, *Agrostis curtisii* (bristle bent) is absent and *Molinia caerulea* (purple moor-grass) less common (Rodwell, 1992a; H8). This is perhaps the most common kind of lowland heath in England and Wales, dominated by the three species *Calluna vulgaris* (heather), *Ulex gallii* (western gorse), and *Erica cinerea* (bell heather).

Two distinct heath assemblages are confined to the Lizard peninsula in Cornwall (Rodwell, 1992a; H5 and H6) (Figure 2.30). Both have *Erica vagans* (Cornish heath) and other heath species, either *Erica cinerea* (bell heather) or *E. tetralix* (cross-leaved heath).

In many parts of the world it is possible to recognize a peculiar flora associated with serpentine soils. The most important outcrop of serpentine rock in the British Isles is that on the Lizard peninsula. Serpentine soils generally have high levels of magnesium relative to calcium, and toxic levels of nickel, chrome and cobalt. They are peculiar in being base rich but low in calcium. However, on the Lizard they are largely overlain by superficial deposits so that, except where there has been erosion, the soil is not normally serpentine (Proctor and Woodell, 1971).

Nevertheless, calcifuge species grow on the Lizard on base-rich soils, taking advantage of the lack of calcium. In addition, the low level of major plant nutrients on these shallow, skeletal soils, with nevertheless a high level of bases, also allows the growth of acidofuges, what we would normally call calcicoles, such as *Geranium sanguineum* (bloody cranesbill), *Filipendula vulgaris* (dropwort), and *Hypochaeris radicata* (spotted cat's-ear). The dominance of *Schoenus nigricans* (black bog-rush) is an indicator of the very low phosphorus levels present. In the drier areas *Ulex europaeus* and *U. gallii* (gorse and western gorse) are co-dominants with the heath species.

3.3.3 THE DEVELOPMENT OF MOORLANDS

Podzolization was especially prevalent in the north and west and at high altitude (Figure 3.14). Cold, wet conditions encouraged the development of a thick mor humus and then blanket peat (Merryfield and Moore, 1974; Lamb, 1981).

The uplands may have been fairly intensively utilized in earlier times. Bronze age archaeological remains are more abundant than from any other prehistoric period. In the wetter conditions of the first millennium BC moorland communities became more widespread in the uplands. Human disturbance in these climatically marginal habitats eventually led to the development of blanket peat. The tree canopy, which had intercepted the rain so that some of it evaporated without reaching the soil, and roots, which had dried the soil by transpiration, had been removed. Without the trees the soils became waterlogged. Without tillage to stir the soil and organized manuring there was rapid leaching. Soils acidified and podzolized. In waterlogged conditions there was the development and then the spread of blanket peat (Askew, Payton and Shiel, 1985). The initiation of blanket-peat formation did not happen everywhere at the same time, but from now

on the general trend was for an ever more marked distinction between the upland, with its leached acid soils and blanket peat, its moorland, and the cultivated lowland (Brown, 1977). This change is seen in many pollen diagrams, such as those from the Derbyshire uplands above 300 m (Hicks, 1971). A fairly extensive *Calluna*-dominated vegetation was present already by 1500 BC in the Bronze Age.

In different areas, under different climatic regimes, different moorland vegetation arose. In the southern Pennines and North Yorkshire Moors, and also scattered throughout the Midlands, a *Calluna vulgaris–Deschampsia flexuosa* community (Rodwell, 1992a; H9) has been described. This is related to the southern heathlands but has a poorer flora. The mosses *Pohlia nutans* and *Hypnum cupressiforme* are particularly common. *Vaccinium myrtilus* (bilberry) or *Galium saxatile* (heath bedstraw) are common in different subcommunities. Towards the north and west, in Wales, the Lake District and Scotland, there is an 'Atlantic heather moor' (Rodwell, 1992a; H10). Here the moorland is dominated by *Calluna vulgaris* and *Erica cinerea* (bell heather). *Potentilla erecta* (tormentil) is particularly conspicuous and *Galium saxatile* (heath bedstraw) is common. On higher ground and away from oceanic conditions, in the most widespread kind of moorland, *Calluna vulgaris* shares dominance with *Vaccinium myrtilus* (bilberry) and *V. vitis-idaea* (cowberry) (Rodwell, 1992a; H12). This community can be found from the uplands of Dartmoor to the Grampians. In central parts of England, around the industrial towns of south Lancashire, Yorkshire and the Midlands its distribution may be limited today by pollution. This community type is a late development in some areas, arising perhaps with the greater use of the uplands for sheep grazing in the Middle Ages. Its structure and variation is greatly influenced by burning to improve grazing for sheep and, more recently, to create grouse moors.

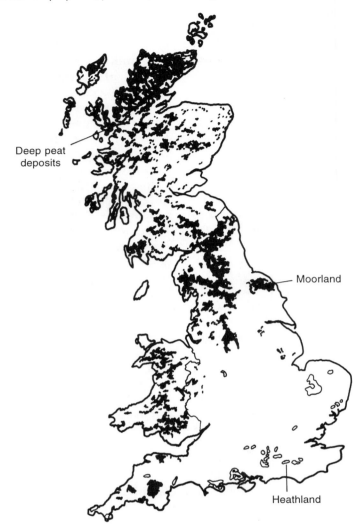

Figure 3.14 Distribution of lowland heath (outline areas) and moorland (black areas) in England, and areas of major deep peat deposits (black) in Scotland and Wales. Areas of moorland and bog are much more extensive in Wales and Scotland than the areas indicated. (Adapted from the Nature Conservancy Council, 1984a; Ratcliffe and Oswald, 1988.)

The different species of heather show different patterns of ecological preference (Grime, Hodgson and Hunt, 1988). Several are confined in the British Isles to small areas in the west and south-west. Of the two widespread species of *Erica* (Figure 3.11), *E. tetralix* (cross-leaved heath) is not as badly affected by waterlogging as *E. cinerea* (bell heather) but it lacks flexibility to other environmental variables. It does not compete very effectively on the drier soils. *Erica cinerea* suffers from iron toxicity in waterlogged soils but, because it is more deeply rooted, it can cope with drier conditions better. In comparison, *Calluna vulgaris* is so widely successful because it has a greater physiological potential. For example, it can continue to photosynthesize effectively, maintaining its stomata open at much lower cell turgidities, down to 65–75% turgidity, in much drier conditions than other Ericas. Equally important is the relative performance of seeds and seedlings. *Calluna vulgaris* seeds germinate very readily and grow rapidly. The seeds of *Erica tetralix* germinate only poorly except in wet conditions, while those of *E. cinerea* show enhanced germination after the heat treatment of a heath fire.

A very widespread community in the wet uplands of the west and north is that dominated by the tussocks of *Juncus effusus* (soft rush) with the mosses *Polytrichum commune* and *Sphagnum* (Richards and Clapham, 1941a). It favours base-deficient mineral soils or thin peat, especially where there is a fluctuating water table, and it can also be found in the lowlands beside ponds and streams. It is abundant on trampled or drained peat. Two other rushes that closely resemble *J. effusus* may be present. They are *J. inflexus* (hard rush) and *J. conglomeratus* (compact rush). *Juncus conglomeratus*, which differs from *J. effusus* in forming smaller, less-dense tussocks, and having a dull, ridged stem, and a flat, rather than rolled bract above the inflorescence, is found in similar habitats (Richards and Clapham, 1941b). *Juncus inflexus*, with its sharp leaves and interrupted pith, is more common on heavy neutral or basic soils in the lowlands (Richards and Clapham, 1941c).

3.3.4 PEATLANDS

The wetter areas of western Britain were especially liable to the development of blanket peat (Figure 3.14) (Moore and Bellamy, 1974). Peatlands were especially widespread in Ireland, despite a widespread underlying geology of Carboniferous limestone because most of this is overlain by glacial drift deposits. Podzolic soils developed on these. Blanket bogs developed in the western highlands of Ireland and raised bogs in the central lowlands on the waterlogged soils of the central plain. The central lowlands are marked by eskers, gravel ridges laid down by water issuing from the front of glaciers. Large areas have many drumlins, little hills of glacial drift moulded by the ice.

Peat soils have at least 40 cm of organic deposits within the top 80 cm, or have 30 cm of peat on bedrock. Organic material in peat is humified so that it is converted from raw organic material via a semifibrous state to a more homogeneous humified matrix. The peat either develops *in situ* in wet conditions or, occasionally, it is deposited as sediment in lake bottoms. The extent of peat soils is probably much reduced from what existed in the past. Today they form only 3% of soils in England and Wales, 16% in Ireland, but more in Scotland, which has extensive areas of blanket bog.

The complex history of bog development is exemplified by the pollen record in one area of western Mayo (Moore, Dowding and Healy, 1975). The woodland of *Pinus sylvestris* (Scots pine) and *Quercus* (oak) was drastically reduced, perhaps by burning, during the fourth and fifth millennia BC. Peat growth ensued but eventually pine forest colonized the peat. This second forest was cleared about 2300 BC by burning, and subsequently there was the marked spread of blanket peat even on to better-drained areas. The whole area has remained bog and largely uninhabited until today, although there was some clearance in the eighteenth and nineteenth centuries. In recent centuries peat has accumulated at a rate of 10 cm/century.

Mesolithic burning of woodland is also implicated in the development of blanket peat on high moorland in Dartmoor (Caseldine and Hatton, 1993). In the Early Holocene a fern-rich hazel woodland developed which was converted to acid grassland by a phase of enhanced burning 7700–6300 years BP. The early transitional community, like a woodland margin vegetation, included *Calluna vulgaris* (heather) and *Melampyrum* (cow-wheat) an association which often indicates burnt ground. *Melampyrum* is interesting as a hemiparasite, especially of grasses; growing on the roots of its host and gaining some nutrients from it. The acid grassland, similar to the heath-derived grassland that can be found on Dartmoor today, with *Lotus corniculatus* (common bird's-foot trefoil) and *Rumex acetosa/acetosella* (sorrels), lasted for 600–1000 years but eventually gave way to sedge-rich and blanket-bog assemblages, which included *Potentilla erecta* (tormentil), *Narthecium ossifragum* (bog asphodel) and *Drosera* (sundews) growing amongst the *Sphagnum* (bog moss). A similar history is recorded at Derry-inver Hill in County Galway, with grassy heath developed before 2400 years BP and the development of blanket bog delayed to 1700 years BP by continued land use (Molloy and O'Connnell, 1993).

The most extensive area of blanket peat and bog in Britain is the Flow Country of Caithness and Sutherland, which, in the 1950s, covered 4000 km^2 scattered over a wide area (Ratcliffe and Oswald, 1988). The Flow land has a diverse origin, from woodland or from more open herb and scrub communities. In the west a pollen diagram from Scourie shows that a woodland of *Betula* (birch), *Pinus sylvestris* (Scots pine) and *Quercus* (oak) disappeared rapidly between 4500 and 4000 years ago. Pine stumps buried in the peat date from this period (Geike's Upper Forestian). Woodland was comprehensively lost relatively synchronously across north-western Scotland, so it has been suggested that there was a climatic cause for woodland decline at this time. Woodlands in this part of northern Scotland were evidently already in a marginal situation prior to climatic decline. A pollen diagram from central Caithness at Loch Winless shows little evidence of there ever being much woodland in this area, only *Corylus avellana* (hazel) and *Betula* scrub in sheltered places. The Early Postglacial tall-herb and birch and willow scrub communities, which were quickly eliminated elsewhere in Britain, survived for several thousands of years in this area, only to be suppressed by the spread of bog and blanket-peat communities, starting about 4500 years ago (Moar, 1969; Peglar, 1979).

3.3.5 THE CELTIC LANDSCAPE

Population levels started to increase again towards the middle of the first millennium BC (Figure 3.15).

About the eighth century BC both iron-working and horse-riding were introduced into Europe by the Cimmerians from the

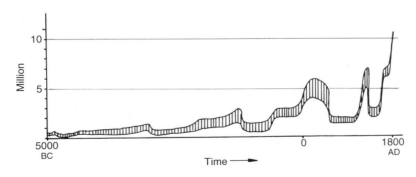

Figure 3.15 Human population levels, showing likely maxima and minima between *c.* 7000 and 200 BP. (From Burgess, 1985.)

Caucasus. First the Hallstatt and then the La Tène Celtic cultures arose, centred on different areas of central Europe (Laing, 1979). The first Celts arrived in Britain during the seventh century BC. Their arrival is marked by a hoard of objects discovered on draining Lyn Fawr in Glamorganshire. The bronze objects are like those of the Hallstatt Celts of Europe, and include various parts of the harness for a pony, but there are also iron objects – a spear-head, a sickle and part of a sword.

At first the Celts were more influential through trade and by cultural innovation rather than by the influx of large numbers of people. The iron axe made it easier to clear woodland and also to wage war. There was renewed construction of hill-forts, leading to the development of huge embanked sites such as Maiden Castle in the second and first centuries BC. The construction of these hilltop defences coincided with the late second century BC influx of people of the La Tène culture, followed a little later by the Belgae, from an area just to the south of Belgium. The Belgae were Celts, sophisticated by their contact with the classical world. The Belgae paved the way for the Roman conquest of south-eastern England. For the first time we have reports of what it was actually like to be in Britain 2000 years ago. Its as if talkies have been invented and we have the first eye-witness reports. Caesar wrote of Britain (Handford, 1951):

> The population is exceedingly large, and the ground thickly studded with homesteads, closely resembling those of the Gauls, and the cattle very numerous . . . There is timber of every kind, as in Gaul, except beech and fir. Hares, fowl and geese they think it unlawful to eat, but rear them for pleasure and amusement. The climate is more temperate than Gaul, the cold being less severe . . . Most of the tribes of the interior do not grow corn, but live on milk and meat and wear skins.

So starts the historic period. Even though it is inevitably very biased, Caesar's account gives us the best picture of Celtic culture to come from Classical writers.

By the time of the advent of the Romans many of the aspects of our living landscape had been established. There was a distinction between upland and lowland (Chambers, 1988). Close contact with the more advanced cultures of the continent exaggerated the division between the lowlands of the southeast and the uplands of the north-west. The lowlands were well populated and agriculturally well developed with a patchwork of fields and woods. In the uplands arable farming was relatively unimportant. These areas had entered a cycle of decline to moorland and bog where even pastoral farming was limited. There were grasslands, heaths and moorland. Much of the woodland had been cleared in southern Britain. In the Anglo-Scottish border region massive deforestation accompanied the large-scale developments of Hadrian's and the Antonine walls and their associated forts, roads and settlements (Barber, Dumayne and Stoneman, 1993). From the midland valley of Scotland the landscape was a quilt of natural seminatural and managed plant communities.

The density of cattle might reflect that methods for maintaining cattle over winter by the harvesting of hay had been developed. Before the Roman period the wet pastures and seminatural grasslands of the upper Thames Valley were of the *Cynosurus cristatus/ Centaurea nigra* (crested dog's-tail/common knapweed) sort (Greig, 1988; Rodwell, 1992b; MG5). They were occupied, perhaps seasonally, by herdsmen exploiting the rich summer pasture. In the Roman period the flood plain was used for hay meadows which had become a valuable resource. They were converted to *Alopecurus geniculatus/Sanguisorba officinalis* (marsh foxtail/great burnet) grasslands (Rodwell, 1992b; MG4) which are sometimes enriched by shows of *Fritillaria meleagris* (snake's-head fritillary).

By the end of the Iron Age the weed/arable flora showed many similarities to most later pre-herbicide weed floras (Jones, 1988). In the Iron Age bread wheat was only a minor crop and emmer and the more robust spelt wheat were more commonly cultivated. Six-row barley was quite extensively cultivated.

The Roman period did not bring a sudden change but more a continuation and development of trends already obvious in the Celtic period. There are some indications to suggest that Roman fields were relatively water-logged and nutrient poor. Weeds, such as *Chenopodium* spp. (goosefoot), which prefer high levels of nutrients declined in importance. In southern England *Eleocharis palustris* (common spike-rush), common in mire and still aquatic communities, where it spreads by fragmentation, was present in crop/weed communities. This is perhaps an indication that valley-bottom flood plains, which were periodically flooded, were being cultivated more intensively (Lambrick and Robinson, 1988). Ground-water gley soils were brought into agricultural usage by drainage. Stagnogleys, which have restricted downward movement of water so that it stagnates, proved difficult to drain but were commonly used for the production of grass. In northern England *Danthonia decumbens* (heath grass) seems to have been behaving as a weed. Today it is confined to nutrient-poor grasslands because it is unable to compete in more fertile soils with more productive species. The presence of these species indicates low-intensity arable farming and the problems faced in controlling weeds.

The older Celtic society survived intact in Ireland (Aalen, 1978) and in parts of western and northern Britain. Tribal and overwhelmingly rural, it survived into the Middle Ages. Settlements, notably 'raths', were scattered over the landscape. Each rath was a circular enclosure, a 'ring fort', bounded by an earthen bank and ditch about 15–46 m across. Crannogs, tiny settlements built on artificial islands, provided an alternative place of security. Settlement patterns were fluid as cattle herders moved herds to new areas. The rath was the centre of a mainly pastoral farm. Fields were preserved for cattle pasture, and tillage seems to have been carried out irregularly in patches on an outfield. The landscape was rich in degraded secondary woodland, mainly *Fraxinus excelsior* (ash), which had colonized after the decline of elm. Forest clearance became more widespread in the middle of the first millennium AD.

3.3.6 ROMAN EXPLOITATION

The Romans changed the landscape, not necessarily by clearing much greater areas of woodland or introducing woodland management on a large scale but by encouraging trade and industry. For the most part, Celtic field systems survived the Roman period (Wacher, 1982; Hodder and Millett, 1990). New military roads, like the Pye Road in East Anglia, were sometimes cut across them. The cultivation of those fields did not change much. A coulter to break the soil ahead of the plough-share helped, but a mould-board to turn the soil and so help control weeds was a late introduction. In some areas, like the Dengie peninsula in Essex, very regular rectangular Roman fields were marked out. In the Roman period there was a marked increase in the size of livestock. Sheep were more favoured than they had been previously. The close-cropped grass sheep walks of Sussex and Wiltshire chalklands may date from the third century AD. Sheep with a continuously growing fleece, which were regularly shorn, were introduced.

The growth of Roman towns provided a market for agricultural produce. Britain was divided into civitates, at the centre of each was a capital. In addition, London, as the major trading centre, became the provincial capital. The urban population never amounted to more than 10% of the total population. The countryside was relatively populous. The total population of Roman Britain rose to perhaps 3–4 million, some estimates say even 6 million. On the basis of recent surveys there were at least 92 000 rural settlements. Perhaps about 700 of these were the villa estates of the aristocracy. Some were very large, covering over 800 ha.

There were still some large areas of natural vegetation, which often lay in marginal areas between different civitates. There were the great fens and marshes and great woodlands like the Weald. But the mark of Rome was nearly everywhere. Parts of marsh and fen, as in the Romney marsh, were drained. The rivers of the East Anglian Fenland were used to transport grain to the north. The Car Dyke, a great drainage canal, was cut around the western margin of the fen between the Cam and the Witham to divert water into the main rivers which flowed across Fenland. The silt fens were cultivated, protected behind a seabank. The islands in the inner peat fens had settlements. Substantial efforts to reclaim the marshes for more profitable use were made by, for example, embanking the Medway marshes in Kent.

The woods were coppiced to provide a continuous supply of wood for Roman industry. Especially important were the iron workings of the Weald and the Forest of Dean. The extent of the Roman workings in the Weald suggest that about 15% of the Weald would have supplied their needs. Only 23 000 acres of coppice wood would have provided a similar need. The distinction between timber and wood is recognized in the Latin words, *meremium* (= timber) and *boscus* (= wood), to describe it. One species which coppices well is *Castanea sativa* (chestnut) (Rackham, 1976). The Romans probably introduced it, perhaps not for coppice but for its nuts. *Carpinus betulus* (hornbeam) also began to be an important tree during the Roman period. Roman towns had an insatiable requirement for fuel. Hornbeam was one of the best woods for this purpose. A trade in fuel-wood was established. Coppices of hornbeam were still supplying London with much of its fuel in Tudor times.

By the middle of Roman times many areas, especially in East Anglia, were cleared permanently. Along the great river valleys and on the chalk, woodland was cleared. Villa estates abutted on the domain of rural settlements for mile after mile. The villa estates were the rural seats of a landed aristocracy, each managed like a small industry. They became more elaborate and luxurious by the third century AD and as rural sites of industry developed, the towns declined in importance. Meanwhile, the pattern of rural settlement became more concentrated. A study in areas bordering the Fenland reports that at first homesteads were dispersed but as the population grew the number of habitations increased and new homesteads were established between the old, so there was a tendency for greater agglomeration in large settlements.

With the spread of towns in the Roman period, a new flora of plants associated with human habitations can be recognized. Perhaps included here are *Aegopodium podagaria* (ground elder), *Atropa belladonna* (deadly nightshade), *Lavatera arborea* (tree-mallow), *Verbena officinalis* (vervain) and others. *Lavatera arborea*, which is probably native on coastal cliffs in the west of Britain, became more widespread elsewhere. *Atropa belladonna* was very common in the Roman period. It was used as a cosmetic, atropine, for enlarging the pupils. These species were in addition to the longer associates of settlements. *Urtica urens* (small nettle), an annual, is strongly associated with archaeological sites in the Holocene and it can be an abundant weed under crops repeatedly cultivated (Greig-Smith, 1948). *Urtica dioica* (common nettle) a strongly rhizomatous perennial, grows naturally in fen carr, but it, too, must have taken advantage of the nutrient-rich habitats created by man. Soil loosened by cultivation or disturbance is more easily penetrated by its rhizomes. Nettle seedlings show a very marked enhanced growth response to the addition of phosphate and nitrate. A pair of species with a similar nutrient requirement are the elders *Sambucus nigra* (elder) (Figure 3.16) and *S. ebulus* (dwarf elder).

The latter is a Roman introduction.

Figure 3.16 Plant species associated with human habitation: (a) *Urtica dioica* (nettles); (b) *Atropa belladonna* (deadly nightshade); (c) *Hyoscyamus niger* (henbane); (d) *Sambucus nigra* (elder). (Copyright Jon B. Wilson.)

Sambucus nigra is strongly associated with archaeological sites going back to the Mesolithic, which may be partly the result of the collection of its berries for food.

Most produce of the countryside or towns was consumed locally, but there was also more widespread trade. Although most trade probably took place by river, the construction of roads opened up Britain. The roads provided routes by which introduced exotic plants could spread. The roads were constructed on an embankment called an *agger* which must have provided a suitable open habitat for colonization, just as railway embankments and cuttings were to do in a much later period. We can imagine the roads bordered with flowers such as *Chrysanthemum segetum* (corn marigold) and *Agrostemma githago* (corn cockle). Trade and transport made Britain open territory for plant colonization.

Britain was an important source of grain for the Roman Empire (Johnson, 1982). In AD 362, 800 shiploads of grain were sent to Gaul. Barley was so abundant it was used to fatten pigs. Britain was so important to the Empire that it was worth stationing large numbers of soldiers here. Several emperors and pretenders to that position had a strong-enough power base in Britain to mount their *coups d'état* from here. This was to bring the downfall of Roman Britain. In response to invasions of Gallic provinces by Germanic tribes, and despairing of help from Rome, Britons elected a series of their own emperors. A soldier, Constantine, elected in AD 407, crossed with most of the British garrison to head off the Germanic threat and re-establish contact with Rome. After a succession of unlikely events, including the rebellion of Constantine's lieutenant Gerontius, and open rebellion by the Britons left in Britain to protect their own towns from the barbarians, Constantine attempted to invade Italy. Perhaps in pique, and certainly in no position to re-establish the Roman administration in Britain, the rightful emperor, Honorius,

wrote to the civitates in Britain in AD 410 to look to their own defences.

3.3.7 TURMOIL AND DISTURBANCE

Many years of unrest followed the collapse of Roman organization (Laing and Laing, 1982b). Gildas records what happened: 'the fire of the Saxons burned across the island until it licked the western ocean with its red and savage tongue'. The Saxon 'conquest' of Britain was probably greatly aided by a major disruption of the native British population by plague. As if things weren't bad enough already, the weather took a turn for the worse. In the late sixth century there was a sharp climatic downturn to colder summers and wetter winters. Conditions became more continental during the eighth century AD with warmer summers but also colder winters. On average the temperatures were lower. The population plummeted, down perhaps to 2 million (Jones, 1990) (Figure 3.15). All this had a great effect on the landscape as fields and towns were abandoned. It was in this devastated landscape that the Anglo-Saxon kingdoms were established.

Despite the centuries of turmoil, at first, in many areas, a Romano-British pattern of land usage was maintained and later a pattern of land ownership and usage emerged which was to continue for many centuries.

It is traditional to consider the Anglo-Saxons as the main clearers of our virgin woodland, the wildwood, but as far as they did clear any woodland they only continued in a long tradition. They cleared mostly secondary woodland and scrub. They did not have to contend with the huge trees of the wildwood. The surviving areas of virgin woodland continued to change as they were utilized as a source of fuel-wood, timber or for grazing. King Alfred recorded (Whitelock, 1979):

We wonder not that men should work timber-felling and in carrying and building, for a man hopes that if he has built a cottage on laenland of his lord, with his lord's help, he may be allowed to lie there awhile, and hunt and fish and fowl and occupy laenland as he likes, until through his lord's grace he may perhaps obtain some day boc-land and permanent inheritance.

The use of woodland was advised by King Alfred,

And I gathered for myself staves and props and bars, and handles for all the tools I knew how to use, and crossbars and beams for all the structures which I knew how to build, the fairest pieces of timber, as many as I could carry . . . nor did it suit me to bring home all the wood, even if I could have carried it. In each tree I saw something that I required a home. For I advise each of those who is strong and has many wagons, to plan to go . . . and load his wagon with fair rods, so he can plait many a fine wall, and put up many a peerless building, and build a fair enclosure with them.

In Epping Forest, east of London, the decline of *Tilia cordata* (small-leaved lime) started at about AD 600 (Baker, Moxey and Oxford, 1978), one of the latest radiocarbon-dated clearances for lime, the majority of which are recorded in the Neolithic and Bronze Ages. In Epping the decline is probably associated with the greater use for grazing of the ridge on which Epping Forest lies. The decline is accompanied by a decline in fern spores and a patchy increase in grasses.

In Epping, *Fagus sylvatica* (beech) took advantage of the more open ground and a secondary woodland dominated by beech arose. A late expansion of beech was general throughout Britain. Today, it is at present probably the natural climax tree in southern Britain below a line from Weymouth to Swansea to Kings Lynn (Rackham, 1980). It owes its dominance to its great arching canopy, so uniform that it can effectively outshade other plants. Light levels can be cut to 2% below it. Beech also creates a thick mor humus which prevents the development of seedlings. It is widely tolerant of a range of acid, neutral and base-rich soils. It prefers dry, porous soils and is replaced by *Quercus* (oak) on deeper, wetter, richer soils, but it is also restricted by summer drought. Rackham (1980) noted its peculiarly patchy distribution, perhaps evidence of how it has relied on the accidents of disturbance to establish.

3.4 OWNERSHIP, RIGHTS AND DUTIES

In the eighth century AD, following the Celtic and Roman pattern, the land was farmed from scattered hamlets and farmsteads, although these were spread much more thinly than before. Towards the end of the Anglo-Saxon period a revolution in land usage took place in part of England, especially the Midlands. By the tenth century, village England had been established, houses clustered together beside a village green and huge, open, shared fields, 'held in severalty' (Rowley, 1981). This change has been called the 'Anglo-Saxon shuffle'. It took place in the context of increasing Viking raids, but it was probably not a response of villagers coming together for protection. It also took place in the context of the coming of Christianity to England. St. Augustine had arrived in Kent in AD 597. By AD 670 every English king had been converted to Christianity. The change may have been encouraged by the Church and a Christian nobility, anxious to settle villagers around a parish church.

Most importantly, the change to open-field farming took place in the context of the confirmation of Anglo-Saxon dominion and the feudal hierarchy of the monarch and his nobility, a nobility whose loyalty was rewarded by lease of land. This dominion over land and the people it contained is seen in the descriptions of lands granted by the kings. For example, one from about AD 685 (Whitelock, 1979) records:

> Wherefore I, Caedwalla, by the dispensation of the Lord, king of the Saxons, confer on you into possession for the construction of a monastery the land whose name is Farnham, of 60 hides . . . with everything belonging to them, fields, woods, meadows, pastures, fisheries, rivers, springs.

By the later Anglo-Saxon period the estates granted by the king were smaller. There was an increased necessity of effectively managing the manor to maximize the wealth of the lord. Village organization permitted greater control by the lord. There was a rigid hierarchy of rights and duties.

The change to open-field farming is also the expression of the expansion of arable land into areas which had reverted to grassland or scrub. Villagers came together to combine their efforts to clear or keep clear an area of common pasture or arable field. Another element of an agricultural revolution was already in place, the deep plough with a mould-board, which helped maintain fertility and control weeds, was introduced in late Roman times. Co-operation between villagers was necessary because of the necessity of maintaining a team of plough oxen, which was impossibly expensive for any one peasant household.

Because of the parcelling out of land, access to distant upland grasslands became relatively more restricted. A more rigorous organization of available land to allow periods of ley and fallow became a necessity. The difficulties of controlling cattle rustling are hinted at in laws restricting or taxing cattle movement. King Edgar's code of the late tenth century instructs at length how to detect stolen cattle on common land and punish the rustler. The open-field system established a system of fair distribution of land and minimized conflict by dispersing ownership. At Bleadon in Avon the Celtic

fields were overploughed to establish open fields.

Whatever the reason for their establishment, it is clear from the Anglo-Saxon charters, especially the geographical distribution of terms associated with open-field systems like *furh*, *Æcer* and *heafod*, that open fields were well in place by the time of the Norman conquest. They represent an intensification of land usage which was to allow a great expansion of population (Figure 3.15).

However, the fruit of these changes was postponed. The centuries of strife and calamity of the Dark Ages continued. There were years of conflict involved in the amalgamation of the smaller English kingdoms, and the taking over of marginal British ones, which led to the establishment of the large English kingdoms of Wessex, Mercia and Northumbria. There were Welsh and Scottish raids. Then there were years of conflict with the Vikings, Danes in the east and Norwegians in the north-west. Coastal areas were especially vulnerable to attack. Portland was attacked in AD 787, Lindisfarne in AD 793 and Jarrow in AD 794. A prayer was added to the church litany, 'from the fury of the Northmen good Lord deliver us'.

Then after years of raids in AD 865 a great heathen army arrived in East Anglia. By AD 877 they shared Northumbria and had carved away part of Mercia. In AD 886 Alfred recaptured London and the Danelaw was established. The establishment of England came with the recovery of Danish Mercia and East Anglia in AD 917 and AD 918 and the defeat of the kingdom of York in AD 954. But the troubles were not over because there were fresh raids from AD 980 onwards, London attacked, Oxford, Thetford and Cambridge burned in AD 1009 and London captured in AD 1013.

Before the Viking era most rural land was organized as 'multiple estates', grouping together areas with different natural resources. For example, the monastery of Lindisfarne owned extensive areas of northern England, and an estate in Kent owned woodland, salt-making rights and fishing rights. This did not last. In a time of strife loyalty was bought by the liberal granting of estates to followers. Large estates were fragmented. Manors granted in the tenth century were smaller than those granted in the eighth century. In the Danelaw land was parcelled out in return for military service. Each manor became a parish with its own parish church and defended centre, a fortified enclosure, a castle or defended manor house. In the heart of the Danelaw in Lincolnshire and the East Riding of Yorkshire there are very many small parishes, the only large ones to survive were royal or ecclesiastical estates. The Viking division of land may have paved the way for more buying and selling of land and the people on it.

The enforcement of lordship led to changes in land usage. The lord was keen to exploit his small manor as efficiently as possible. This encouraged the development of the nucleated villages and the open-field system in the Midlands. On the South Downs it seems that it encouraged, from the seventh century onwards, a shift from the hill tops down into the dry valleys, leaving the hilltop grasslands for sheep pasture. There were variations in land usage. In the Lake District the Norwegian Vikings followed a more dispersed pattern of settlement, familiar to them from their homeland, as they recolonized the uplands. The large, multiple estates survived longer in the old Celtic regions of Wales, Cumbria, Northumbria and also in the East Midlands. In the Celtic lands slavery also lasted longer.

Throughout this period a wilder vegetation must have re-established itself in many areas. Areas were recovered for cultivation only to be lost again. Each disturbance brought the potential for change, for the establishment of new species, for the rise to dominance of a different species and for more permanent changes in soil type. The Norman conquest was not the last disturbance, but it was swift and, in the south, brought more than 200 years of peace and stability. North of a line from the Humber the story was different. Northern England was laid to waste in the conquest. It took years to recover, a recovery not helped by continued strife in Wales and on the Scottish border.

3.4.1 THE HOT SUMMER, THE OPEN FIELD

There was a remarkable flowering of medieval culture in the twelfth and thirteenth centuries while the British Isles enjoyed its mildest climate for 2000 years. We are fortunate in having in the Domesday Book an unsurpassed record of the medieval landscape in England at the beginning of this period. Land is recorded by its usage, and although the interpretation of records is not straightforward because of variations in the terms used, the Domesday Book gives an excellent picture of that time a millennium ago (Darby, 1976).

The population was probably recorded as heads of households or families but a reasonable estimate can be obtained of actual total population. The population was about 2 million people, about half of the population of Roman Britain. However, there were concentrations of people, especially in East Anglia and the coastlands of Sussex, and in pockets elsewhere on richer soils, like southeast Kent, parts of Lincolnshire and south Cambridgeshire. Only here did the population rise to more than 15 people/square mile. Elsewhere there were large areas which were very sparsely populated. The most important low-population area was everywhere north of the Trent. In the south, the marshes and fens, and great areas of woodland like the Weald, often had fewer than 5 people/square mile.

There is only one direct record for the cultivation of scattered strips in the Domesday Book, that for Garsington in Oxfordshire.

Whether this is because it was so common as to be unremarkable or because it was actually rare is difficult to know. Nevertheless, the cultivation of strips within a vast open field had become the normal pattern of cultivation by the thirteenth century. The system was so successful that it provided food for the burgeoning population. The population rose rapidly so that by AD 1300, at the height of medieval development, it was 5–6 million, concentrated in a great swathe across the Midlands into Lincolnshire and East Anglia (Donkin, 1976). Since by far the majority of the population was working the land, and comparing this with fewer than 300 000 people working on the land today, the countryside must have been teeming with people (Campbell, 1990). The area with the greatest population, the Midlands and East Anglia, was also the area where there was the greatest development of the open-field system. In the Domesday Book arable land was measured by the number of plough teams, and already in AD 1086 these were concentrated in the Midlands and East Anglia where there were more than 3.5 plough teams/square mile. The number of plough teams quadrupled in the next 150 years.

There was not one system, open-field cultivation that did not conform to one pattern, even within the area where its development was greatest. Generally there were three vast open fields, there could be two or four, each divided into rectangular blocks or 'furlongs' which were subdivided into individually owned strips. A typical strip had an area of about 1 acre, about 200 × 20 m (1 furlong × 1 chain = 220 yd × 22 yd) and could be ploughed in 1 day. It has been suggested that strips originally arose as square fields, about a furlong, which were divided between the sons. A strip often consisted of several adjacent plough ridges, selions, about 5 m wide. These may have initially been raised with spades to aid drainage, but the method of driving the plough clockwise around the ridge turned earth towards its centre. The

ridge was about 1 m higher than the furrow. The ridge was oriented down the slope to improve drainage.

The strips were tenanted, though most work was done communally. There were generally three classes of peasants: freemen or franklins; villeins, husbonds or neats, who were the most numerous; and farmhands called cottars, cotterells, bordars or undersettle. The holding of a freeman or villein was about 12 hectares composed of scattered strips and proportional shares of meadowland and grazing rights on the common. The open-field system was a sophisticated kind of land management which depended upon a balance of fallow, manuring and rotation of leguminous crops which maintained fertility and controlled weeds.

Strips were not separated by fencing or hedges, except perhaps at the boundary of the field. Laws were laid down for the compensation of those who lost produce through the failure of someone to maintain his part of the common fence. The system was governed with by-laws, for example from the village of Newton Longville in Buckinghamshire, they included: 'no one shall gather beans, peas or vetches in the fields who holds land of the lord, except from the land which he himself has sown' (Myers, 1969).

There are good descriptions of how the open-field system worked at the very end of its lifetime in the county reports to The Board of Agriculture and Internal Improvement at the beginning of the nineteenth century.

3.4.2 A WEED FLORA

The rotation of crops in close vicinity provided ample opportunity for different kinds of weeds. As well as fallow land there were extensive headlands, where the plough turned round, which were raised from the accumulation of soil pushed ahead of the plough-share. There were also gores, awkward angles of land between strips.

These areas, which were frequently disturbed, must have had a diverse ruderal and weed flora as well as stray crop plants. These areas could also be harvested. One by-law of Newton Longville allowed 'that a pauper shall gather beans not inside, but at the head or alongside the selions. And if they do otherwise they are to surrender whatever they have gathered and not be allowed into the fields to gather beans again' (Rothwell, 1975).

The introduction of regular rotation favoured some weeds over others. Rotation of crops was practised, but in different ways in different places. It could occur by the whole field, so that only one kind of crop was growing in the fields at a time. Alternatively, the field would have some furlongs under winter wheat or rye and others with spring-sown barley, oats, peas, beans or vetch. Each time weeds could re-establish from the soil seed-bank.

A seed bank of weeds and grassland species can persist for many years in a soil cultivated continuously. For example, in one study of an arable field taken into cultivation from grassland many weed species were continuing to appear for 4 or more years. Seedlings of *Trifolium repens* (white clover) were still appearing after 20 years. When left fallow an arable field can soon produce a rich herb and grass cover. We can see this on today's farms in many examples of EEC-sponsored 'set-aside' land. Some agricultural weeds are first recorded in the Middle Ages. Each successive culture had its own set of associated agricultural weeds. Many were probably imported with cereals.

The brome grasses, including *Bromus secalinus* (rye brome), also called chess or cheat, and *Bromus arvensis* (field brome), were weeds that dominated wheat fields, but which were also harvested as a famine food. The brome grasses are a good example of weeds closely associated with a particular crop. Some species were pre-adapted to be agricultural plants and weeds. Soft brome

has a coastal variant, *Bromus hordaceus* subsp. *thominii*, with low stature and a procumbent growth form which adapted it to inland pastures. A different variant, *Bromus hordaceus* subsp. *pseudothominii*, shares many similar characteristics but probably evolved convergently. Both subspecies are seed contaminants of *Festuca pratensis* (meadow fescue), a valuable hay and meadow grass, and neither can be easily cleaned from the fescue seed. *Bromus interruptus* (interrupted brome) evolved as a weed of *Onobrychis viciifolia* (sainfoin), *Lolium perenne* (rye grass) and *Trifolium* (clover). It is an interesting weed because it is not an alien introduction but one of our endemic species. It is a weed of scattered sporadic occurrence with a distinctly 'mutant' morphology. It probably arose relatively recently in Britain. Sainfoin was introduced as a crop only in the seventeenth century. *Bromus commutatus* (meadow brome) and *Sanguisorba minor* ssp. *muricata* (fodder burnet) were also adapted to growing in sainfoin. All these sainfoin specialists became rare when sainfoin ceased to be cultivated. *Bromus interruptus* has become very rare in the wild.

With the cultivation of the deep plough with a mould-board, some weeds of earlier importance, such as *Artemisia vulgaris* (mugwort), declined and others, like *Vicia sativa* (common vetch), *Agrostemma githago* (corn cockle) (Figure 3.17), *Anthemis cotula* (stinking chamomile) and *Centaurea cyanus* (cornflower) became more common.

Weed infestation must have been a very serious problem in the open field. Henry II issued an ordinance against 'Guilde Weed' *Chrysanthemum segetum* (corn marigold) (Figure 3.18) and legislation was also enacted against it in Scotland (Howarth and Williams, 1972).

In 1597 Gerard described corn cockle with these words, 'What hurt it doth among the corne, the spoile of bread, as well as in colour, taste, and wholesomeness, is better known than desired' (Woodward, 1972). An

(a)

(b)

Figure 3.17 Arable weeds once common and now very rare: (a) *Agrostemma githago* (corn cockle); (b) *Althaea hirsuta* (hairy mallow); (c) *Adonis annua* (pheasant's-eye); (d) *Bupleurum rotundifolium* (thorow-wax). (Copyright Jon B. Wilson.)

(c)

(d)

Figure 3.18 Field infested with *Chrysanthemum segetum* (corn marigold).

alternative name for cornflower was hurt-sicle. Gerard described the way cornflower 'hindereth and annoyeth the reapers, by dulling and turning the edges of their sicles in reaping corne'.

A particularly problematic weed was *Anthemis cotula* (stinking chamomile). An alternative name for it is stinking mayweed from the Anglo-Saxon 'maegthe' (Kay, 1971). It became so common in some areas that it gave its name to villages like Mayfield in Sussex and Maytham in Kent. Fitzherbert in his *Boke of Husbandrie* (1523) described 'mathes' or 'doggefenell' as 'the worst weed that is except terre'. 'Terre' is probably *Vicia sativa*. Stinking chamomile gained its bad reputation in part because its ripe achenes could cause the skin to blister. It flourished in the open field because of its genetic diversity and plasticity in growth. It could produce vigorous new shoots from the surviving basal part of the stem after the top was scythed off and it could also grow as a winter annual.

One way to control weeds was to allow grazing on the field after harvest. However, *Anthemis cotula* was relatively trampling-resistant and unpalatable to grazing animals. This method of weed control could give rise to problems on neighbouring strips. A by-law of Newton Longville said (Myers, 1969) 'no one shall cause his beasts to graze in any piece of cultivated land before the crop of at least one acre adjacent is wholly removed'. In addition, at any time, one whole field was left to fallow. Cattle were grazed on the fallow field. A period of fallow helped to control weeds. The folding of grazing animals on the fallow field allowed the trampling of weeds and manuring to be concentrated in particular areas. 'The great intention of a fallow is to pulverise the land and destroy weeds' (Pitt, 1809). In the fallow, weeds were grazed back and trampled, not allowed to flower by being ploughed in green in early summer, and suffered competition from more vigorously growing plants in longer fallow. At first, fallow was allowed to re-grass naturally, but later the fallow was sown with seed from hay. Grasses such as *Lolium perenne* (perennial rye-grass) and *Cynosurus cristatus* (crested dog's-tail) provided the most important grazing.

3.4.3 MEADOWS

An important part of the open-field system was the availability of other areas which provided grazing or fodder for animals. A distinction between grazing on the fallow field and semiwooded common or heath and the production of hay from permanent meadows had long been maintained. Common land was converted into arable but the meadows were preserved. Meadowland, which was never ploughed, could be worth twice as much as ploughland. Those harvested for winter hay were greatly prized because they allowed livestock to be kept over winter. They are recorded in many place names -meade, -mede, -ley, and -ham. The meadows were usually situated in low-lying alluvial soils beside rivers or marshes.

A herb-rich sward was subject to regular inundation. Common species were *Filipendula ulmaria* (meadowsweet) and *Alopecurus pratensis* (meadow foxtail). Waterlogged soils favoured *Deschampsia cespitosa* (tufted hair grass) and *Juncus inflexus* (hard rush). In better-drained areas there was *Juncus effusus* (soft rush) and *Poa trivialis* (rough meadow-grass). Meadows were subdivided into doles or strips marked out by stones. Some portions were permanent possessions but

others were shared by lot. Grass was allowed to grow ungrazed from Candlemass (2 February) until harvest by Lammas (1 August). Then the 'aftermath' was grazed and manured until the next Candlemass. Baker (1937) contrasted the floras of two different grasslands near Oxford. One kind had been grazed by cattle, horses and geese for nearly 900 years and only cut for hay in the Civil War between 1643 and 1645. The other had been kept for hay and only grazed after the hay was cut. The grazed pasture had 56 species of vascular plants and the hay meadow 69, but these grasslands only shared 30 species. Both grasslands had several perennial grasses but the hay meadow had a higher proportion of annual grasses and short-lived perennials, such as *Bromus* spp. (brome) and *Anthoxanthum odoratum* (sweet vernal grass). Dicots included many upright species with stem leaves, like *Lychnis flos-cuculi* (ragged robin) and *Rhinanthus minor* (yellow rattle). In contrast, the dicots in the grazed pasture were laterally spreading ones, such as *Ranunculus repens* (creeping buttercup), *Bellis perennis* (daisy) and *Prunella vulgaris* (selfheal).

Surviving herb-rich meadows are rare because of drainage and ploughing. The application of fertilizers, either inadvertently because of drainage from surrounding arable fields, or on purpose, has been especially damaging. It has encouraged coarse, highly competitive grasses, resulting in reduced floristic diversity. Luckily, some meadows have been protected as local nature reserves. Notable alluvial meadows still regularly flooded are North Meadow near Cricklade in Wiltshire, which is noted for its mass flowering of *Fritillaria meleagris* (fritillary) and the Sibson-Caistor Meadows on the River Nene in East Anglia. The Derwent Ings in Yorkshire are a larger area of flood meadows and pastures managed in the traditional way which are flooded for longer periods, up to 8 weeks. The 2350 ha of the Ouse Washes may be flooded from November to March. The

Ken-Dee marshes in Kirkudbrightshire are the largest area of alluvial and permanent meadows in Scotland. The most important area of alluvial grasslands still managed in the traditional way, are the River Shannon Callows in Ireland, which amount to over 3500 ha (Heery, 1991). There is a mosaic of grazing and cutting meadow. Hay meadows are often owned in long, unfenced strips by many farmers.

The great value of meadows led to the development of water-meadows, a mainly sixteenth and seventeenth century development. They were constructed in the chalk valleys of southern England. Water was distributed evenly over the meadow by a series of 'carriers' or 'carriages' and drawn off by another network of interpolating 'drains' or 'drawns'. 'Floating' the meadow with warm calcium-rich water in spring encouraged the early growth of grasses and could quadruple the yield of hay for cattle and sheep raised for mutton. Common plants were *Cardamine pratensis* (cuckoo flower) and *Festuca arundinacea* (tall fescue). The remains of one seventeenth century water-meadow can be seen at Lower Woodford in Wiltshire.

3.4.4 MANAGING THE WOODLANDS

By 1300 much of central England was an open sea of arable fields. There was intense pressure on other kinds of land. Areas of 'waste' declined. The wood became an important possession so that its boundary was marked by a woodbank and ditch and protected by law. In the Domesday Book only 3.4% of the land is recorded as woodland, almost entirely as either coppice woodland or as wood pasture. The use of woods as a particular kind of pasture, especially for pigs but also for deer and cattle, was well established. Pannage, the right to fatten pigs in the woodland, was one way in which the extent of the woodland was measured, as a wood for so many swine. In pasture woods the trees were pollarded so that the tender

branches which provided poles and rods were protected from grazing by growing on a short trunk above grazing height, being cut every 10–20 years. Gradually pasture woods had a tendency to be converted to open grassland as woodland regeneration was prevented. A sophisticated management of the resource was maintained by tradition backed up by law. Grazing rights and wood rights were jealously protected. An important way of rewarding faithful servants of the crown was the granting of woodland rights.

By the thirteenth century much of the lowland wood pasture had disappeared because of too much grazing. It was replaced by open common land. It is only relatively recently, in the past 150 years, as grazing pressure declined that many of these commons became partially wooded again. In a few areas there was too little grazing and secondary woodland arose. Burnham Beeches in Buckinghamshire arose in this way. These kind of secondary woodlands can be recognized by their concave outline. Areas where the roads crossed or entered the wood pasture were not recolonized by trees so the roads taper into the wood. There may also be ancient pollards. In contrast, abandoned coppice woods have a more convex but rather irregular outline since they had an obvious boundary at one time.

Both parks and commons tend to have a less rich flora than coppice or old coppice woods. However, the presence of some very old trees provided sites for a rich epiphytic lichen and bryophyte vegetation. *Tilia cordata* (small-leaved lime) survived in some parklands.

Changes in medieval woodland came from a number of causes. Population was generally rising, though with temporary setbacks from phases of the Black Death (Baker, 1976). By 1600 the population of England was 4.5 million. The population of the countryside had doubled and 25% of the population now lived in towns. In particular, London had a population of 250 000, providing an important market for the woodland products of Essex and Hertfordshire especially. One enterprising man bought the landowner's right to cut wood in Monk Wood, Loughton in Essex for £20. He spent £35 felling and transporting the wood and sold it for £120.

Woodland was cleared, assarted and surviving patches more intensively managed. Penalties for assarting, clear felling and enclosure, varied and could be severe but it is unlikely that the wood was ever restored once it was felled. In Epping the commoners had the right to lop wood between 12 November and 23 April. In some cases fines for the illegal collection of wood or chopping down of trees provided a useful source of income for the crown, a kind of VAT. In time fines were replaced by regular annual rents.

Some woodland was found in the specially protected areas called Forests. In these the king, or some great nobleman, had the right to keep deer and there were laws to protect and encourage them. Most Forests were established in the early Norman period, later in Ireland and Scotland. In some areas, like Essex and Cornwall, there were very large Royal Forests. In Epping Forest there was a hierarchy of officials headed by a steward with verderers and underkeepers. Enclosure of the king's forest was not allowed. It was illegal to erect a fence high enough to keep deer out. Grazing animals had to be allowed to wander freely in case by herding they were encouraged to occupy the best areas to the detriment of the deer. During fence month (21 June–21 July) grazing of any sort was not allowed, so that the newborn fawns were not disturbed. Peasants could have only as many animals as they could maintain in that month on the village common.

Forests included not just woods but heaths, moors and marsh. Some hunting took place for sport, but more importantly was the provision of venison for feasting and as a gift to subordinates. The Forest law was a way in which the king exerted rights over those of the local landowner and commoners. Their imposition was resented by the landowners and in the Magna Carta there are these articles (Rothwell, 1975): 'All Forests that have been made Forest in our time shall be immediately de-afforested . . . All evil customs connected with forests . . . foresters . . . shall immediately be inquired into . . . and . . . shall be utterly abolished'.

Wood pasture survived longer in some areas like the Royal Forests than in unprotected woodlands, but the Forests were also subject to assarting. Fines for assarting the Forest provided another kind of income for the crown. The extent of Royal Forests declined as land was granted by the crown as reward for service. The best-surviving example of a Forest is Hatfield Forest (Rackham, 1992). Here there is a mixture of fenced coppices and open plains with pollarded trees. The plains are open to grazing by cattle and deer at all times. The Forest is 'compartmented'. The coppices around the central plain are protected by a wood bank and fenced to keep out deer and cattle for the first 9 years of the 18 year coppice cycle (there were originally 17 coppices).

The coppice woods, recorded in the Domesday Book, were to survive relatively unchanged until coppicing declined greatly in the early Victorian era. It survived in a few woods and it has been reintroduced in several others. Outstanding examples of coppiced woodlands can be seen at the Bradfield Woods and Hayley Wood (Figure 3.19a) (Rackham, 1975).

Coppices were intensively managed to provide underwood for fuel, poles, rods and withies. Amongst the coppice some standard trees were maintained to provide larger timber. Coppicing was carried out on a short 5–7 year cycle in the Middle Ages. Longer coppicing cycles of 10–17 years became more common from the sixteenth century onwards. A woodland might be compartmented into coppices at different ages to provide a continuous supply of wood. Each compartment

Figure 3.19 Coppice wood: (a) Hayley wood; (b) the scene of devastation' after coppicing at Brinwells, Sussex, but necessary to maintain a flourishing herb flora. ((b) Copyright Jon B. Wilson.)

was protected from animals by a fence or a steep woodbank and ditch. After coppicing the coppice looks as if a disaster has hit it (Figure 3.19b). However, coppicing maintains a high species diversity. The Bradfield Woods in Suffolk have more than 300 species of herbs. The coppice wood provides a wide variety of structural and climatic conditions which change on coppicing. Changes include a twentyfold increase in summer light and a three- to fourfold increase in spring and winter light. There can be an increase of temperature of 6–9°C near the ground. Increased exposure to wind and sun and increased transpiration can lead to greater drought.

Coppicing leads to an increase in the flowers of herbs due to their greater vigour, though this increase may be delayed until the second year. There can be a twenty- to fortyfold increase in spring flowers such as *Primula vulgaris* (primrose) and *Anemone nemorosa* (wood anemone). *Anemone nemorosa* is favoured because it is more tolerant of the disturbance and trampling that occurs through coppicing than either *Hyacinthoides non-scripta* (bluebell), *Allium ursinum* (ramsons) or *Mercurialis perennis* (dog's mercury) (Shirreffs, 1985). Light-demanding species from the wood margin or clearings, such as *Juncus* spp. (rushes), *Euphorbia amygdaloides* (wood spurge), *Teucrium scorodonia* (wood sage), *Lythrum salicaria* (purple loosestrife), *Lamiastrum galeobdolon* (yellow archangel), *Silene dioica* (red campion) and *Malva* spp. (mallow), establish. Weedy species such as *Carduus* and *Cirsium* (thistles), *Epilobium* (willowherb) and *Digitalis purpurea* (foxglove), some from further afield, also appear, and grasses invade. The number of species present rises to a maximum after 5 years. At first there is a decline of the few shade-tolerant species like *Mercurialis perennis* (dog's mercury), but gradually they increase to dominance. After about 5 years shade-intolerant species begin to decline under the canopy of shade-tolerant herbs and the grow-

ing coppice. *Hyacinthoides non-scripta* (blue-bell) is not very much affected by coppicing but flourishes especially after 7 years when the canopy closes.

One tree that is common in old coppices in south-east England is *Carpinus betulus* (hornbeam). *Carpinus* first started to become a more important element of woodland in the Anglo-Saxon period, for example in Epping Forest (Baker, Moxey and Oxford, 1978). Coppicing and wood pasture encouraged *Carpinus* over *Fagus*, although it does not provide especially good underwood. Both are species of the acidic, nutrient-poor soils which developed late in interglacial periods. *Carpinus betulus* is resistant to browsing, because although it will be eaten by deer, it is not particularly palatable. It has not been a source of fodder for domesticated animals. *Carpinus* is only useful as firewood since it provides only poor timber, nuts or fibre.

The limited distribution of *Carpinus betulus*, its restriction to coppice woods, gives a false impression about its potential as the dominant tree of wet woodlands in southern England. Here it is confined to rather well-defined areas in the south-east, especially in Essex–Hertfordshire–Middlesex and parts of Kent–Sussex. However, English hornbeam woods are an extension of those on the continent, where *Quercus/Carpinus* woods are characteristic mature woodland of central Europe between 45°N and 55°N (Polunin and Walters, 1985). Several types can be recognized. A north-western type has *Lonicera periclymenum* (honeysuckle), *Fragaria vesca* (barren strawberry), *Anemone nemorosa* (wood anemone) and *Stellaria holostea* (greater stitchwort). In south-west England and north-west France *Narcissus pseudonarcissus* (daffodil), *Hyacinthoides non-scripta* (bluebell), *Luzula sylvatica* (wood rush) and *Ruscus aculeatus* (butcher's broom) are found.

In the Middle Ages it became necessary to obtain timber from standard trees left to grow amongst the coppice or in fields and along field margins. The wildwood could no longer provide them. Trees of great age had disappeared from England. When a great construction project was taking place, like the construction of a cathedral, the countryside was scoured for suitable trees. The octagon of Ely cathedral required 16 struts 40 ft (12 m) long and 13 inches (32 cm) square but only 10 could be found. The design had to be modified to include some shorter ones. Large timbers were also required in windmills (for the post) and prices were high, probably several thousand pounds in today's prices, because of scarcity. The importance of timber for mankind before the industrial revolution cannot be overemphasized. A medium-sized timber-framed farmhouse from Suffolk built in 1500 used 330 trees (Rackham, 1976). Half of these were less than 9 inches (22 cm) in diameter and only three exceeded 18 inches (44 cm) in diameter. Most trees were felled after 25–75 years. Trees older than 75 years were difficult anyway to fell, transport and cut up into usable timber. In 1483 the first statutes were introduced to protect woodland. Woodland had to contain a minimum number of standard trees. Fencing had to be maintained to prevent the development of wood pasture.

In some parts of the country woodland became a scarce resource, surviving mainly as hedgerows. Some were remnants, 'ghosts' of woodlands (Rackham, 1976, 1980). They have a sinuous or curving line representing the margin of the original wood. They have a rich flora, including *Tilia cordata* (small-leaved lime), *Anemone nemorosa* (wood anemone) and *Mercurialis perennis* (dog's mercury). Already in Saxon times fields were being enclosed by hedgerows, either to protect arable fields or to enclose livestock. If planted, they were gradually colonized by woodland species. *Ulmus* (elm) became a more important woodland tree in eastern England. Its timber was valued because of its durability under water and its great strength. Lineage elms (*Ulmus procera, U. minor, U. plotii*) were planted in hedgerows in the Stuart period, especially in the Midlands. Different variants of elm were favoured

in different areas. Some spread into existing woodland or, by suckering, led to the development of secondary elm woods on farmland.

The increase of industry had localized effects on the woodland. The iron-working industries of the Forest of Dean and elsewhere were based on extensive coppice systems. The requirement for charcoal maintained many coppices. Even in 1905 a third of all woodland in England was coppice. Charcoal was produced by 'colyers' working in the forest. Wood was cut to the correct length and then stacked to dry for several months. Then it was arranged in a circle about 7 m in diameter around a central pole and piled to make a tall cone. The outside was covered with ash and bracken to make an airtight seal. Then the central pole was removed and glowing embers added to ignite the cone. It was left for 2–4 days. *Carpinus betulus* (hornbeam) was the preferred wood for charcoal but *Quercus, Fagus* and *Betula* were also used. Coppicing for charcoal became less economic after the railway network made available cheap coal from the north. Coal was brought by ship to London from Newcastle in the Tudor era.

3.4.5 COLONIZING AND ASSARTING THE MARGINAL LANDS

In the Middle Ages overpopulation spilled over into the marginal uplands or lowland wetlands (Parry, 1985). The strict social stratification of established estates prevented personal betterment and the enterprising sought out new areas. Some landowners in marginal areas encouraged settlement, with the ready granting of rights and greater freedom of tenure. The first areas to be colonized were the abandoned lands of the north which had been laid waste by William the Conqueror. In addition, the better climate allowed renewed occupation of the uplands (Figure 3.20).

The village of Farnacre was established on

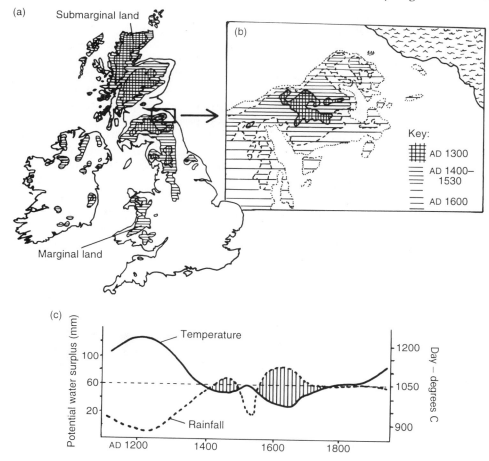

Figure 3.20 Climatically limited, agriculturally marginal and submarginal land in the British Isles: (a) marginal and submarginal land; (b) submarginal land south-east Scotland AD 1300–1600; (c) accumulated temperature and potential water surplus, with shaded periods showing when large areas became submarginal. (After Parry, 1985.)

Bodmin Moor at an altitude of 305 m. The great monastic orders, especially the Cistercians, were important in this process, although their promotion of sheep grazing could also lead to the destruction of settlements. Monastic granges, worked by a lay brotherhood, could be either arable or pastoral.

In the retaken marginal lands in poorer parts of the countryside, such as the valleys of northern and western Britain, where there was restricted availability of good arable land, a different agricultural system developed to the open-field system of the lowlands. Called the rundale or run-rig system, it may have been introduced by the Norwegian Vikings. An infield was divided into strips and allocated according to shares. At Wasdale Head, in the Lake District, the infield of 300 acres was shared between 18 farmers. The strips were narrow and parallel, known as run-rig. They were rotated, for example with oats–oats–barley, but not fallowed. Instead, the field was regularly manured with dung collected from byres in which cattle were kept over winter. In spring the dung was spread over the meadow and livestock let out to graze. The infield amounted to about a fifth of cultivable land. There was also a meadow, closed to grazing in May and hay harvested in late July or August.

Cattle and sheep grazed the stubble after harvest but in summer stock were dispersed to the 'outfield' (D.S.F. Williams, 1963). This was divided into 'breaks', about a third of which were cultivated each year for oats. Otherwise this provided rough pasture. Breaks were cultivated continuously for about 5 years, not fertilized but had livestock folded on them to graze the stubble after harvest. Between periods of cultivation the outfield was left for a long fallow. In addition, at higher altitude there were other areas of permanent pasture, used only seasonally.

Areas outside England were not immune from colonization. Lowland Scotland was Normanized by people from the south. Ireland was invaded by Norman 'marcher' lords, mainly from Wales, in 1167. In 100 years about two-thirds of Ireland came under the control of the Anglo-Norman lords who only paid a superficial fealty to the English king. Cistercian and Augustine monasteries provided another organizing influence. In south and east Ireland a feudal landscape of open fields surrounding villages was established. Deforestation was widespread. Areas in the west on the poorly drained soils remained Irish and followed older patterns of land use.

3.4.6 USING THE WETLANDS

In some areas the woodland was not sufficient to provide fuel for the burgeoning population. Other fuels, mostly peat, but also furze (*Ulex europaea*) and fern (*Pteridium aquilinum*) had to be exploited. In Norfolk, which had some of the densest populations, huge areas were dug for peat in the Broads (Lambert *et al.*, 1960). Peat digging may have been begun by the Danes in the tenth century. The peat has layers, 'lenses', of reed and sedge and moss peat, representing different phases of peatland development, in the Ant and Bure valleys. Deep down, at depth of 6 m in Ranworth Broad for example, there is brushwood peat, representing the earlier development of a fen carr of alder, oak and birch. This brushwood peat was particularly favoured for fuel – 900 million cubic feet (27 million m³) were removed. In 1287 there was a tidal surge when the Zuider Zee was flooded in The Netherlands, the Broads and parts of Romney Marsh were flooded. Outside the Broads much of the flooded land was later recovered but the Broads remained flooded. Here the margins of the peat cuttings were colonized by a natural fen and marsh community. In places there has been terrestrialization and development of large areas of alder carr. At Buckenham, in a neutral fen, the moss community has alternated between one rich in hypnoid mosses and bog mosses; from being dominated by *Homalothecium nitens* with *Drepanocladus revolvens*, *Calliergon giganteum*, *Campylium stellatum* and *Cratoneuron commutatum* to *Sphagnum teres* and *S. palustre*. The origin of the Broads is revealed by their steep sides, step-like profile, and parallel lines of peat islands which mark the boundaries between diggings.

The low-lying marshlands represented one of the last surviving areas of uncolonized land (Godwin, 1978). It was not really virgin land because, like the woodlands, the marshes were actively used. The wetlands provided rich summer grazing. The name Somerset is said to derive from the Anglo-Saxon Sumorsaetum, meaning perhaps summer-dwellers, where people from the uplands drove their cattle down into the Levels (Purseglove, 1989). Winter flood, rich in alluvium, fertilized the grazing-lands.

The wet climate of the Dark Ages had undone much of the earlier efforts to drain the wetlands. Fenland was described (Godwin, 1978) as 'a hideous fen of huge bigness which, beginning at the banks of the river Gronte, extends itself from the south to the north even to the sea'. It is no accident that Alfred launched his resistance against the Danes from his hide-out in the island of Athelney in the Somerset Levels. Hereward the Wake in Ely managed to resist the Normans until 1071 in the Fens. The poor regard that wetlands had was due in part to the 'ague' they caused, attributed erroneously to the miasma, the damp and mists, of the wetlands. The ague was malaria, less virulent than tropical malaria, but also spread by mosquitoes.

In some areas patterns of ancient farming of the wetlands survived. Romanus or Rumen, the seventh century priest to the wife of King Oswy, gave his name to Romney Marsh, much of which he owned and probably farmed. There was piecemeal drainage in some areas of Worcestershire and in parts of the Fenland. The silt lands of the East Anglian Fens had been re-occupied in the late Dark Ages by Angles and Vikings. A ring of settlements was protected to the seaward side by the 'Roman Bank' and to the peat-fen side by the 'Fenbank' perhaps remade from the Roman times. However, only the re-introduction of Roman-like organization in post-conquest times, with the establishment of the great monasteries, were large-scale drainage and embankment attempted again.

The driving force for change was, as in the uplands, the Church. Monasteries and churches established on islands in the marshes or on the surrounding uplands now sought to increase their holdings. Much of the Romney Marsh, the Essex marshes, the Somerset Levels and the Fens were dissected by banks and ditches enclosing rich sheep pastures. New settlements came into being; Emneth and Tilney in south Lincolnshire, and four different Wiggenhalls (W. St. Germans, W. St. Mary Virgin, W. St. Peter, W. St. Mary Magdalene) grew out of one. In the Somerset Levels Glastonbury and Muchelney abbeys organized the straightening of the rivers. The silt lands of the Fens and Levels were transformed into two of the most prosperous areas in the country (Figure 3.8c, 3.21).

The silt lands became very rich and prosperous despite many setbacks. Flooding was a perennial problem but on occasion it was disastrous. In 1320 the monks of Christchurch in Sussex lost half their 10 000 sheep in a flood. In places there was impressive co-operation between neighbouring estates and settlements to maintain the dikes and ditches. In time, the responsibility for drainage and land reclamation devolved upon commissioners of sewers. In 1532 Henry VIII passed a Statute of Sewers.

Inland, in the peat fen, winter flooding provided excellent summer pasture for the large estates. Areas on the drier uplands could rely on the Fenland for grazing and hay and fuel and so could dispense with pasture, meadow and woodland. Within the peatlands the peasants lived on the few gravel islands. They seemed poorer than the inhabitants of the silt fens. They subsisted on fishing, fowling, turf cutting and reed gathering but they were not necessarily poor because they had cash crops of eels and fowl, as well as reeds and sedge. Fenland produce was widely traded. As well as cattle, geese were herded on the Fens. There was rich fishing and fowling on the meres of East Anglia. In the eighteenth century live fish were sent to London in water butts. Most importantly in determining the landscape was the way the vegetation was harvested. Cutting the reeds and sedges halted the succession and maintained an open community rich in flowers. Fen sedge was sent to Cambridge for kindling and sedge was used for farmyard litter. An Italian visitor to London in the fifteenth century described the muddy streets and the way in which the ground floors of buildings were covered by a litter of straw for people to clean their feet

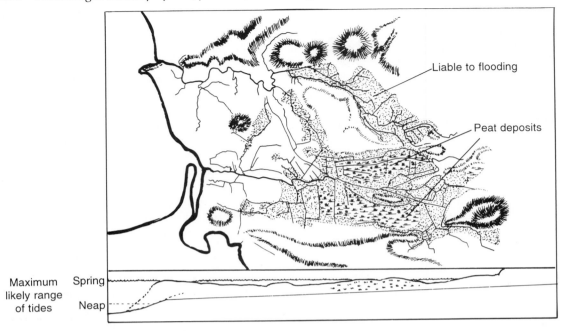

Figure 3.21 Somerset Levels: plan 1794, showing rivers, drains and areas liable to flooding; profile of transect with vertical scale exaggerated.

upon. Reeds were used for thatching. Willows were cut for basketry and hurdles. Brushwood from fen carr, mostly alder, was also utilized but as fuel, and the cutting of peat for fuel was also a very important activity.

The cutting of fen reeds, rushes and sedges maintained a mixed open community, halting the fen succession and preventing the development of a mature oak–alder carr. Behind the fringe of reeds or sedges a more diverse fen, or mixed sedge, community develops. The community is particularly rich with both tall and smaller herbaceous species (Wheeler, 1980). For example, at Wicken Fen over 300 vascular plants have been identified from the fen, with others on surrounding dry ground. Rooted in the sedge tussocks there may be tall herbs like *Eupatorium cannabinum* (hemp agrimony) and *Lythrum salicaria* (purple loosestrife). In particularly fertile conditions *Urtica dioica* (nettles) is very common, even suppressing other vegetation.

In the spaces between tussocks there are other species such as *Iris pseudacorus* (yellow iris). In calcareous conditions there are many different species of small sedges like *Carex pulicaris* (flea sedge), rushes like *Juncus subnodulosus* (blunt-flowered rush) and orchids like *Epipactis palustris* (marsh helleborine).

Cladium mariscus (sword sedge) is an important species of the mixed sedge community. Cutting on a 3–5 year cycle encourages the mixed sedge community with herbaceous plants. *Cladium* was regularly cropped to be used as thatch or as a capping for *Phragmites* thatch because of the durability of the leaves, but it was never grazed (Conway, 1942). An annual or biennial cutting reduces the vigour of *Cladium* so that a mixed *Molinia caerulea* (purple moor-grass) community with many 'dwarf' species, such as *Hydrocotyle vulgaris* (marsh pennywort), *Succisa pratensis* (devil's bit scabious) and fen orchids arises. This could provide

Figure 3.22 View of cut fen, one of the droves through Wicken Fen.

grazing and could also be cut to provide farmyard litter. The droveways at Wicken, which are cut and trampled more frequently, have their own characteristic flora (Figure 3.22).

Without cutting a full succession occurs. The temporal sequence is recorded in the spatial arrangement of plants leading back from the open water. It is also recorded in the pollen stratigraphy.

3.4.7 A COMMERCIAL LANDSCAPE

With the rise of towns during the Middle Ages the countryside became part of the general economy. Although much agricultural production was for local use, grain and wool were an important part of trade. By 1300 London, with a population of 100 000, was the second largest city north of the Alps.

It was to dominate trade over the succeeding centuries, but many other market centres arose, dominating their hinterland. The rise of trade encouraged a process of specialization in the agricultural economy which was to mark the countryside indelibly.

Intensive farming was concentrated in East Anglia and the south-east. The intensive systems had ready markets. Outside the common-field area, especially in parts of East Anglia and in the Home Counties, there were mixed farming systems with an emphasis on dairying to provide cheese and butter for the market. Arable land was less intensively developed and relied on low rates of fallowing but required the availability of manure from dairying stock to maintain soil fertility.

At the height of rural development at the beginning of the fourteenth century, away from ready markets, agriculture was less intense. Durable products, grain, especially wheat, or wool were produced for sale, the former mostly in the common-field area and the latter from the south and south-west. Arable productivity was limited by soil fertility and the poor availability of manure. There was a greater reliance on fallow. In Somerset, the Welsh borders, the West Midlands and the north-east, the systems were geared towards self-sufficiency. In the north and west, and on uplands elsewhere, low-density livestock-rearing on permanent pasture was the main economic activity, with an emphasis on breeding stock for the lowlands.

Wool was the most important raw material of the international trade. The highest quality came from England and Wales. The trade was dominated at first by foreigners, especially Italians, but gradually English merchants, centred on London, gained economic strength. The importance of wool as the foundation of the national economy is recognized even today by the Lord Chancellor sitting on the Woolsack in the House of Lords. From the fourteenth century the export of woollen cloth became more important than export of fleeces. The availability of cash crops encouraged the trend away from managed estates to tenant farming where the landowner reaped the benefits of ownership without any expense. The emphasis on commercial return encouraged the move away from co-operation, an erosion of commoners' rights, towards enclosure and individually concentrated holdings occupied by a single tenant farmer.

3.4.8 DISEASE AND CLIMATIC DECLINE

The open-field system was robust and efficient enough to last until the eighteenth century in the lowlands. In the Middle Ages tillage was relatively shallow and crop yields relatively low. The demands on the soil were not necessarily great. There need not necessarily have been much decline in soil fertility despite the short periods of fallow. There was a decline in arable yields between 1315 and 1322 probably because of a series of wet, cool summers (Baker, 1976). This was the beginning of a world-wide climatic decline, the Little Ice Age, which, after an interval of more clement conditions, culminated between the mid-sixteenth and early nineteenth century (Grove, 1988). Underlying a decline in crop yields there may have been a decline in soil fertility, not because of changes in tillage but because of a decline in the ability to manure the soil. At this time there were severe losses of sheep and cattle, probably due to rinderpest and liver-fluke, which were encouraged by the wetter conditions. A decline in population followed, predating the more extreme decline when plague arrived. The decline of population because of the Black Death, the scarcity of labour, a fall in land prices, the rise of a country gentry, and economic changes such as the increasing prices of wool, all added to hasten the eventually to the demise of the open-field system.

The intensive farming of the open-field system required a large supply of cheap labour. Arable farming was disproportionately damaged by the decline of population in the fourteenth century and there was a shift to less intensive systems of agriculture. In Ireland, for example, there was a general shift back to pastoral farming and an expansion of 'waste' and woodland, rough pasture and hazel scrub. This was the beginning of a long shift in land use that continued throughout the following centuries (Yelling, 1990). Generally it favoured the development of a more extensive system of agriculture, a shift towards the keeping of livestock, and sheep rather than oxen or other cattle. Permanent pasture arose on the ridge and furrow ploughland, thereby preserving the corrugated landscape under grass. The corrugations have a marked effect on the distribution of pasture plants, associated in part with differences in drainage.

Open fields in geographical areas outside the main 'champion countryside' of the Midlands were enclosed.

Climatic cooling reached a minimum about 1500, then there was a short-lived improvement followed by a deeper decline to the 'Little Ice Age' of seventeenth century. Sea-ice formed around the coast in winter. In 1684 sea-ice reached nearly 40 km out from the Dutch coast. Between 1530 and 1600 the growing season shortened by 8 days in Scotland. All this meant a very marked retreat from marginal land in the uplands, especially in the period between 1530 and 1700 (D.S.F. Williams, 1963). A speech of Burgundy in Shakespeare's Henry V provides a description of France at the time of the wars, a description of a countryside in decline, perhaps a scene Shakespeare was familiar with:

> . . . her hedges even-pleached,
> Like prisoners wildly overgrown with
>   hair,
> Put forth disordered twigs; her fallow
>   leas
> The darnel, hemlock, and rank fumitory
> Doth root upon, while the coulter rusts
> That should deracinate such savagery
> The even mead, that erst brought sweetly
>   forth
> The freckled cowslip, burnet, and green
>   clover,
> Wanting the scythe, all uncorrected, rank,
> Conceives by idleness, and nothing teems
> But hateful docks, rough thistles, kecksies,
>   burs,
> Losing both beauty and utility . . .

3.4.9 SHEEP AND RABBITS

Meanwhile the number of sheep increased through the Middle Ages. In particular, upland grasslands became more widespread, their composition changing as the intensity of grazing changed. Two kinds of sheep-grazed grasslands have been described from south-east Scotland (Hunter, 1962). On the better soils there is a mull sward, with *Agrostis* spp. (bents) and *Festuca* spp. (fescues), which is utilized intensively in summer. This kind of grassland has a tendency to become infested by *Pteridium aquilinum* (bracken). On mor soils tussocks of *Molinia caerulea* (purple moor-grass) and swards of *Nardus stricta* (mat grass) are often accompanied by *Calluna vulgaris* (heather) and *Eriophorum vaginatum* (hare's-tail cotton-grass). These low-quality grasslands are grazed more intensively in winter. *Molinia* escapes much of the grazing because, unusually for a grass, it is deciduous. It occupies an intermediate ecological position between *Eriophorum vaginatum* on the wetter soils and *Nardus stricta* on drier soils. *Nardus stricta* is favoured by intensive sheep grazing because it is relatively unpalatable to them. It can form small tussocks but it is tolerant of trampling, by growing as a flattened form beside paths. It reproduces rarely by seed and spreads vegetatively by rhizomes. Fragments of rhizome, broken off by trampling, or by decay of older parts of the plant, are very effective at spreading the plant.

In addition, herb-rich hay meadows have provided an important source of grazing in the uplands (D.S.F. Williams, 1963). In the Craven Pennines they are grazed intensively in the lambing period, up to 20 May. This reduces the growth of *Lolium perenne* (rye grass) and *Dactylis glomerata* (cocksfoot), so that later-growing *Agrostis* spp. (bents) and *Cynosurus cristatus* (crested dog's-tail) flourish. The meadows are cut in July and can then be grazed again.

One of the most significant events for the British flora was the introduction of rabbits in the twelfth century, bred for their meat and skins (Sheail, 1971). They needed tender nursing to help them survive. They were protected in fenced warrens or on offshore islands like Lundy which had no predators. Molly-coddling extended even to the construction of burrows for them with huge augers. Pillow mounds were constructed for the burrows. Many warrens were established on common land by the lord of the manor or ecclesiastical lord, sharing grazing with the sheep and cattle of commoners. Warrens with pillow mounds are a widespread feature of our countryside. By the nineteenth century 11% of the Breckland was covered with warrens. Active farming of rabbits on Newborough Warren in Anglesey did not start until after 1608. By the seventeenth century rabbits were very significant in maintaining particular kinds of grassland. In 1 year in the twentieth century 7000 pairs of rabbits were taken off Newborough Warren annually by the Forestry Commission. Rabbit pellets lay thick on the ground and the turf was only 1–2 cm high. In 1551 a Swiss naturalist recorded, 'there are few countries where, coneys do not breed but the most plenty of all is in England' (Williams, 1971).

At all ages abandoned arable fields have reverted to grassland. Rabbit grazing has been very important in recent centuries in preventing the further succession to scrub and woodland. For example, on the Porton ranges, (Wells, Sheail and Ball, 1976) without rabbit grazing, scrub invaded with shrubs separated by an open, tall *Arrhenatherum elatius* (false oat-grass) grassland. Rabbits are precise eaters. Where heavily grazed by rabbits, a *Festuca ovina* (sheep's fescue) and small rosette herb grassland interwoven with lichens survives. The influence of leguminous herbs such as *Lotus corniculatus* (common bird's-foot trefoil) enriches the soil. *Carex flacca* (glaucous sedge), one of the most widespread species of calcareous grassland, is abundant in regularly grazed areas (Taylor, 1956). With less grazing a taller grassland with *Festuca rubra* (red fescue) and herbs such as *Sanguisorba minor* (salad burnet) is maintained. With no grazing this is converted to a tussock grassland of *Festuca/rubra/Helictotrichon pubescens* (red fescue/downy oat-grass) grassland.

The importance of rabbits was noticed particularly after the introduction of myxomatosis from South America in the 1950s. By 1955 rabbits had ceased to graze the vegetation of Newborough Warren significantly (Ranwell, 1960b). The first effect was a greatly increased flowering of most plants, especially the grasses *Festuca rubra* (red fescue) and *Agrostis stolonifera* (creeping bent). The grasses and sedges continued to expand at the expense of low-growing broad-leaved herbs. By 1956 the turf was 5–8 cm high. Shoots of *Calluna vulgaris* (heather), which had previously been grazed back to moss-like stature, were 3–5 cm high and flowered frequently. By the beginning of 1958 the turf was 8–10 cm high. By the end of that year it was 15–20 cm high.

The consequence for floristic diversity of rabbit grazing has depended on the soil and vegetation type. In experiments where rabbits were excluded from areas for several decades in the Breckland it was shown that on shallow calcareous soils there was a rise of species diversity, at least at first (Watt, 1962). *Anthyllis vulneraria* (kidney vetch), *Leontodon saxatilis* (lesser hawkbit), *Linum perenne* (perennial flax), *Bromopsis erecta* (upright brome) and *Silene otites* (Spanish catchfly) all invaded. Other species flowered more freely. The spring and summer annuals were relatively unaffected at first. In contrast, on acid soils *Festuca ovina* (sheep's fescue) rose to dominance and eliminated smaller species. In another experiment Tansley and Adamson (1925) excluded rabbits from floristically rich old chalk grassland. In the first 6 years floristic diversity declined as a tall *Bromopsis erecta* (upright brome) grassland arose.

One extreme community associated with intense rabbit grazing has been described from the Chilterns; a 'grassland' rich in *Sedum* spp. (stonecrop) and *Myosotis* spp. (forget-me-nots) and tufted mosses with *Iberis amara* (wild candytuft), *Crepis capillaris* (smooth hawk's-beard) and *Senecio jacobaea* (common ragwort). *Senecio jacobaea* must have been particularly favoured by the establishment of rabbit-grazed grasslands since they conspicuously avoid eating it (Harper and Wood, 1957). It is one of a range of unpalatable species which include *Atropa belladonna* (deadly nightshade), *Solanum dulcamara* (bittersweet), *Urtica dioica* (nettle) and *Sambucus nigra* (elder), which are particularly associated with rabbit burrows. The disturbed soils allow plant establishment and are also nutrient rich. In contrast to its avoidance by rabbits, *Senecio jacobaea* is readily eaten by sheep, which are only rarely poisoned by it.

3.5 EXPLOITING THE COUNTRYSIDE

An Italian visitor in 1497 described the English landscape (Williams, 1971). 'The farmers so lazy and slow that they do not bother to sow more wheat than is necessary for their own consumption; they prefer to let the ground be transformed into pasture for the use of the sheep that they breed in large numbers.'

Another visitor noted, 'The population of this island does not appear to me to bear any proportion to her fertility and riches. I rode . . . from Dover to London, and from London to Oxford, a distance of more than 200 miles, and it seemed to me to be very thinly inhabited.'

These observations predate a revolution in the countryside which dates back to the Tudor period and from which there was a rise of the country gentry. The country squire exercised a large measure of control over his tenants. He exercised his power to create larger holdings and to encourage improvements. The rise of the gentry was encouraged by the dissolution of the monasteries in the sixteenth century. Crown lands were replenished for a short while, but soon the lands were given away to favoured courtiers. The estates of Church and Crown occupied 25–33% of the land in the early sixteenth century. By 1668 they occupied only 5–10%.

The greatest beneficiaries were middling to lesser manorial lords. This represented the rise of a class for whom the countryside was seen as a capital investment which required a financial return. The most important improvement was enclosure of open field and common land. There were good arguments for enclosure, including the provision of fuel. A submission to the Commission on Enclosures in 1517 stated (Williams, 1971):

. . . where there is no wood nor timber growing within twelve to fourteen mile of the same lordship the said John Spencer hath there set trees and sown acorns for timber and wood, and double diked and set with all manner of wood both in the hedgerows, and also betwixt the hedges adjoining to the old hedges that William Coope made before . . . for in those parts there is no wood, so that the poor men of the country are fain to burn the straw that their cattle should live by, therefore it were a great loss to destroy those hedges for it is a greater commodity than either corn or grass in those parts.

In 1698 Celia Fiennes noticed pats of cow dung drying on the walls of houses and outhouses in the vicinity of Peterborough for use as fuel, such was the shortage of firewood (Browning, 1953). However, the effects of enclosure on the poor could be severe. Another return of the Commission on Enclosures reads (Browning, 1953):

. . . he [William Coope] enclosed those tenements with hedges and ditches on all sides . . . and he willingly allowed those messuages and cottages aforesaid to be wasted and fall into decay and ruin . . . And so he converted the aforesaid arable land from cultivation and arable into pasture for animals . . . whereby twelve ploughs which were fully engaged in the cultivation of these lands are completely idle, and sixty persons who lived in the aforesaid houses . . . were compelled tearfully to depart, to wander and be brought to idleness and so presumably perished from want.

In another place 40 people were replaced by one man and his shepherd. At Fawsley near Daventry in Northamptonshire, two villages were depopulated in the Tudor period and the estate given over to 2500 sheep.

In Ireland the transformation of the medieval landscape dates back to the 'plantations' which started in the Tudor period (Aalen, 1978). Protestant settlements were planted for the political reason, among others, of establishing a loyal population. Ulster received its first settlers in 1609–10. There was renewed clearance of forest and expansion of farmland, and the older villages and open fields were transformed into estates of enclosed fields. In military campaigns woodland was cleared for tactical reasons to remove refuges. Confiscation and military oppression turned the Irish into refugees, pushing them out from the lowlands to the hills with their cattle herds, so that the surviving upland woodlands came under increased grazing pressure. While in 1600 about one-eighth of Ireland was wooded, by the beginning of the eighteenth century it was virtually devoid of woodland.

Enclosure of fields took place at first against the will of Parliament. Enclosure represented the possibility of consolidating holdings so that the individual farmer could undertake improvements like drainage or manuring but it also represented a change in the social structure of the country, and at first Acts of Parliament were passed to prevent it. Nevertheless, land was enclosed by common agreement. By 1700 about half of the arable land in the country remained in open fields but from then private Acts of Parliament were passed for enclosure (Walton, 1990). Between 1802 and 1844 2000 enclosure acts were passed, enclosing 1 000 000 hectares. The straight lines of the parliamentary enclosures, often superimposed on older features can be seen in many places. While

earlier ancient hedgerows may be 'ghosts of woodlands' and have diverse woodland species like *Mercurialis perennis* (dog's mercury), *Hyacinthoides non-scripta* (bluebells) and *Primula vulgaris* (primroses), the later planted hedgerows are straight, and rather species-poor. Many hedges used to be rich in elm trees, before the recent resurgence of elm disease. The planting of elm was encouraged to provide timber for ship-building. Many eighteenth and nineteenth century hedgerows are primarily *Crataegus monogyna* (hawthorn) and they contain non-woodland species such as (*Anthriscus sylvestris*) (cow parsley).

Hooper (Pollard, Hooper and Moore, 1974) has recorded a rough correlation in southern Britain between the number of shrub species in a 30 yd (27 m) length and the age of the hedgerow in centuries. However, there is a great deal of regional variation and probably even local variation, as well as a broad statistical variation. In a local survey on the Northamptonshire–Huntingdonshire border the relationship is about 99× the woody species −16. This kind of accuracy is spurious in most cases because many hedges were planted as mixed plantations of saplings collected from a local woodland to begin with. As well as *Crataegus monogyna* (hawthorn), *Malus sylvestris* (crabapple), *Ilex aquifolium* (holly), *Corylus avellana* (hazel), *Quercus* (oak), *Ulmus* (elm) and *Fraxinus excelsior* (ash) were all used. In addition, as well as increasing with time the number of species can go down with time as the highly competitive suckering of *Prunus spinosa* (blackthorn) and *Ulmus* (elm) ousted other species. Nevertheless, older hedges are richer in woody species, herbs and animals.

Hedgerows provide 160 000 ha of woodland linking otherwise isolated woodlands (Figure 3.23).

The importance of hedgerows comes from the range of habitats associated with them. There is a tree and a well-developed shrub-layer. Hedgerows are often raised up so that

Figure 3.23 Hedgerow with *Crataegus monogyna* (hawthorn) in Kent. (Copyright Jon B. Wilson.)

there are differences in drainage, and there is often a ditch as well which provides for a tall-herb layer, a low-herb layer and a wetland community. The management of hedges has been different in different parts of the country. In East Anglia the hedge was coppiced. Elsewhere plashing or pleaching was carried out so that twiggy growth was cut away and main stems slashed, bent and fastened down to posts hammered into the hedge at every few feet. Heathering may be carried out, where pliable willow twigs are woven between the posts to maintain the fence until regrowth.

Since the Second World War there have been large losses of small fields and the hedges which enclosed them. In 1986 the annual rate of loss was 1% a year, amounting to 5000 miles/yr (8000 km/yr). Surviving hedges have not been maintained in the traditional way. The use of a 'hedge-chopper' damages the hedge bottom and leaves a mulch of dead vegetation which suppresses the normal perennial herb flora. In addition, intensive cultivation right up to the hedge bottom results in mechanical damage or the application of defoliating herbicides. As a result, an opening is provided for troublesome weedy species, such as *Galium aparine* (cleavers) and *Anisantha sterilis* (barren brome), which can then use this refuge to invade the crop.

3.5.1 DRAINING OF THE WETLANDS

One example of the growing commercialization of the countryside in the Elizabethan period was the renewed interest in draining the wetlands (Darby, 1956). This was part of a general move towards enclosure of common land and as such it was fiercely opposed by commoners anxious to preserve their rights. The dissolution of the monasteries in the 1530s, the dissolution of monastic rights to the fens and wetlands, presented an opportunity for capitalists to profit. Fenland life had continued relatively successfully throughout the Middle Ages, surviving a period of higher sea level and greater rainfall. Now it was threatened by venture capitalists from outside who wanted to turn Fenland, with its rich patchwork of shared ownership, rights and co-operative exploitation, into the private profit of arable fields on drained land. Venture capital paid for Dutch engineers and workmen experienced in drainage. The foremost engineer was Cormelius Vermuyden. He arrived in Britain in 1621 and was occupied in several projects, apparently leaving the Thames flood defences at Dagenham in a poor state, and provoking riot and revolt because of drainage work at Hatfield Chase west of Scunthorpe, before moving on to the Fens.

In 1631 Russel the Earl of Bedford formed the Bedford Level Company with 13 other business adventurers to drain large parts of Cambridgeshire, Huntingdonshire, Norfolk and Lincolnshire. Vermuyden was employed and there followed a period of intense activity. The first major canal, the Old Bedford River was finished in 1637. The early efforts were not entirely successful, and in 1638 the commoners paid Cromwell, a resident of Ely, to champion their opposition. With the Civil War works went into abeyance but Cromwell switched sides and became one of the commissioners of a new Act for the Draining of the Fens. He sent a major of his own regiment to suppress the commoners' riots. In 1651 the New Bedford Hundred Foot River was completed. Running parallel to the Old Bedford river the system was designed to allow the area between to flood in winter, thereby protecting surrounding areas. The Ouse washes are flooded from November to March in wet winters. *Juncus* spp. (rushes),

Oenanthe spp. (water dropworts) and, low down, even some coastal elements such as *Aster tripolaeum* (sea aster), *Apium gravolens* (wild celery) and *Bolboschoenus maritimus* (sea club-rush) are found. The ditches are rich in aquatic plants like *Potamogeton* spp. (pondweeds), *Nymphoides peltata* (fringed water-lily) and *Hydrocharis morsus-ranae* (frogbit).

Drained lands provided rich farming. In 1670 John Evelyn visited Soham Mere where there were (DeBeer, 1955) 'hopes of a rich harvest of hemp and col-seed'. Evelyn described the

> . . . Engines and Mills, both for wind and water, drawing it through two rivers or grafts cut by hand . . . discharging water into the Sea, such as this Spot had been the former winter which was now so drie and so exuberant and rich as even astonished me to see the increase there was; Weedes grew as high as horse and man almost upon the bankes.

It was not all advance. There were periods of flooding. Celia Fiennes (Browning, 1953) described the Fenlands of 1698:

> Thence on the fen banks, on the top of which I rode at least two miles with fens on both sides, which now were mostly under water, a vast tract of such grounds which are divided by the dikes without trees a those I observed before. And these banks are made to drain and fence out water from the lower grounds, and so from one bank to another, which are once in many acres of land 100, so that at length it does bear off the water. But in the winter it returns, so a they are forced to watch, and be always in repairing those banks.

Silting up and shrinkage of peat made drainage more difficult, but the widespread use of wind pumps from the seventeenth century onwards had transformed the situation, and the introduction of steam pumps later resulted, at long last, in the permanent 'recovery' of wetlands. As a result the wetland flora almost disappeared. It is now preserved in a few relatively tiny nature reserves. Today, because of shrinkage of the peat in the surrounding lands, Wicken Fen stands out above the surrounding farmland. In periods of drought it is only preserved by water being pumped up into it.

3.5.2 IMPROVING THE COUNTRYSIDE

In the seventeenth century the landscape of England was beginning to be recognizably that of today, although large parts were still open fields. John Evelyn noted in 1654 that Leicestershire was 'much in Commune'. In the summer of 1654 he visited three contrasting areas (De Beer, 1955), 'part of Huntingdonshire . . . the Country about it so abounding in Wheate that when any King of England pass thro it, they have a costome to meet his Majestie with a hundred plows'; then there was mixed countryside between Pontefract and York, '. . . is goodly, fertile, well water'd and wooded country, with pasture and plenty of all provisions'; Salisbury Plain was 'a goodly plaine or rather a Sea of Carpet, which I think for evennesse, extent, Verdure, innumerable flocks, to be one of the most delightful prospects in nature'.

Even while the national population was rising, that part of it living and working in the countryside was decreasing. By 1700 cereal cultivation had become concentrated in the Midlands and east England. Heavier rainfall in the west encouraged pasturage on deep, neutral fertile loam. Grazing was reduced in spring to encourage a dense grass sward to develop.

The spectrum of weeds has changed with patterns of cultivation. In the eighteenth and nineteenth centuries there was an influx of new weeds from the New World. They included *Conyza canadensis* (Canadian fleabane), *Coronopus didymus* (lesser swine-cress), *Galinsoga parviflora* (gallant soldier), *Matricaria* *discoidea* (pineapple weed), and *Claytonia perfoliata* (Springbeauty).

On his various tours of the country Defoe found barren uplands were present in Derbyshire, Cumberland, Westmoreland, Northumberland, West Durham, West Yorkshire and Lancashire (Defoe, 1971). One estimate has 28% arable, 26% pasture and meadow, 8% woods and coppices, 8% forests parks and commons, 26% barren and 2% built upon or wetland at this time. However, in the far north of Scotland Defoe described Strathnaver as 'the remotest part of all the island, though not the most barren or unfruitful, for here as well as on the eastern shore is good corn produced, and sufficient at least for the inhabitants'. The crofting communities, which Defoe saw in the Highlands, represented some of the last people living in an ancient pattern of subsistence mixed farming. It did not last. These communities were the latest and last to suffer at the hands of landlords bent on 'improving' their land. Crofts were cleared to make space for sheep despite the questionable economic advantage of sheep rearing in these marginal areas where soils without tillage became quickly podzolized or turned to bog. Some of the clearances were particularly vicious, with crofts burnt and homeless, starving Highlanders forced to become refugees.

At about the same time there was a paradoxical concern about preserving the strength of the countryside, mainly represented by its trees. Evelyn in his *Silva* (1664), concerned about the shortage of timber, encouraged 'planting his Majesties Forest of Deane with Oake so much exhausted of the choicest ship-timber in the World'. The Forest of Dean was one of the Crown lands which had survived from piecemeal 'improvement'. Others were not so lucky. The great Essex Forest (Waltham Forest) declined to the remnants of Epping Forest and Hainault Forest. The need for shipbuilding has long been blamed for the disappearance of our woodlands. In fact there was always a good supply of timber (Rackham, 1976). Paranoia about the state of readiness of the navy translated itself to a concern about the trees with which it was built. Oak, which could be shaped into the peculiar timbers necessary in a ship came from the oaks of parkland and hedgerows. Even in the ship-building areas of Suffolk and Essex prices of timber were scarcely inflated. In 1706 Defoe (1971), noted that Kent, Sussex and Hampshire were 'one inexhaustible storehouse of timber never to be destroyed'. The psychological importance of oak probably preserved many oakwoods, but in the period of greatest requirement between 1780 and 1860 imported softwoods began to replace local hardwoods. In Scotland Defoe wrote,

> On the most inland parts of this country, especially in the shire of Ross, they have vast woods of fir trees, not planted and set by men's hands as I have described in the southern part of Scotland, but growing wild and undirected, otherwise than as nature planted and nourished them up, by additional help of time, nay of ages. Here are woods reaching from ten, to fifteen and twenty miles in length, and proportioned in breadth, in which there are firs, if we may believe the inhabitants, large enough to make masts for the bigger ships in the Royal Navy.

This may have been the last surviving remnant of the truly wild wildwood, but most of it did not last much longer. Defoe advised, 'If they cannot fell the timber . . . having no navigation; they may yet burn it, and draw from it vast quantities of pitch, tar, resin, turpentine etc. which is of easier carriage.' In 1773 Dr Johnson claimed that much of the rest of Scotland was treeless, much to the chagrin of the Scot, Boswell (Birkbeck, 1950), 'from the bank of the Tweed to St Andrews I had never seen a single tree, which I did not believe to have grown up so

far within the present century. The variety of sun and shade is utterly unknown'.

At the end of the eighteenth and beginning of the nineteenth century the Board of Agriculture published a series of General Views or Surveys of the Agriculture of most counties. They provide a clear picture of the state of the countryside and our flora. The one for Leicestershire by William Pitt, published in 1809, gives a view of what had become of the champion open-field countryside in the middle of the agricultural revolution (Pitt, 1809). In the whole of the county there were only six or eight open-field systems left, amounting to 10 000 acres (4000 ha). The surviving system at the parish of Glenfield is described:

> That this parish produces more sustenance and employment for mankind than the average of enclosed parishes in this county, of equal staple of soil, I have not the least doubt; but respecting net profit to the proprietor and occupier, I believe the balance to be in favour of enclosure.

In contrast one farmer noted,

> . . . this land in its open state was very unprofitable to the occupier, though rented at from 10 to 12s per acre; the great expense of cultivation, and collecting crops from patches of land dispersed over the whole lordship, the trespasses from stock getting loose, and loss from disorders in sheep.

In the process of enclosure the poor tenant had little power over how it was carried out. Often enclosed open fields were laid down to permanent pasture, allowing the better fattening of animals for market, although the total number of sheep kept did not necessarily increase. The decline of arable acreage was compensated in part by the enclosure and 'improvement' of the common waste land. It was in this way that the poorer tenant lost out; access to the common grazing of the wastelands had been a vital part of his economy.

3.5.3 AN AGRICULTURAL REVOLUTION

By the beginning of the nineteenth century much of the pattern of the countryside had been established. It was a rich and complex mosaic. There was a clear difference between the open, planned or 'champion' countryside of central and eastern England and the older ancient countryside elsewhere (Figure 3.24).

However, the pressure for improvement continued. In *A General View of Agriculture of the County of Dorset to the Board of Agriculture and Internal Improvement* (Stevenson, 1815) there is this recommendation,

> There is much unprofitable pasture-land in Dorset, that ought to be broken up by paring and burning, with the addition of chalk and lime where the soil requires them. It might be made to bear corn but once in three or four years, with green crops and artificial grass between. That by thus converting a part of the rough pasture and down land of this county into tillage, there would be an advantage, I am well convinced . . . The practice is spreading in most counties.

The new scientific attitude to farming was accompanied from 1700 by an improved climate. There followed what has been called an agricultural revolution (Reed, 1983).

In 1730 Tull published *The New Horse-houghing Husbandry*. He invented or re-invented a range of tilling implements, hoes, harrows and the seed drill. Ploughing, harrowing and rolling the land was to destroy weeds, break up soil aggregates and make a fine tilth, thereby increasing contact between the roots and soil; the remnants of

Figure 3.24 View of an old field system, East Dony Valley, Shropshire.

the crop, stubble and weeds were to be chopped up and buried; and drainage and rooting conditions improved. Instead of broadcasting the seed, sowing the seed in rows by the use of seed drills, allowed the crop to be hoed between the rows to control weeds. A distinction was made between seed weeds and root weeds (Pitt, 1809), 'In all cases root weeds are best destroyed in dry weather, and seedlings in showers.' It was recommended (Pitt, 1809),

> To prevent their growth [i.e. seedling weeds] in crops, the turnip fallow should be pulverised early by ploughing and harrowing, and then let to lay for showers to vegetate the seeds; then harrow again repeatedly, and at length plough to expose a fresh surface, and fresh soil, to the sun, air and showers; when this begins to vegetate, harrow and pulverise, and let it lay for showers, and by degrees, and repeating the operations, most of the seeds in the soil may be expected to vegetate, and may then be destroyed before sowing the crop.

While couch grass, 'the harrow teeth fetched out, which should be spread abroad . . . in hot dry weather it will perish when loosened from the soil by spreading in the sun and air.' Tull broke the ground. Other innovators sowed it.

'Up and down' husbandry was an innovation of the seventeenth century. It was a regular rotation of grassland with arable. In the eighteenth century 'Turnip' Townshend introduced on his estate at Raynham the 'Norfolk' four-course rotation, consisting of (1) 'roots', either turnips or swedes; (2) spring barley, undersown with clover or rye, or sometimes sainfoin and trefoil; (3) 'Seeds', a hay crop of grass and red clover; and (4) winter wheat. The turnips provided fodder for sheep, either folded directly on them or kept in stock-yards. The rotation helped to control weeds and so replaced the fallow period. On heavy clay soils the livestock was fed in stock yards and the manure spread over the field.

In 1776 Thomas Coke, later the Earl of Leicester, took over the estate at Holkham. In his home farm he introduced a four-course rotation, introduced swedes instead of turnips, and marled the soil with lime to increase fertility. His tenancies were based on the strict observance of his rules of husbandry. In doing so they prospered. As a result, the rent-roll increased from £2200 to £20 000 over 40 years. At sheep shearing he opened his house to interested agriculturists. In 1836 there were 7000 guests. The yield of wheat increased to 1.7–2.0 onnes/ha nationally, and up to 3.6 tonnes/ha on the fertile Fenland soils. These yields were scarcely exceeded up to the 1930s when a satisfactory yield was just 3.6 tonnes/ha. Today, with the high level of input of chemical fertilizers, yields of 5 tonnes/ha are normal.

In the early nineteenth century six-course 'Midland' rotations became more common, with turnips, swedes, barley, two courses of rye grass, and wheat (Whitlock, 1983). As well as the swede, other new crops were tried. The mangold was introduced from France. It cropped heavily but was not as nutritious or as hardy as the turnip or swede. Potatoes were grown increasingly for human consumption. Rape was grown for fodder and for oil-seed. Some older crops, especially flax and hemp (*Cannabis sativa*) which were very labour intensive were grown less widely.

Grass and *Trifolium* spp. (clover) or *Onobrychis viciifolia* (sainfoin) seed mixes became available commercially. Grasslands were sown with monocultures of rye grass or mixtures with clover to eliminate useless weeds. *Trifolium repens* L. (white clover) was first grown as a separate fodder crop in The Netherlands in the seventeenth century (Burdon, 1983). By the 1750s it was well established in Britain as a crop. Seed was imported from The Netherlands. It was realized that a grass–clover mix was very productive. Kentish wild *Trifolium repens* (white clover), raised in Romney Marsh and native *Lolium perenne* (perennial rye-grass) formed a famous mixture. In some areas Devonshire rye-grass was favoured. The first authenticated introduction of *Lolium multi-florum* (Italian rye-grass) was in 1831 (Beddows, 1973). By 1850 2500 bushels (91 000 litres) were being imported annually by one among many seed firms. It was used in seed mixtures to provide a quick grazing sward or, after spring grazing, allowed to grow for hay or silage. Sown in high density it produced a sward capable of carrying a high density of stock. Diverse variants are found, many of which have arisen from hybridization with *Lolium perenne* (perennial rye-grass).

3.5.4 DIVERSITY IN DECLINE

So much of our countryside was finally established in the eighteenth and nineteenth century. It is a landscape familiar to us from the paintings of landscape artists. The landscape was at its most diverse. It was at a time when improvement was not universal and so remnants of older landscape, vegetation and flora survived. For example, a major difference to the present was the floral diversity of the agricultural landscape. A list of agricultural weeds of Leicestershire is provided by Pitt (1809),

> The couch grasses are the bane of arable crops . . . Dog's Couch grass (Triticum repens), Benty Couch grass (Agrostis elatior), Hard fescue (Festuca duriuscula) and Creeping Soft Grass. The other root weeds to be destroyed in a fallow are common or curled dock . . . common thistle, and spear thistle, The seedling weeds . . . chickweed, ivy-leaved chickweed (Veronica hedere folia), fat hen or wild spinach (Chenopodium viride) willow-weed (Polygonum parsicaria), Bird's lake-weed (Polygonum aviculare),

> shepherd's purse (Thlaspi alpestre bursa-pastoris). The other most pernicious were . . . Chadlock; in three distinct plants, wild mustard; wild rape; wild radish. Corn chamomile, in corn and bean crops, Corn marigold, goldings in barley and turnip crops. Corn crowfoot, hungerweed (Ranunculus arvensis), sow thistles with white and yellow blossoms . . . Other corn weeds of this county of less importare nettle hemp, white dead nettle, bindweed, bearbind (Polygonum convolvulus), shepherd's needle, corn scabious, blue button (Scabiosa arvensis), knapweed or blue bottles (Centaurea cyanus, scabiosa) coltsfoot . . . I have observed in corn fields and fallows, the greater daisy or white marigold, groundsell, scorpion-grass (Myosotis scorpiodes) goose tansy (Potentilla anserina) . . . in addition . . . corn mint (Mentha arvensis) tares, two and four seeded (ervums hirsutum & tetraspermum) . . . also corn goosegrass, or cleavers, nettles, poppy, cockle.

What a contrast to today when there are scarcely any broad-leaved arable weeds at all. At the beginning of the nineteenth century perhaps our native vegetation had achieved its maximum diversity. Five thousand years of human interference had created a thousand different kinds of habitats. The landscape was a rich mosaic of fields, wetlands, woodlands, heaths, moorlands, meadows and downlands. But already it was marked for decline. The passion for 'improvement' was leading to drainage of wetlands, liming of heathlands, ploughing of downlands. All these changes were motivated by the ever-greater influence of the marketplace on the countryside and by the increase in population.

Notably, changes in cereal prices have markedly changed the acreage of arable land (Walton, 1990). Changes in grain prices have led to changes in arable practice. The cultivation of wheat was encouraged by rising prices; there was a rising population which created demand and the Napoleonic Wars restricted supplies from elsewhere. The Corn Laws maintained prices thereafter. The area of arable land increased to a peak in the 1840s.

Repeal of the Corn Laws in 1846 did not at first spell disaster for home agriculture. The tremendous increase in the population of the towns and cities kept demand high. The population of Britain had risen from 11 000 000 in 1801 to 16 500 000 in 1831. By the end of the century it had reached 32 500 000. By 1851 over half the population lived in towns. However, by the 1870s food imported from the Empire, the USA and Argentina was much greater. Wheat acreage fell by 1 000 000 acres between 1875 and 1885, and by the same amount again by the end of the century. Large areas were converted to permanent pasture. In Ireland the potato famine led to massive depopulation of the countryside and a shift to pastoralism.

The areas that went out of cereal cultivation first were the marginal ones, so that there was a decline in mixed farming and increased specialization. Farms became either mostly arable or mostly pastoral. As a result the ley (grass) or fodder crops of rotations, such as clover, sainfoin, coleseed, rape and lucerne were much less commonly grown. Where an arable rotation was maintained, a high-value break crop, such as potatoes, sugar beet, oil-seed rape or peas, was substituted. The cost of farm produce fell by 25%. Another result was a depopulation of the countryside. Farmer after farmer became bankrupt. Millions of acres were left to the rabbits.

Cereal cultivation picked up in the First World War. Corn acreage increased by 2 million acres (0.8 million hectares) but the prosperity did not last and there was another rapid decline in the 1920s. This was the kind of countryside depicted in the novel *Cold Comfort Farm* (Gibbons, 1977). In 1939 less than 30% of the country's food was being produced on British farms. Britain only produced 46% of its requirement of the seven major staples: wheat, barley, oats, milk, eggs, pork and poultry.

3.5.5 THE TAMED COUNTRYSIDE

Agriculture was kick-started by the Second World War, and since then has shown an inexorable rise (Nature Conservancy Council, 1990). It has been a revival led by the development of arable farming. The rise of arable has continued, either despite or because of EU (EEC) interference, to this day. The cost of this has been great. In the period since 1950 livestock of all sorts, especially chickens and pigs, and to a lesser extent cattle, have greatly increased in number, but more and more they have been kept inside in intensive rearing systems. The amount of arable land increased from 3.8 to 5.5 million hectares at the expense of rough grazing. Remaining grasslands were improved to provide silage and hay. The change of breed of cattle from slow-growing Galloway and Welsh Black to fast-growing, nutrient-demanding Charolais and Limousin has necessitated a high-nutrient food. This includes cereal fodder from greater arable acreages. By 1974 nearly 31% of all agricultural acreage was devoted to cereals. Upland grasslands were left to more and more sheep: the sheep stocking rate rose from 2.9/ha to 6.0/ha between 1951 and 1981. Encouraged by European Community intervention minimum prices, and increased market stability, there has been a shift away from mixed farming towards intensive systems, especially arable. Production increased by 2% a year between 1973 and 1988. Britain became self-sufficient in most of the major staple foods by 1983.

The intensification of agriculture has had a wide range of costs. Between 1935 and 1975 the number of people working on the land declined by more than half, to about 300 000. The intensive sowing of 'winter' crops has proved particularly damaging because cultivation in autumn exposes the soil to erosion. Much of the topsoil of Britain, unprotected by grassland or stubble, is being washed away. Hedgerows have been ripped up to enable farm machinery to work more efficiently, removing one possible kind of protection (Barr *et al.*, 1992). The net loss of hedgerows was 121 000 km in Britain between 1984 and 1990, and those lost have been mostly the species-rich mixed hawthorn ones. Old meadows and grasslands, which were maintained by low-intensity grazing and one annual cutting, have not fitted in with new regimes. Drainage has been improved to allow easier access to farm machinery.

Our countryside has been pruned and cleansed. In some areas every trace of wildness has been expunged and every patch of wilderness destroyed or tamed to earn its keep.

3.5.6 THE UNTAMED FLORA

The weed flora represents that last buccaneering element of our flora, the part which refused to be tamed. Throughout the centuries it has added a wild splash of colour to our landscape, softening the straight-edged realities of farmland. The older countryside and older floras have been destroyed. There has been wholesale loss of floras and habitats to 'development'; to prairie-like arable fields, to monocultures of forestry conifers and by drainage of wetlands. But one of the most profound changes in our flora this century has been the decline, even the disappearance, of the weed flora. Some arable weeds such as *Agrostemma githago* (corn cockle) and *Centaurea cyanus* (cornflower) had already become relatively rare by the beginning of the nineteenth century as modern agricultural practice produced cleaner 'seed corn'. *Chrysanthemum segetum* (corn marigold) and *Spergula arvensis* (corn spurrey) became rarer because of the liming of acid soils (Howarth and Williams,

1972). Species such as *Missopates orontium* (weasel's-snout), *Kickxia* spp. (the fluellens) and *Silene noctiflora* (night-flowering catchfly) used to flower in the stubble and were destroyed by stubble burning.

The most important change was the introduction of herbicides in the 1930s. Early herbicides were dinitro compounds related to 2,4-dinitrophenol. They were general biocides, such as glyphosate, which gained their specificity in that they had poor wetting ability and tended to drip off the narrow upright leaves of grasses but stayed on the broad, flat leaves of dicotyledonous weeds. Weeds were most susceptible close to flowering when they have achieved maximum leaf area. A few dicotyledons with dissected leaves, like *Torilis arvensis* (spreading hedge-parsley), escaped. *Convolvulus arvensis* (field bindweed) has a shiny surface which is difficult to wet with herbicide sprays. Later, more specific 'hormone weedkillers', such as MCPA (2-methyl-4-chlorophenoxyacetic acid) and 2,4-D (2,4-dichlorophenoxyacetic acid), which caused abnormal growth patterns in some species, were introduced. Today more than 90% of arable land is treated with weedkiller each year, and 40% more than once a year. Even non-arable areas may be treated. 'Asulam' (sulphonyl carbamate) has been used to control *Pteridium aquilinum* (bracken). Applied on the mature plant it attacks the rhizome which rots away.

Generally it has been relatively easy to target dicotyledons, 'broad-leaved' weeds, in a grain crop, so that as a result they have become relatively less important in comparison to grass weeds. Some broad-leaved species which were once common, like *Ranunculus arvensis* (corn buttercup), are now rare. Even *Papaver rhoeas* (common poppy) is not commonly seen in great profusion as it once was. *Galium aparine* (cleavers) is the only serious dicotyledonous weed because it has proved insensitive to the commonly used phenoxy-herbicides. It can grow in the tall-herb community of the field margin or in ephemeral assemblages on disturbed arable land. A few weed species, such as *Arnoseris minima* (lamb's succory) and *Bromus interruptus* (interrupted brome), have become more or less extinct in the wild in Britain.

It has not all been loss. The weed flora has been constantly changing with changing agricultural practice. The lack of deep ploughing with the introduction of direct drilling has allowed *Anisantha sterilis* (barren brome), a common weed in the Iron Age to become important again. *Capsella bursa-pastoris* (shepherd's purse) became rare as an arable weed, but in recent years the cessation of stubble burning has allowed it to spread again, choking barley and contaminating the corn. A few plants, such as *Artemisia vulgaris* (mugwort), *Heracleum sphondylium* (hogweed) and *Anthriscus sylvestris* (cow parsley), survive in the headland and field margin (Roebuck, 1987). Others colonize from the uncultivated areas of the hedgerow and pathside. These refuges are narrower as cultivation has been pushed closer and closer to the field margin.

The most important weeds today are grasses (Figure 3.25). They include *Alopecurus myosuroides* (black grass), *Avena fatua* (wild oat), *Bromus commutatus* (meadow brome), *Phalaris canariensis* (canary grass) and *Apera spica-venti* (loose silky bent). *Avena fatua* and *Alopecurus myosuroides*, which are resistant to some herbicides, have become relatively more important. *Avena fatua* and *A. sterilis* (winter wild-oat) seemed for a while to be a more of a growing problem compared to *Alopecurus myosuroides* (black grass) but a concerted campaign of intensive herbicide application has led to a relative decline compared to the latter. Weeds show a pattern of geographical importance. For example *Galium aparine* (cleavers) is important in the east of England and *Viola arvensis* (field pansy), *Stellaria media* (common chickweed) and *Polygonum aviculare* (knotgrass) are more important in the west.

Successful arable weeds share a syndrome

Figure 3.25   Important weeds in today's landscapes: (a) *Avena strigosa*; (b) *A. fatua*; (c) *Apera spica-venti*; (d) *Bromus commutatus*; (e) *Phalaris canariensis*; and (f) *Galium aparine*.

of characteristics. Weeds have great plasticity in development, allowing 'adaptation' to different crops. Weeds have a rapid growth rate. They favour highly fertile soils and their high growth rate is associated with the ability to show a quick response to fertilizer addition. Rapid growth in the seedling phase is associated with having a relatively large seed. Reproduction is precocious. Seed production occurs within 5 weeks from germination in *Stellaria media* (common chickweed). Some are potentially tall, but height is often associated with a twining or scrambling habit, as in *Convolvulus arvensis* (field bindweed), *Fallopia convolvulus* (black bindweed) and *Galium aparine* (cleavers), thereby minimizing their requirement for structural tissue and allowing rapid elongation growth. For example,

Galium aparine uses its backward-pointing prickles on the leaves to clamber up over other plants. Even at a density of 5–10 plants/m² it can reduce crop tiller number, the number of grains per ear and grain size, but it is also a nuisance because of the way it tangles farm machinery.

Most weeds are at least facultative annuals. The annual grasses *Alopecurus myosuroides*, *Avena fatua* and *A. sterilis* are the most widespread weeds of winter-sown crops, wheat, beans and oilseed rape. *Anisantha sterilis* (barren brome) behaves as a winter or summer annual, rapidly colonizing bare ground. Most of the annual grasses have little innate dormancy so that they germinate whenever conditions, such as moisture or sufficient contact with the soil, are favourable. Some have a weak vernalization requirement. The first large caryopsis of *Avena sterilis* has shorter dormancy than the second and germinates in the autumn, infesting winter crops. In contrast, most of the seed of *A. fatua* germinates in spring.

The seed of *Alopecurus myosuroides* only has dormancy if it is induced by waterlogging or burial below 50 mm, though dormant seed suffers a high mortality. Rotational agriculture controlled it and it is now favoured by continuous cultivation of cereals and also by the use of combine harvesters, which allow the crop to stand longer in the field before harvesting (Naylor, 1972). It is widespread in winter-sown crops such as winter wheat and oilseed rape in southern, eastern and midland England, though it is declining. In 1972 it occurred on 300 000 ha and seriously affected 100 000 ha. It has an effect on wheat yield from as low as 10 plants/m² and when it reaches a density of 100 plants/m² it reduces yield by 1.0 tonne/ha. Winter barley is less affected than winter wheat because it is sown earlier and achieves a dense competitive canopy in autumn.

Other weed species produce a persistent seed bank with a pattern of dormancy suiting a particular crop, like sleepers, secret agents, they wait for the correct signal to become active. *Galium aparine* (cleavers) has a pattern of seed dormancy which allows it to germinate over an extended period in autumn and spring. Germination of the seed of *G. aparine* is promoted by light. Dormancy is lost in the autumn but regained in spring. High summer temperatures break the dormancy but germination does not occur until autumn when winter cereals are sown (Roebuck, 1987).

Weeds have very high rates of seed production in favourable conditions and good dispersal. The success of *Alopecurus myosuroides* is related to its high potential seed productivity for a grass of 500 per plant and its long flowering period of May–August. Up to 13 000 seeds/plant are produced by *Stellaria media* (common chickweed), 6000 by *Veronica persica* (common field-speedwell), and up to 40 000 by *Rumex obtusifolius* (broad-leaved dock). Weeds have effective seed dispersal. Commonly, weed seeds hide as an impurity in the grain, but they are also transported in mud attached to machinery or implements. *Agrostis* spp. (bents) have light seed so that wind dispersal can be important in establishment. The sticky seeds of *Galium aparine* contaminate cloth and sacking. Some weed seeds survive ingestion by farm animals.

Weeds have genetic variability, with different ecotypes adapted to particular crops, germinating either in the autumn or spring or genetically adapted in other ways. They are the perfect undercover agents, mimicking the crop they infest, like the brome grasses, described in section 3.4.2. Some have survived the demise of their cover. *Stellaria media* (common chickweed), is thought to have been an important weed of flax in the past (Sobey, 1981). Perhaps it will enjoy a resurgence as a weed of linseed, which has suddenly become popular again. Most weeds also grow in more natural communities such as hedge and field margin, and some have ecotypes adapted to these environments, but potentially providing sources of weed seed. The corn spurrey has a variety, *Spergula arvensis* var. *arvensis*, with papillate seeds, and a variety, var. *sativa*, with smooth seeds, more common in the north-west. *Galium aparine* (cleavers) has a hedgerow ecotype with a lower growth rate and a different pattern of dormancy to the weed ecotype (Roebuck, 1987). *Avena fatua* (wild oat) is able to grow and set seed in hedgerows and verges, and though this is usually at low density it nevertheless acts as a reservoir for infestation. In contrast, *Alopecurus myosuroides* (black grass) is largely confined to cultivated soils and is rarely found in hedgerows.

Reproduction by seed is universal amongst important arable weeds and only a few broad-leaved arable weeds, such as *Cirsium arvense* (creeping thistle), *Rumex obtusifolius* (broad-leaved dock) and *Convolvulus arvensis* (field bindweed), have a high capacity to regenerate from root or rhizome fragments. Perennial weeds have ceased to be a very important problem in arable fields because of current agricultural practice. Repeated cultivation generally destroys the rhizomes and stolons of *Elytrigia repens* (common couch) and *Agrostis stolonifera* and *A. gigantea* (creeping and black bent), which can invade from the field margin. However, the rhizomes of *Elytrigia repens* (common couch) can grow at a depth of 400 mm, thereby resisting control by frequent cultivation. Cultivation can act to spread the infestation rather than destroying it. *Elytrigia repens* is one of the world's worst perennial weeds. It is recorded from 32 crops in 40 countries. *Agrostis stolonifera* can remain a problem because it regenerates from stolon fragments as well as from stolons.

False oat-grass has a bulbil-producing variety, *Arrhenatherum elatius* var. *bulbosum*, the onion couch. In western Britain it is a plant of rough grasslands, hedgebanks and roadsides, while it is mainly an arable weed in central England. The bulbils are actually basal, short-culm internodes which are

swollen and corm-like. The bulbils were collected to be eaten in the Neolithic and Bronze Ages but they also provide a very effective means of propagation on deep fertile soils. Without control, the number of bulbils in soil can increase by a factor of 34 over 2 years. Late December ploughing is an effective control. Rhizomes and stolons need to be damaged and buried deeply or exposed to winter frosts.

Plants of *Alopecurus myosuroides* (black grass) surviving in stubble go on to produce four times as many fertile tillers as plants growing newly from seed, thereby multiplying potential contamination. Regular cultivation can limit vegetative regeneration, but in the absence of a herbicide it can also spread propagating fragments very effectively. Burning of the stubble kills 40–80% of the seed and encourages an early germination so that seedlings can be killed with herbicide. Surviving plants can be killed by ploughing but 95% control is necessary to reduce numbers.

In an agricultural landscape, wild plants have been pushed more and more to the ever-narrowing marginal land. Ones that have not declined are those that can withstand the frequent disturbance of trackways and paths. In the soils churned by wheels of farm machinery *Matricaria discoidea* (pineapple mayweed), *Polygonum aviculare* (knotgrass) or *Persicaria maculosa* (redshank) grow. On trampled paths there are *Poa annua* (meadow grass), *Lolium perenne* (perennial rye-grass), *Plantago major* (greater plantain) and *Trifolium repens* (white clover).

3.5.7 PLANTS AT THE MARGIN

Changes in agricultural practice have been the most important reason for the increased rarity of some of our native species. A by-product of agricultural intensification has been the loss of any areas considered marginal to agricultural productivity. Natural or seminatural habitats have been trimmed to near extinction in many lowland areas. Losses since the Second World War reported in 1984 (Nature Conservancy Council, 1984a) were 40% lowland heath, 40% ancient lowland woods, 50% lowland fens and 60% lowland raised mires. In addition, it is estimated that there was a decline of 30% upland heath or bogs. Seminatural communities have continued to decline, saltmarshes 1% per year between 1984 and 1990, but the arable landscape has suffered the most, with the loss of marginal habitats. The number of species recorded in the arable landscape has dropped by 30% between 1978 and 1990 (Institute of Terrestrial and Freshwater Ecology, 1993).

After arable weeds, the most important decline has been in wetland plants, largely because of improved drainage to 'improve' grasslands and convert land to arable use. Together changes in arable farming and drainage of wetlands account for at least 50% of the extinctions or decline of flowering plants and ferns in Britain in the twentieth century.

Drainage and conversion to arable land has occurred widely in the East Anglian Fenlands. Lowering of water tables has resulted in a shrinkage and desiccation of the peat and peat decay. Since the mid-nineteenth century and widespread introduction of steam pumps the East Anglian Fens have been lowered by 3 m. The proportion of Fenland under cultivation is 95%, compared with only 10% 50 years ago. Most of the Somerset Levels are still under pasture with numerous small fields, King's Sedgemoor had only 10% of land under cultivation in 1987, but here, too, there has been improvement by drainage. A more direct threat to peatlands than drainage, but now probably less significant overall, is peat cutting. One-fifth of British output of peat for horticulture is from the Somerset Levels (Figure 3.26).

It also occurs in Cheshire. Peat extraction increased in the past 50 years for garden peat. Coir is championed as an alternative material.

A number of lowland fen and marsh species have become very rare. Most are confined to tiny patches of peatland in a few nature reserves. *Viola persicifolia* (fen violet) was extinct for a while, but was rediscovered in 1980 when soil samples were taken from Wicken Fen and monitored for the appearance of seedlings breaking dormancy. In the nineteenth century it had been abundant there but it had not been seen for 60 years. Later it reappeared in the fen in two different areas where the vegetation was cut and the peat surface was disturbed (Rowell, 1984). *Senecio paludosus* (fen ragwort) was rediscovered in the wild in 1972. It is still confined to a single site, though it is in cultivation at Cambridge Botanic Gardens. It had died out at Wicken Fen at the end of the nineteenth century. Continued survival is difficult because of low seed set due to self-incompatibility. Both these species only survived because of the long-term viability of seeds in wet marsh soils. There are plans to re-establish populations in former sites. Two other very rare species are aquatics. *Alisma gramineum* (ribbon-leaved water-plantain) is confined to a single site in Worcestershire. It appears to be very sensitive to water quality. *Corrigiola litoralis* (strapwort) is found on the margins of water bodies which have a fluctuating water level. In Britain it is now found regularly only on the shores of Slapton Ley.

Peatlands of different sorts have declined very greatly in extent. One estimate suggests that 87% of raised peat bogs have been lost to forestry or agriculture. In Ireland, they have been exploited for fuel for power stations. The heather moorlands and blanket mires have also declined due to overgrazing, drainage or afforestation. Grampian region lost 26% (702 km^2) of its heather moorland between 1940 and 1970, and Dumfries and Galloway 63% (476 km^2), two-thirds to afforestation, in the same period. Many peatlands are rather species-poor. That is normal. The Flowlands of Sutherland and Caithness may seem dreary, monotonous habitats but

Figure 3.26 Peat cutting in the Somerset Levels.

they represent the largest continuous expanse of a very special type of vegetation. The increasing floristic diversity recorded from moorlands indicates their decline because of disturbance and pollution. The loss of peatlands is particularly important because Britain has between one-seventh and one-tenth of the world's total of this kind of vegetation. In international terms, peatlands are our national speciality.

Agricultural improvement of woodland, scrubland and lowland grassland has also made some species rarer. At various times cereal cultivation has been encouraged with a corresponding decline in grassland. Permanent pasture has declined, especially in the past few decades. There are important regional differences. In Cumbria 82% of enclosed farmland is grassland but only 11% unimproved or semi-improved (Fuller, 1987; Morris, 1991). In Hampshire only 2% of chalk downland was under grass in 1980. There has been a shift to arable or rotational grassland or to forestry. These changes account for 95% of loss of unimproved 'natural' grasslands in the twentieth century. The rich, unimproved grasslands of North Devon have undergone a spectacular decline, with 65% outside Sites of Special Scientific Interest 'improved' in the 5 years up to 1990. Pastures have been 'improved' by drainage and reseeding. Other marginal grasslands, on slopes too steep to plough in arable areas, have been abandoned, so that without grazing they have experienced a scrub invasion. Similarly, commons on the urban fringe have suffered because of the decline of grazing and because of the threat of vandalism and litter. Pastures lost 14% of their diversity between 1978 and 1990 (Institute of Terrestrial and Freshwater Ecology, 1993).

Established woodlands have undergone

many changes over the past 150 years. Most importantly was the cessation of coppicing. In grazed woods such as Epping Forest and the New Forest there were also striking changes in grazing pressure. In Epping in the late nineteenth century there was concern about the close-cropping of the vegetation by deer and cattle which led to attempts at fencing. As a result the woodland canopy closed, young seedlings failed to establish and birch colonized the clearings. With the myxomatosis of the late 1950s, scrub colonized the grassy plains and the short turf became coarse, tussocky grass and thorn scrub. This kind of change has occurred in glades and rides of woodland and also in the chalk grasslands.

Many processes have acted together to allow the spread of scrub. Boggy areas have been drained. Scrub has not been cleared or kept clear by fire. The marginal increase in broad-leaved woodland, 1% between 1984 and 1990, has to be viewed in this light as not necessarily a healthy sign. Conifer plantations increased 6% in the same period, while old woodlands continued to be felled (Institute of Terrestrial and Freshwater Ecology, 1993).

The importance of the maintenance of open communities for maintaining diversity was illustrated by the accidental burning of the clifftop grasslands of Rickham Common in Devon in 1975. A patchy grassland was established which included *Lotus subbiflorus* and *L. angustissimus* (hairy- and slender bird's-foot trefoil), both rare species. By 1983 gorse scrub had spread and they only survived at the edge of the clifftop path. In the Avon Gorge the cessation of sheep grazing in the mid-1920s and the spread of exotic trees has threatened the rarities. In the Lizard the decline in rough grazing, changing use of paths, mostly overuse, tourist pressure on the coast and uncontrolled heathland burning have all taken their toll. Lowland heath has been particularly vulnerable to improvement because so much of it was

common land. Lowland heathlands have been set to grass for pasture by ploughing and reseeding, or ploughed for continued cultivation, or used for forestry. West Penwith in Cornwall had 7882 ha of heathland in 1840 but only 4217 ha in 1980 (Johnson and Rose, 1983). South-east Dorset had 40 000 ha in 1750 and was down to 6100 ha by 1983 (Figure 3.27) (Waton, 1983). The Nature Conservancy Council estimated a 40% loss between 1950 and 1984 (Nature Conservancy Council, 1984b).

A number of recent changes have made the Breckland rarities rarer. The presence of the Breckland species on the margin of their natural climatic range makes them susceptible to other changes, especially those in land management. Increased enclosure of fields and the sowing of hay meadows with *Medicago sativa* (lucerne) and *Dactylis glomerata* (cocksfoot) for cattle has reduced the brecks. On one estate the number of cattle went from 400 in 1920 to 3695 in 1972. Myxomatosis was introduced in 1956. The effect on the rabbit population can be gauged from the bag of gamekeepers. On one estate the bag fell from 128 000 rabbits in 1921 to 1700 in 1970. There were fewer rabbit scrapes to provide open ground for the establishment of rarities. *Pinus sylvestris* (Scots pine) wind breaks were planted to stabilize soils. In addition, large areas have been covered with Forestry Commission plantations. Other areas have been taken over for airfields. The deep ploughing of arable fields and the use of heavy-tracked vehicles on parts of the land by the Ministry of Defence have destroyed the soil stratification and patterning. Today many Breckland rarities can be found only on the roadside verges, at the margins of forestry plantations or on relict areas of heath.

Urban spread has played its part in the decline of heath. Bournemouth was built on Poole Heath. Urban development and the depredations of holiday-makers are responsible for the increased rarity of some coastal species. Several of the rarities are beach and

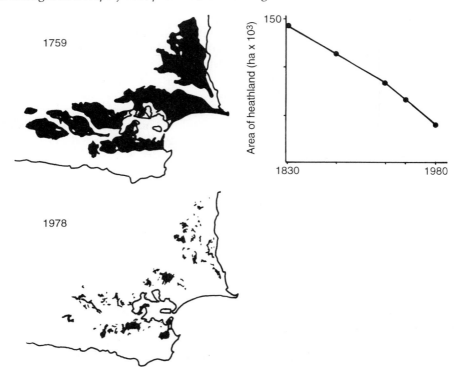

Figure 3.27 The decline of heath: (a) distribution of heath in Dorset *c.* 1759 and *c.* 1978 (Haskins, 1978; Webb and Haskins, 1980); (b) loss of heath 1830–1980 (Nature Conservancy Council, 1984b).

shingle species. These species are important as a remnant of an early postglacial flora which was widespread inland. *Euphorbia peplis* (purple spurge) was once present at 28 sites but may now be extinct in Britain. Cliffs are mostly very safe except for holiday development and coastal protection schemes.

With 10% of England now classified as 'urban' the spread of concrete and tarmac has been relentless. In England 110 km^2 of countryside is being lost per year (Institute of Terrestrial and Freshwater Ecology, 1993). The decline of the flora around London has been very marked. Extinctions of native species in Surrey have been calculated. Before 1900, 17 species disappeared. Since then, especially with suburban spread in the 1920s, a further 34 species became extinct up to the 1970s. It is estimated that 7% of the

extinction or decline has been due to destruction of habitats by building. However, not just the urban fringe has been at risk. The development of an industrial nation with its transport links has exposed the countryside to the depredations of 'plant lovers'. Bunches of *Fritillaria meleagris* (fritillary) from Thames Valley meadows were once sold in florists. The Great Western Railway annually organized a daffodil express on Easter Sunday down to Gloucestershire to pick wild daffodils. In the Victorian era and after rare native plants were widely collected by botanical enthusiasts to be pressed and mounted in personal herbaria. Pressed specimens were exchanged through the post like stamps. Ferns were particularly vulnerable, because they were collected for fern gardens and glasshouses. More recently, our mosses have suffered greatly as people have collected them indiscriminately to line hanging baskets.

The decline of our rarest species *Cypripedium calceolus* (lady's slipper orchid), which survived in the wild as one plant and a few seedlings, is a sorry tale in which notable botanists are some of the worst culprits (Holliwell, 1981). It was collected from the wild for the gardens of John Gerard, John Parkinson, John Rea (the flower enthusiast and gardener, not John Ray the taxonomist) and the Tradescants among others. It never thrived and repeated collections were necessary. Market gardeners pillaged the woods in the Arncliffe Valley. A price was placed on surviving plants. One old vicar protected the survivors by picking the flowers so that they could not be found. The plants survived as inconspicuous leafy rosettes. Then one year he became ill and the plants flowered. Each plant was betrayed by its own flower, a red and yellow flag marking the spot to dig, and soon they were all grubbed up.

Some losses are simply disgraceful, like the 45% of limestone pavement which has been damaged or destroyed to provide stone for fake rockeries.

3.5.8 REASONS OF RARITY

Humanity is not to blame for the rarity of all native species. There are also a number of species which it seems are naturally rare. About 26% of extinctions or decline of our native species have been due to 'natural' causes. These include the alpine species confined to high-altitude low-competition habitats. Several species of orchid and parasitic plants have exhibited strange changes in abundance. *Himantoglossum hircinum* (lizard orchid) was known in 1900 from a single stable population near Dartford in Kent, and had a sporadic occurrence in 20 other sites. Between then and 1933 another 108 locations were spotted. Since 1933 there has been a decline so that only about nine localities are known today.

The orchids and parasites have tiny seeds. When properly pollinated they produce thousands of tiny seeds. This seed is dispersed both in space and in time, because of seed dormancy, surviving in the seed bank to wait for the possibility of the establishment of a successful mycorrhizal relationship. Orchids have a tendency to be spotted by botanists and so changes in distribution patterns probably do reflect real changes of distribution rather than differences in recording. Their ability for long-distance dispersal is shown by the spread of *Spiranthes romanzoffiana* (Irish lady's tresses) after its arrival from America. It was first recorded in western Cork in 1810, then found by Lough Neagh in 1892 and, more recently, it has spread to Scotland and South Devon. *Neotinea maculata* (dense flowered orchid) seems to have spread to the Isle of Man from western Ireland. Another example was a swarm *Dactyloriza* spp. (marsh orchids) which established, out of the blue, on the sterile soil of fly ash pumped into a gravel pit beside a power station in Essex. The swarm flourished for several years until overcome by willow carr.

Another of our rarest species is a spiny shrub with pinkish-white flowers, *Pyrus cordata* (Plymouth pear). It is confined to the western edge of Europe and is found in Britain in a few sites, old hedgebanks around Plymouth and Truro in the west country. Its limited distribution is a bit of a mystery. Some plants are restricted to a few sites but are abundant there. For example, *Helianthemum appeninum* (white rock-rose) grows only in two areas, north Somerset and south Devon. Found in only eight sites, it is abundant in five.

3.6 EVOLUTION IN OUR FLORA

Some of our rarest species are the endemic species, treasured because they grow nowhere else. Unfortunately our endemic flora is not rich. In fact it is very small despite the fact of the British Isles being an archipelago of islands. Islands are sometimes marked by having high levels of endemism with a flora rich in species which grow nowhere else. These species are either relict species, called palaeoendemics, which have been preserved on islands while they have been eliminated elsewhere, or they are neoendemics which have arisen *in situ*. The process of speciation is encouraged by isolation, so that islands which have been isolated by great distances over very long periods of time have the highest levels of endemism. By either of these criteria the islands of the British Isles would not be expected to have large numbers of endemic species. Britain lies very close to the continent and has been insulated only very recently. Also 15 000 years ago, at the height of the last glacial stage it hardly had a flora of any sort, let alone one rich in palaeoendemics. Also there has not been enough time for a great range of neoendemic species to arise. If the history of the evolution of the world's land flora, 400 million years long, is represented by a single 24 hour day, then our Holocene history only represents the last couple of seconds of that day.

Even though there has been little time in the British flora for the degree of isolation necessary for the evolution of new endemic species here, in fact several have arisen because they possess a breeding system which strengthens genetic isolation. Peculiar variants have arisen, sometimes following hybridization, and have been preserved and stabilized by self-pollination and inbreeding. One of our most beautiful endemic species, *Primula scotica* (Scottish primrose), is like this (Figure 3.28a) (Ritchie, 1954).

It is found in the northernmost part of Scotland and Orkney. A single seed has been discovered in Cambridgeshire from the full glacial. It is related to *Primula scandinavica* from Norway. Both species are very similar to, and derived from, *Primula farinosa* (bird's-eye primrose) (see Figure 2.14C) but differ from it in being able to produce seed by self-fertilization, a characteristic which may have been important in their ability to establish isolated populations.

A similar process of hybridization and inbreeding has given rise to endemic species of eyebright. *Euphrasia rivularis* is probably a stabilized hybrid segregate of *E. rostkoviana* × *E. micrantha*, and *Euphrasia virgursii* is one of a cross between *E. anglica* and *E. micrantha*. Eyebrights regularly self-pollinate, so that distinct lines can be stabilized from hybrids by inbreeding. Four other endemic eyebrights, *E. campbelliae*, *E. marshallii*, *E. pseudo-kerneri* and *E. rotundifolia* have probably also originated in this way.

Another example is found in the fumitories, which often produce closed self-pollinating flowers late in the flowering season. There are two endemic species of fumitory, which probably have a hybrid origin: *Fumaria occidentalis* and *F. purpurea* (western ramping-fumitory and purple ramping-fumitory).

Isolation of small populations can also speed up evolution and lead to the evolution of locally different populations, even new species. *Gentianella anglica* (early gentian) is another endemic species where a hybrid origin may be implicated. There is extensive hybridization in the genus. Patterns of variation are complicated by some species producing annual and biennial plants which differ in leaf shape and habit. There is also extensive local differentiation. The early gentian has two subspecies, *Gentianella anglica* subsp. *anglica* and subsp. *cornubiensis*, which differ in their habit and also in their time of flowering. The latter is confined to clifftop grassland and dunes in north Devon and north-west Cornwall. Another gentian, the autumn gentian, is found outside the British Isles but has evolved an endemic subspecies *Gentianella amarella* subsp. *hibernica* found in Ireland, and another, *G. amarella* subsp. *septentrionalis*, found in Scotland.

(a)

(b)

(c)

Figure 3.28 Endemic species: (a) *Primula scotica* (Scottish primrose); (b) *Coincya wrightii* (Lundy cabbage); (c) *Bromus interruptus* (interrupted brome). ((a) Copyright Jon B. Wilson.)

The isolation of small populations is significant in the evolution of the endemic *Coincya wrightii* (Lundy cabbage) (Figure 3.28b), which was only described as distinct in 1936. It is only found on Lundy and in a small part of north Devon. Its genus, *Coincya*, is a complex small group of species confined to mountain ranges or the coast. There are 10 species recorded for Europe in *Flora Europaea*, including four other narrow endemics. One other native species of *Coincya* is the Isle of Man cabbage, which is an endemic subspecies called *C. monensis* subsp. *monensis*.

Some plants have the ability to produce seed by agamospermy, without the normal processes of pollination and fertilization. Each seed is genetically identical to its parent and as a result large numbers of identical plants are produced. Each group of identical plants has often been recognized as a species, sometimes called a microspecies. Genera that include endemic microspecies are *Taraxacum* agg. (dandelions), *Rubus fruticosa* agg. (blackberries), *Calamagrostis purpurea* (small reeds), *Alchemilla* spp. (lady's mantles), *Limonium binervosum* agg. (rock sea-lavenders), *Ranunculus auricomus* (goldilocks) and *Sorbus aria* (whitebeams).

Examples where evolution has gone so far and fast, natural selection has been strong so that it is possible to recognize newly evolved species taxonomically, are rare in the British flora. *Bromus interruptus* (interrupted brome) described above is one (Figure 3.28c). However, endemic subspecies have been recorded more frequently. One notable example is the population of the Irish saxifrage which grows only on Arranmore Island off the western coast of Ireland. It has been taxonomically recognized as *Saxifraga rosacea* subsp. *hartii*. It has more robust plants than normal, with hairs that are all very short and glandular rather than the normal mix of glandular or non-glandular long ones. Another endemic subspecies is *Linum perenne* ssp. *anglicum* (perennial flax) (Ockendon, 1968).

Considering the short time scale, it comes as a surprise that there is in fact a lot to interest the student of genetic evolution in the present British flora. In fact, genetic evolutionary change, if not speciation, can occur very rapidly, depending on the processes of migration and colonization, the establishment and survival of small populations. Our flora has for so long been in transit, unstable and unsettled, that it has a rich history of chance events.

One common way of studying genetic evolution is to record patterns of variation within and between species. Natural selection for and against particular genetic variants because of edaphic or climatic factors is recorded in the different pattern of distribution of the variants. There may be the gradual change in some plant characteristic across an environmental gradient, or the frequency of a particular variant in populations may change, or there may be an abrupt change so that different morphs are present in different areas.

There are many examples of these kinds of patterns within species in the British flora. *Alnus glutinosa* (alder) varies in leaf and catkin size from north-west to south-east Britain (McVean, 1953). In most cases these patterns represent the sorting out of already pre-existing variation in the process of the colonization of the British Isles. The patterns do not halt at the Channel. *Geranium sanguineum* (bloody cranesbill) has a more finely dissected leaf in south-east Europe than in north-west Europe. *Silene dioica* (red campion) has an extensive pattern of variation in flower, capsule and seed characters across Europe. *Tripleurospermum maritimum* (sea mayweed) has contrasting leaf types; more or less dissected or succulent (Figure 3.29). *Lotus corniculatus* has a variant with a dark keel which is commoner in eastern England and the Pennines than the west (Crawford and Jones, 1986) (Figure 3.30).

One well-studied example has been the variation in *Plantago maritima* (sea plantain). It exhibits extensive geographical variation across the northern hemisphere. There is also local variation across ecological zones in saltmarshes. One hairy variant is common in very exposed situations on the edges of cliffs. In this case the distribution of variants, as in the distribution of pollution-tolerant plants, can be easily attributed to natural selection. Many times we do not understand the adaptive significance of variation patterns though we assume that differences have been selected. The process of natural selection and genetic adaptation is clearer in those variants that have adapted to the new polluted environments made by industrialized humans.

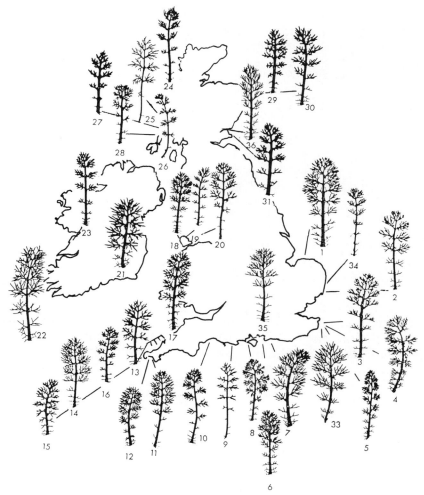

Figure 3.29 Geographical variation in leaf shape in *Tripleurospermum maritimum*, plants grown in simultaneous cultivation (Kay, 1972).

3.6.1 NEW INDUSTRIAL HABITATS

Most soils in the British Isles show the human influence in one way or another. However, some very recent soils are particularly peculiar. These include cultivated soils which have been artificially thickened by the application of organic material, and various soils that are the result of industrial development, like those on open-cast mines and spoil heaps. In the past 200 years there has been a very great increase in the area of polluted soils as the by-product of mining or some other industrial activity. These include colliery spoil heaps and slag heaps from iron-smelting works, present especially in the Black Country of the Midlands, around Manchester, around Barnsley in West Yorkshire and in central Scotland. In Cornwall there are polluted soils around old tin and copper mines and large areas of china clay pits and heaps. Elsewhere there are quarries

(a)

Figure 3.30 *Lotus corniculatus*: (a) normal variant; (b) cline in the proportion of dark-keeled variant across northern England. ((a) Copyright Jon B. Wilson; (b) after Crawford and Jones, 1986.)

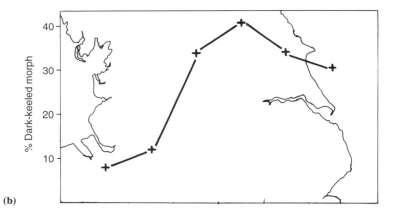

(b)

of all sorts for brickearth, limestone, slate and sand and gravel works. There are large areas of railway sidings and airfields.

The soils of such sites are very variable, but they are often inimical to plant growth. The original topsoil is usually missing, stripped away by quarrying activities or covered by spoil. Paradoxically some of these open habitats are open because they are inimical to tree growth. In the past few centuries there has been a renewed diversity of open habitats, such as quarries, sand-pits and wastelands of all sorts, which have provided habitats for herbs. Many man-made soils are nutrient poor, unconsolidated and have poor drainage. They are sometimes polluted with toxic constituents. Many are polluted with toxic elements, such as the heavy metals; lead, zinc and copper. Lead, for example, decreases the rate of root cell division.

An interesting example of a species distribution determined by spoil type is that of *Thlaspi caerulescens* (alpine penny-cress) which is confined to soils with high levels of lead and zinc (Figure 3.31).

It does not require lead or zinc but grows only where it suffers less competition from other plants because they find lead and zinc highly toxic (Ingrouille and Smirnoff, 1986). It is one of a group of species that have colonized old lead mine sites, but it is also very rare on natural outcrops of lead ore.

Other species that have evolved ecotypes tolerant of the heavy metals; copper, lead and zinc are the grasses *Deschampsia flexuosa* (wavy hair grass), *Anthoxanthum odoratum* (sweet vernal-grass) and the herbs *Armeria maritima* (thrift) and *Minuartia verna* (spring sandwort) (Bradshaw and McNeilly, 1981). Most of these species have populations growing on non-contaminated soils where there is a low frequency of tolerant individuals,

perhaps 1 in 10 000 or less. These odd tolerant individuals are selected on the contaminated soils. For example, the level of tolerance to copper in *Agrostis capillaris* (creeping bent) in grasslands around a copper refinery near Liverpool is greater in 70-year-old lawns than those 4–14 years old (Wu, Bradshaw and Thurman, 1975). The younger lawns were patchy with poor growth, while the older ones had a good turf. The natural selection of a copper-tolerant population had occurred within 70 years, through selection of the few tolerant individuals in a young turf. Different species tend to have different natural tolerances: *Dactylis glomerata* (cocksfoot) and *Agrostis capillaris* (common bent) to copper and *Anthoxanthum odoratum* (sweet vernal-grass) and *Plantago lanceolata* (ribwort plantain) to zinc and lead. Long-lived woody plants tend not to have tolerant variants. Quite often root growth is limited because the soils are physically poor, with a limited range of particle size and no organic material. They lack aggregated crumbs which allow drainage and soil aeration, and can be tightly packed.

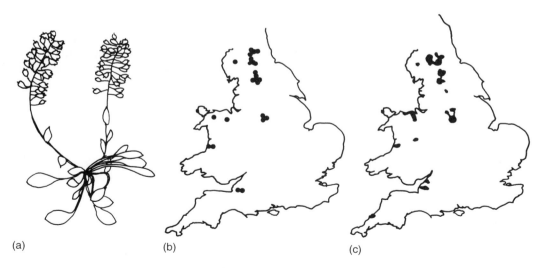

(a) (b) (c)

Figure 3.31 *Thlaspi caerulescens* (alpine penny-cress): (a) plant in fruit; (b) distribution in England and Wales; (c) distribution of regions where the lead concentration in stream sediments >160 mg/g. (From Ingrouille and Smirnoff, 1986.)

They often lack nitrogen and phosphorus, and some even lack calcium, in forms available to plants.

An important part of genetic adaptation is the evolution of reproductive isolation between differently adapted populations. Without isolation the adapted gene pool is diluted by an influx of unadapted genes in pollen coming into the population from outside. It is fascinating that mine-site plants have been found to flower earlier than non-tolerant plants in adjacent grasslands, reproductively isolating them (McNeilly and Antonovics, 1968). At the Drws-y-Coed mine in North Wales the earlier flowering was greatest in the area of most likely contact between different populations at the margin of the mine.

Some industrial soils are very peculiar because they have very high concentrations of particular minerals. Some are very acidic because of the oxidation of pyrite. The reclamation of industrially derived soils has taken place on a large scale. At best the topsoil has been removed before exploitation and is replaced carefully in a way preventing too much mixing of top and lower layers and compression by vehicles. Mechanical treatment may include ripping and harrowing and the inclusion of a great deal of farmyard manure or sewage sludge. However, soil microflora are relatively sensitive to heavy metals, which is why the much higher levels found in sewage, up to 2000–8000 p.p.m., can be a serious problem for the exploitation of treated sludge as fertilizer. Nevertheless, organic material can reduce the toxicity of soil by complexing with heavy metal ions. It also improves both water retention and drainage. Sowing with earthworms can also help to build soil structure. Slow-release fertilizer may be used, but a lot is lost by leaching so that much higher levels have to be applied than on agricultural soil.

If the soil is polluted with heavy metals, they are sown with a mixture of tolerant grasses and nitrogen-fixing plants, such as

Trifolium (clover) and *Alnus glutinosa* (alder). On heavy-metal-polluted soils a variety of metal-tolerant grass has to be used like the variety *Festuca rubra* (red fescue) called 'Merlin', which has been sown on calcareous lead/zinc wastes, or 'Goginan' (*Agrostis capillaris*, common bent) on acidic lead/zinc wastes and 'Parys' on copper wastes. The names of the latter varieties come from the mine sites where they were discovered. The Parys Mountain copper on Anglesey has an ancient history and was heavily exploited in the Roman period. These tolerant varieties are examples of the remarkable way some species have evolved in the British flora recently. In these circumstances natural selection has been very powerful, so that evolution has occurred very rapidly. Without genes for heavy-metal tolerance plants will not grow. Natural toxic soils do exist where metalliferous veins have been exposed by erosion, but a small range of plant species took advantage of the spread of toxic soils.

3.6.2 ATMOSPHERIC POLLUTION

The general level of pollutants in the environment has increased over the past 200 years because of atmospheric pollution. Mosses collected in Sweden between 1860 and 1875 had about 20 p.p.m. lead, but this then doubled between 1875 and 1900 to 40 p.p.m. The increase was a result of increased coal consumption. There was another increase to 80–90 p.p.m. post-1950 because of petrol consumption. These low levels of heavy metals are not a problem to higher plants but other atmospheric pollutants, such as sulphur dioxide and ozone, have been seriously damaging to plants.

Some plants have adapted to atmospheric pollution. In the 1960s Mudd noticed that in the environs of Manchester commercial seed of *Lolium perenne* (perennial rye-grass) looked much worse after winter than rye grass in natural pastures. He demonstrated that the native material was distinctly more tolerant to SO$_2$ pollution. It had been exposed to 100 years of high levels of SO$_2$ (mean levels of 500 µg/m^3 were commonplace in the winters of the past). Seed from populations closest to urban and industrial centres had the highest levels of tolerance. Pollution levels are less than a third of that now. We export most of it in the atmosphere! However, nitric acid pollution from traffic is not declining like the other pollutants and ozone levels exceed safe levels at many times.

The organisms that have shown the most marked change in distribution in relation to atmospheric pollution are the lichens (Figure 3.32).

Their sensitivity is related to their growth form, with the most finely dissected filamentous *Usnea* species or 'shrubby' *Evernia* most sensitive and the crustose *Lecanora* species most tolerant. Foliose *Xanthoria* and *Parmelia* species have intermediate levels of tolerance.

In the most polluted areas lichens are completely absent and only the green alga, *Desmacoccus viridis*, is found on tree bark.

Broad-leaved trees are more tolerant to pollution than conifers. The pine, for example, can keep its leaves for up to 5 years and so is exposed to pollution for 10 times the length of time of deciduous trees. Occult precipitation, tiny droplets condensing from the atmosphere, containing pollutants is much more damaging than rainfall which runs off the plant. In the 1900s acid rain damage was first noticed in *Abies* (fir) as a stork's-nest crown die-back, yellowing of old leaves and the formation of adventitious shoots. Acid rain damage is a complex phenomenon. The presence of ozone is significant. In the 1970s a new type of damage, canopy loss, leaf shedding and yellowing of second-year leaves in the upper crown was noted; this is related to ozone damage. Acid

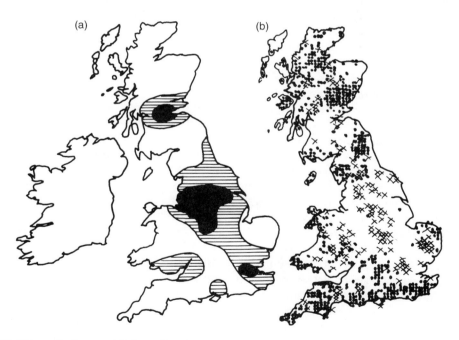

Figure 3.32 SO$_2$ pollution and lichen distribution. (a) Average amount in the air in winter in the early 1950s. Shaded areas, 40–80 µg/m^3; dark areas, 80–160 µg/m^3. (After Meethan, in Bowen, 1965.) (b) Distribution of an *Evernia* species. x, Extinct populations; ●, healthy populations. (From Seaward and Hitch, 1982.)

rain may predispose plants to ozone damage by making the plant surface more permeable. Lesions, small depressions on the leaf surface, are created which collect polluted droplets. Acid rain also acidifies the soil. Large plantations of conifers have trapped acid rain so that soil water has been acidified.

3.6.3 THE EROSION OF PEATLANDS

Throughout the history of the British flora, climate change has been a natural cause for changes in the abundance of species. Many areas where there is blanket peat are now considered to have less annual rainfall than the minimum required to maintain the cover of *Sphagnum* (bog moss) (Tallis, 1965; Pearsall, 1971). Rainfall has declined from that 5000–4000 years ago when there was the most rapid spread of peatlands. Drying leads to aeration of the peat and wastage by oxidative decomposition.

Pollution has exacerbated the decline of upland peatlands (Ferguson, Lee and Bell, 1978). This is especially obvious in the southern Pennines in the vicinity of the industrial areas of the Midlands, South Yorkshire and Greater Manchester, but it is a widespread phenomenon of the high-level plateau peatlands as far north as Sutherland. A black rain fell on the peatlands in the first half of this century. On average 0.4 g/m^2 of soot fell per month on the moors near Sheffield in 1948, marking a dark band in the peat. *Sphagnum imbricatum* has completely disappeared from the southern Pennines in the past 150 years. *Sphagnum recurvum*, a species with high sulphate tolerance, is the most common species now. Acid deposition from acid rain, and especially the increased input of nitrate through rainfall and occult precipitation, has stressed the ecosystem so that it takes longer to revert from other stresses like burning, grazing or periods of drought. A decrease in heather pollen compared to grass pollen has been correlated with a rise in lead, zinc and copper in lake sediments of Loch Enoch in Scotland (Battarbee *et al.*, 1985).

Once the surface of the peat has dried and cracked, it becomes exposed to erosion through a combination of rain, ice and wind. The peat loses its ability to absorb water and is washed or blown away. Burning to manage the vegetation for grazing, and the trampling of grazing sheep or deer are implicated in the beginning of erosion on many sites. A slow headward extension of streams into the blanket peat may have started as long as 5000 years ago, but after about 1770 there was very rapid erosion. At this time the expansion of the Yorkshire and Lancashire woollen industry put increased grazing pressure on the moorlands. Once started the process is irreversible and deep gullies between towering peat haggs are formed. Erosion can be very rapid in summer storms after long periods of dry weather. More than half the sites examined in a survey of Caithness and Sutherland were eroding. Rates of erosion of 10–25 mm/year have been recorded from the Peak District.

3.6.4 POLLUTED WATERS

One of the most serious effects of twentieth century agriculture has been the increased input of fertilizers into natural and semi-natural systems, a pollution as serious as that of any industrial effluent. Ground-water supplies have been contaminated by high levels of nutrients from chemical fertilizers. The amount of nitrogen applied as fertilizer has, on average, doubled each decade from the 1940s, although the rate of increase is lessening. The amount of phosphate applied increased nearly threefold in the 1940s and then stabilized. Between 36–55% of the yield of the four main crops, wheat, barley, potatoes and sugar beet, is estimated to be due to fertilizer input. Fertilizer application has been most concentrated where there has been intensive cultivation of vegetables, like brassicas on the Isle of Thanet in Kent.

Ploughing up of old grassland, deep ploughing generally, also leads to a flush of nitrate in drainage waters. However, the most important source of nitrate has been from livestock residues, farmyard slurry and dung when applied in large quantity as manure. With more and more cattle kept indoors, large quantities of slurry and solid waste have been created. Reported farm pollution incidents trebled in the 1980s. Most of these came from washing yards and parlours. However, spillage of silage effluent, which is very damaging, is less common. In addition to all this farmyard pollution, there has been a higher input of phosphates, as well as other pollutants, from sewage effluent into rivers and waterways, which have suffered eutrophication.

The change has been recorded in the sediments deposited in the Broads (Moss, 1987). The deep sediments from the fourteenth century to the end of the nineteenth century are full of snails, the remains of the water-weed, *Chara*, and epiphytic diatoms. Above this sediment *Chara* is replaced by the aquatic moss *Fontinalis*. The increased input of nutrients stimulated the growth of epiphytes and filamentous algae so that *Chara* was shaded out. Meanwhile, a taller, ranker vegetation of *Potamogeton* (pondweeds), *Myriophyllum* (water-milfoils) and *Ceratophyllum* (hornworts) is established. In Hickling Broad this change came later than some other areas. The cause was also eutrophication but here it was associated with the expansion of a local black-headed gull colony to tens of thousands of birds whose droppings fertilized the water. In the past few decades the aquatic life of the Broads has changed again. The large, rooted aquatic plants have disappeared and have been replaced by planktonic algae and cyanophytes. Rates of sedimentation have increased tenfold so that the Broads are filling up at a rate of 1–2% per year.

The effects of eutrophication are widespread. In dry years water must be pumped up from ditches into Wicken Fen to keep the fen moist. Unfortunately it comes polluted by nutrients from the surrounding fields so that coarse weeds are established in the area of the inlet.

3.6.5 CLIMATE CHANGE

The most significant result of industrial pollution, especially the greenhouse effect of extra atmospheric carbon dioxide from the burning of fossil fuels, seemed to be global climate change. This could cause the most dramatic change in our flora since the introduction of agriculture. Plants seem to have responded already. Dried specimens of a range of species, collected before the industrial era, have more stomata than recent ones (Woodward, 1993). The stomata are pores through which the plant obtains atmospheric carbon dioxide for photosynthesis. If more CO_2 is available, they need fewer stomata.

Some idea of what might happen to the vegetation has been suggested by some of the mild winters experienced recently. Though grass grows throughout the winter it is not necessarily good quality for grazing and there has also been a greater threat from pests and disease. The competitive relationships between species are altered so that potentially new successions can take place. Climatic warming would make some species with a northern or high-altitude distribution much rarer. Regular summer drought would reduce the potential for the grass growth which makes our islands green. A change in the climate has to be seen in the context of a greatly increased potential for change in our flora because of the astonishing range of exotic species introduced relatively recently in Britain. Many have been introduced from warmer climates and, though they are tender to the present climate, they could become the future dominants in our vegetation.

However, the direction of climate change is far from clear (Courtney, 1992). Global temperature declined for the three decades before 1970, leading to the prediction then that we were entering an ice age! It is far from clear how carbon dioxide pollution might cause global warming (Marston *et al.*, 1991). Carbon dioxide, deemed to be the main culprit of global warming, is actually produced in far greater quantities by natural processes, such as forest fires and termites, than by emission from burning fossil fuels. Increased carbon dioxide concentration in the atmosphere may actually be a consequence of global warming rather than a cause of it, because higher ocean temperatures lead to the release of large quantities of carbon dioxide from the ocean to the atmosphere. Instead, recent changes in global temperatures follow most closely natural variation in the activity of the sun (Kuo, Lindberg and Thomson, 1990; Friis-Christensen and Lassen, 1991). Changes in the vegetation over many millennia provide ample evidence that human activity is not necessary for climate change to occur.

3.7 THE EXOTIC FLORA

The *New Flora of the British Isles* (Stace, 1991) records 2990 species: 35 species are conifers with only three native; 81 species are ferns, clubmosses and horsetails, mostly native. The 2909 species of flowering plants include 1737, nearly 60%, naturalized exotic species. This is tremendously exciting. Our flora has gone through a revolution as great as the colonization of these islands after the Ice Age. The change has been very sudden; the majority of these naturalized species have been introduced in the past 200 or 300 years. Two outstandingly successful introductions have been *Conyza canadensis* (Canadian flea-bane), common in south-east England, and *Veronica persica* (common field-speedwell), first recorded in 1825, from western Asia but now the commonest speedwell on cultivated land. *BSBI News*, the newsletter of the Botanical Society of the British Isles makes interesting reading. Every issue includes

reports of new aliens or newly naturalized species. Not since the post-glacial colonization of the British Isles has the vegetation been exposed to such potential for change. We live in exciting times.

Whole floras have been imported in two particular ways. Hayward and Druce (1919) drew attention to the aliens which were associated with wool shoddy, the waste from scouring the wool. The shoddy was used as manure. In 1960, 529 wool-shoddy aliens were counted (Lousley, 1960). Many others have since turned up. They come from anywhere fleeces have been imported: Australia, South America and the Mediterranean. Almost all of them are very transient. *Solanum sarachoides* (leafy-fruited nightshade), an alien from Brazil, is one which turns up very regularly on rubbish tips and is now naturalized on a tip at Dagenham. *Trifolium subterraneum* (subterranean clover) was introduced to Australia from Europe but has made the return journey. It has come back with a different accent: called *Trifolium subterraneum* var. *oxaloides*, it is a more robust and frost-sensitive variety than the native variety.

Another set of species is associated with cage-bird seed or pheasant seed mixtures. Many species are found apart from *Phalaris canariensis* (canary grass) and several species of *Panicum* (millet). Species commonly associated with areas around pheasant coverts are *Helianthus annuus* (sunflower), *Fagopyrum esculentum* (buckwheat) and *F. tataricum* (green buckwheat), *Linum usitatissimum* (flax), *Sinapis alba* (white mustard) and *Vicia sativa* (common vetch).

Other groups of aliens are associated with Tan bark from Turkey, and soyabean waste. Between 1969 and 1979, 220 different aliens were found on tips in the London area alone. Most only lasted a couple of years. Few aliens have yet had an impact on our vegetation. On the banks of the Thames 85 aliens have established from the oil-milling industry. These species mainly have a North American origin and so, being hardier, may at any time begin to make a significant contribution to our flora. Look out for *Artemisia biennis* (slender mugwort) and *Bidens connata* (London bur-marigold). Recently 'wild-flower' seed of native species, but harvested from continental plants, has been identified as a source of exotic variants.

One exceptional alien, which shows what might happen, is *Phalaris paradoxa* (awned canary-grass), which may have established from wool shoddy or from bird seed. It is becoming an important weed of winter crops, even threatening to rival *Alopecurus myosuroides* (black grass). Up to 8000 flowering heads/m² have been counted. It prefers wetter soils and minimal cultivation. Deep ploughing may help to control it.

Many aliens seem to spend some time established only in a small area and then, due to some change in circumstance, undergo a population explosion. *Sisymbrium irio* (London rocket) is an overwintering annual which became abundant after the Great Fire of London in 1666 and then became established quite widely on roadsides and walls and in waste places. Once it was recorded from 24 'counties'. It is rarer now but well naturalized in a few localities and elsewhere appears as a wool-shoddy casual.

The story of the *Senecio squalidus* (Oxford ragwort) is a remarkable example. It was first recorded in 1794 growing on walls in Oxford, as an introduction to the Oxford Botanic Garden from Sicily. With the coming of the railway it spread down the railway lines from 1879, rooting in railway ballast, its parachute weed carried it along in gusts of wind caused by the trains.

Other plants took advantage of the railway lines and the many disturbed soils, burned clearings and other derelict sites of human activity. Notable examples are several species of willow herb, including *Epilobium ciliatum* (American willow herb) and *Chamerion angustifolium* (rosebay willow herb). The former was introduced from America and

first noticed in 1891. It is a fast-growing perennial, regenerating from leafy stolons. Its seedlings can grow, flower and set seed in 10 weeks from germinating. It flowers and sets seed over much of the year and each plant produces >10 000 seeds. *Chamerion angustifolium* is a native species, though it has been behaving like a rapidly expanding introduced species. It was confined to upland habitats as a member of the tall-herb community of cliffs and screes. In the late nineteenth century its parachute seeds rapidly colonized much of lowland Britain. More recently *Epilobium brunnescens* (New Zealand willow herb), which was first noticed in 1908, has been spreading throughout the British Isles.

In a similar way, several species of broom, such as *Cytisus striatus* (hairy-fruited broom), *C. multiflorus* (white broom), *Genista monspessulana* (Montpellier broom) and *G. hispanica* (Spanish gorse), introduced from the Iberian Peninsula or the Mediterranean region, have become naturalized on the well-drained slopes of our motorway banks.

Another fascinating story has unfolded in the past few years of the colonization of our roadsides by salt-tolerant coastal species (Scott, 1985). The application of de-icing salt has opened up a new habitat for species such as *Puccinellia distans* (reflexed saltmarsh-grass) (Figure 3.33). *Cochlearia dancia* (Danish scurvy-grass) and to a lesser extent *C. officinalis* (common scurvy-grass) have begun to favour the central reservation of dual carriageways.

3.7.1 A PROMISCUOUS FLORA

The story of *Senecio squalidus* (Oxford ragwort) does not end with its spread down the railways. It has entered into our native flora in two remarkable ways. It hybridized with our native *S. vulgaris* (groundsel) (Figure 3.34).

The hybrids have the short stature of *S. vulgaris* but there are ray florets, looking like petals, surrounding the floral head. From this

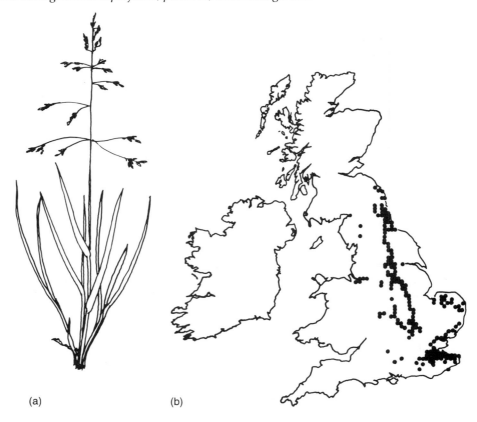

(a)   (b)

Figure 3.33 *Puccinellia distans* (reflexed saltmarsh-grass): (a) plant; (b) distribution along roadsides in Britain up to 1984. (From Scott, 1985.)

sterile hybrid a fertile 'hybrid-species' has arisen. It has been called *S. cambrensis* (Welsh ragwort). It was first found in Flintshire in 1948 and is now spreading out of Wales. It was noticed in Scotland in 1982. Meanwhile, further crosses have given rise to a variant identical to *S. vulgaris* but with the ray florets of *S. squalidus*. A hybrid between *S. squalidus* and the *S. viscosus* (sticky ragwort) is also regularly found, so there is potential for another hybrid species of *Senecio* to arise.

A similar story of the origin of a new hybrid species is the origin of *Spartina anglica* (common cord-grass) following hybridization between our native *S. maritima* (small cord-grass) and *S. alterniflora* (American cord-grass) which arrived as an illegal alien in Southampton Water in the 1870s.

A puzzling story of hybridization is presented by *Platanus × hispanica* (London plane). It is the foremost of our town trees (Wilkinson, 1981). In 1548 William Turner mentions seeing two plane trees, one in Northumberland and one in Cambridge, perhaps brought back by friars, monks or pilgrims from Italy or further east. The London plane has an obscure origin as a hybrid between the American *Platanus occidentalis* and *P. orientalis* from eastern Europe and the Near East. The former is not very hardy in Britain and so rarely grown, but the latter can sometimes be seen in parks

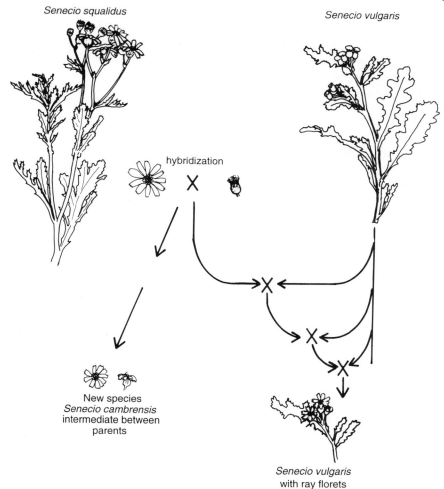

Senecio squalidus

Senecio vulgaris

hybridization

New species
Senecio cambrensis
intermediate between
parents

Senecio vulgaris
with ray florets

Figure 3.34 Hybridization and evolution of *Senecio vulgaris* and *S. squalidus*.

and gardens. The hybrid between them, the London plane, has proved useful because of its tall, straight trunk and its hardiness to pollution, which washes off its shiny leaves. Two of the earliest hybrid trees were planted at Ely in 1680, but where did they originate?

Hybridization has been rampant in the British flora (Stace, 1975). The *New Flora of the British Isles* (Stace, 1991) includes 559 hybrids. Some of these have spread widely or originated in many different places. Those that reproduce vegetatively can often be found in the absence of either parent species. There are many good examples of *Salix* (willow) hybrids like this. They include the *S. × meyeriana*, *S. × ehrhartiana*, *S. × rubens*, *S. × pendulina*, *S. × sepulcralis*, *S. × mollisima*, *S. × rubra*, *S. × laurina*, (shiny-leaved, Ehrhart's, hybrid-crack, weeping-crack, weeping, sharp-stipuled, green-leaved and laurel-leaved willows), and several kinds of osier.

Hybridization between a native species and an introduced exotic species, like that between our native *Hyacinthoides non-scripta* (bluebell) and the exotic *H. hispanica* (Spanish bluebell) is common. *Hyacinthoides hispanica*, with its erect head, bell-shaped flowers and blue anthers, is well naturalized. Hybrids between them with the full range of intermediate characters are common and, being fertile, are often naturalized in the absence of either parent. Hybridization between *Primula vulgaris* (primrose) and *P. veris* (cowslip) can form hybrid swarms (Woodell, 1965) (Figure 3.35), giving a hybrid *P. × polyantha* (false oxlip) which is the ancestor of garden polyanthus.

Geum urbanum and *G. rivale* (wood avens and water avens) hybridize so regularly and with such fertility that when they grow together a hybrid swarm is produced. *Geum urbanum* and *G. rivale* look very different. *Geum urbanum* has yellow, open flowers on erect stalks and *G. rivale* has pinkish-purple hooded flowers which hang down. When

they grow together hybrids are produced which look intermediate in colour and shape and, since these are fertile, they cross with the parental types so that more intermediates are produced. This can take place until there is a complete spectrum of variation between the parental types. However, in a natural vegetation the species would be well separated: *G. urbanum* in shady, dry woods and *G. rivale* in open, marshy conditions. Disturbance of natural vegetation by humans has brought them together, especially the construction of hedgerows, which are strips of dry woodland on the hedgebanks, running alongside marshy areas in the ditches.

3.7.2 UNDESIRABLE ALIENS?

In the long run the most significant aliens may prove to be garden escapees or chuckouts. They cope very well with our climate. Some notable ones, like *Senecio squalidus* (Oxford ragwort), have escaped from botanic gardens. There have been two garden escapees which have become very common, even seeming to threaten our native flora; *Fallopia japonica* (Japanese knotweed) and *Heracleum mantegazzianum* (giant hogweed). *Fallopia japonica* was first found in the wild in Britain in 1886. It is now so widespread that it has already gained a range of vernacular names, and it is so competitive that it is one of only two species of flowering plant scheduled in the Wildlife and Countryside Act as an undesirable alien. *Heracleum mantegazzianum* has been favoured in the recent past as an impressive garden plant. It grows up to 5 m high (Figure 3.36).

This would have been all right but it was widely introduced into nature by enthusiasts. Unfortunately, like many umbellifers it causes a photoallergic reaction in human beings, causing a nasty rash.

A recent escapee in the vicinity of Kew Gardens, *Ceratochloa carinata* (Californian brome) has caused some concern. It escaped in 1919 and is now the most competitive grass

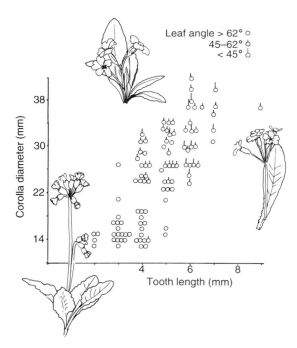

Figure 3.35 Hybridization between *Primula vulgaris* and *P. veris* to give *P. × polyantha*. (After Woodell, 1965.)

Figure 3.36 *Heracleum mantegazzianum* (giant hog-weed). (Copyright Jon B. Wilson.)

along much of the path on both sides of the Thames. It was noticed in Oxford in 1945. It has flowers which self-pollinate and do not even open, so that seed can be produced at any time of year.

Not all successes are flowering plants. The moss *Orthodontium lineare* was introduced from the southern hemisphere in the 1910s. It has replaced our native *O. gracile* as the main species on oak stumps.

However few aliens have proved very damaging to our native flora. Water-weed, *Elodea*, is one alien that has risen to dominance over native competitors in its favoured habitat of calm waters, though it can be suppressed by shading from *Lemna* spp. (duckweed). *Elodea canadensis* (Canadian

water-weed) was first recorded in 1836 in Ireland and then Britain in 1842 (Figure 3.37).

It spread along the canal system, and to other bodies of water, perhaps from aquaria, choking many of them at first. Where introduced into a new area there is a build-up of the colony over 3–4 years, eventually excluding other aquatic plants. The population can then stay in equilibrium for 3–10 years, but often then there is a slow decline over the following 7–15 years so that it eventually disappears or survives only as a small residual population.

Two other species of *Elodea* have been introduced (Simpson, 1986). *Elodea callitrichoides* (South American water-weed), which has red root tips, was first recorded in 1948, but has, as yet, not become very well naturalized. However, *Elodea nuttallii* (Nuttall's water-weed) from North America, which was recorded only in 1966, has spread very rapidly. It is a robust plant and out-competes *Elodea canadensis* (Canadian water-weed) in the nutrient-rich waters common today.

One damaging introduction is not a flowering plant but a fungus. A species of oak mildew, *Microsphaera alphitoides*, was introduced from America at the beginning of this century. Now almost every oak tree has it and, in shaded conditions, it is particularly damaging to oak seedlings and saplings.

What can happen with introduced species is best shown with reference to just two species, the sycamore and the rhododendron. *Acer pseudoplatanus* (sycamore) was rare in the seventeenth century, planted in hedges and gardens (Jones, 1945). It seems to have been introduced into Scotland first in the fifteenth century, where it was called the plane (Morton-Boyd, 1992). In his Herball of 1597 Gerard described the sycamore as 'a stranger in England, only it groweth in the walkes and places of pleasure of noble men, where it is especially planted for the shadow sake'. By 1664 it had been found out as a nuisance. Evelyn wrote: 'the honey-dew leaves, which fall early, like the ash, turn to mucilage and

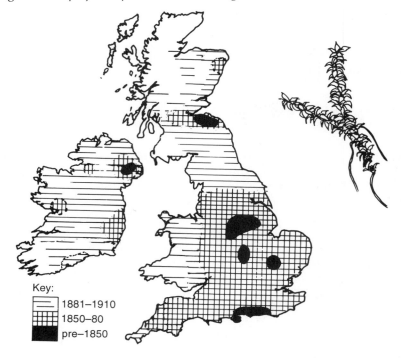

Figure 3.37 The spread of *Elodea canadensis* (Canadian pondweed). (From Simpson, 1986.)

Key:
- 1881–1910
- 1850–80
- pre–1850

noxious insects, and putrifie with the first moisture of the season, so as they contaminate and mar our walks; and are therefore by my consent, to be banish'd from all curious gardens and avenues'. Ray in 1690 still recorded it as growing not wild, but mainly in churchyards. Since that time it has become one of our most common trees by taking advantage of every disturbance of woodlands and scrub to become established. It is relatively frost-tolerant and grows up to 30 cm each year. It can be long-lived, up to 600 years. By 1947 more than 4% of all woodland was sycamore and it was the fifth most common broad-leaved tree. With its numerous wind-dispersed fruits it infiltrates old woods, but its seedlings are rather sensitive to competition. Because of this it is unlikely to cause lasting damage to our native woodlands. Today it has become so much part of our flora that it has 183 different species of epiphytes and parasites. In contrast, *Acer platanoides* (Norway maple) has a distant claim to native status, since it probably grew in Britain in the Hoxnian interglacial. Introduced in 1683 it regenerates freely in many woods but, paradoxically, it is not yet regarded as a nuisance (Jones, 1945).

Another plant that can claim a right to residence, like an immigrant from New Zealand who can claim residence in Britain because his grandfather was born here, is *Rhododendron ponticum*, because of its presence in the Gortian interglacial. Native to south-east Europe and Asia Minor, it was introduced to Britain in 1763. It has been widely planted and is well naturalized, spreading by seeding and suckering. Now it is regarded as a serious problem because it forms dense thickets in western oak woodlands. Another Gortian/Hoxnian aquatic reintroduction, which flourishes at times, is

Azolla filiculoides (water fern). It is from tropical America and suffers a decline after a hard winter.

3.7.3 A GARDEN FLORA

The threat to our native vegetation from alien species is more apparent than real. Aliens have entered our flora very fully already without threatening it. This can be seen by examining the contribution to our vegetation of a single large family, the Brassicaceae or cabbage family. In the *New Flora of the British Isles* (Stace, 1991) there are 127 species of Brassicaceae listed, but only 60 of them are native. The native species include species of almost all the major 'natural' habitats, such as woods (*Sisymbrium officinale*, hedge mustard), wet places (*Rorippa nasturtium aquaticum*, water-cress), mountain ledges and limestone cliffs (*Draba* spp., whitlow-grass species; and *Arabis* spp., rock-cress species), saltmarsh (*Cochlearia*, scurvy grass), dunes (*Matthiola sinuata*, sea stock), and so on. Many native species have also found a home in man-made habitats, such as pastures (*Cardamine pratensis*, cuckoo flower) and hedgerows (*Barbarea vulgaris*, winter cress), or like *Thlaspi caerulescens* (alpine penny-cress) on the polluted soils of lead-mine sites. Several, such as *Capsella bursa-pastoris* (shepherd's purse), are arable weeds, but even so they have such a long presence in our flora, predating the first fields, that they are regarded as native. *Capsella bursa-pastoris*, for example, is recorded from the last full glacial stage in Britain.

The distinction between native and alien is made weaker by the existence of many other ruderals and weedy species which have only a short history in Britain and so must be alien. Weeds of cultivation are *Capsella rubella* (pink shepherd's-purse) from south and central Europe and *Barbarea intermedia* and *B. verna* (medium-flowered and American winter-cress). Others are ruderals or casuals of wasteland. There are bird-seed aliens,

such as *Camelina sativa* (gold-of-pleasure), and wool-shoddy aliens, such as *Sisymbrium erysimoides* (French rocket). There are cultivated species like *Isatis tinctoria* (woad), *Lepidium sativa* (garden cress), *Raphanus sativa* (radish), *Brassica napus* and *B. oleracea* (swede, turnip and cabbage) in all their varieties.

The most interesting, perhaps, are the garden escapees, which have so enriched our townscapes and countryside; *Erysimum cheiri* (wallflower), *Hesperis matronalis* (dame's violet), *Aubrietia deltoidea* (aubretia), *Lunaria annua* (honesty), *Lobularia maritima* (sweet alison), *Malcolmia maritima* (Virginia stock), and so on. Some have not got far away from the gardens into which they were introduced, but can anyone doubt that they will be an established part of our natural vegetation in a few hundred years time.

Our 'native' flora is a robust one. It is essentially a hardy pioneering flora. It has established itself within the past 15 000 years, perhaps most of it in the past 10 000 years. It is not threatened by introductions since it has itself been selected for introduction.

Even our most natural habitats, like those of the coastland, have not suffered appreciably yet. Some introduced species have established but they have generally enriched rather than threatened the 'native' vegetation. For example, the *Acaena* spp. (pirri-pirri-burs) from New Zealand are naturalized in a few places, including some sand dunes, but do not threaten other species. *Acaena novae-zelandiae* first became common as a wool-shoddy alien in the 1900s but is also commonly grown in gardens (Gynn and Richards, 1985). *Tamarix gallica* (tamarisk) is widely planted to stabilize sandy soils and to act as a wind break, but it shows little sign of becoming a serious competitor. It can raise soil salinities by transpiration but it seldom establishes from seed and it is pruned back by rabbit grazing. There is more of a threat perhaps from plantings into new areas of a native species such as *Hippophae rhamnoides* (sea buckthorn). It is native in eastern

England but has been widely planted elsewhere along roadsides and on dunes to bind the soil. It has nitrogen-fixing nodules on its roots and so it changes the nutrient status, and consequently the flora, of the areas where it grows. It can also outshade its potential competitors.

Our flora is immigrant but no less valuable for that, nor should we value recent immigrants less. Our flora has been wonderfully enriched in recent years. Take a species like *Cotoneaster divaricatus* (spreading cotoneaster) for example, which was introduced from western China in 1904. It received a Royal Horticultural Society First Class award of garden merit in 1969, and now it is naturalized in west Kent, a centre for cotoneaster naturalizations. The *New Flora* (Stace, 1991) lists 45 naturalized Cotoneasters! Then there is *Fuchsia magellanica*, planted as hedging and now so beautifully naturalized in the west of Britain and in Ireland. It was introduced from southern Chile and Argentina. Another wonderfully evocative introduction from South America is *Senecio smithii* (Magellan

ragwort), now naturalized in North Caithness, Orkney and the Shetlands. It was introduced into croft gardens by sailors who returned from whaling in the southern seas around Patagonia and the Falklands.

There are so many wonderful new plants. The wonderful yellow splattered with drops of blood-red, of *Mimulus guttatus* (monkey-flower) and *Mimulus luteus* (blood-drop-emlets), or the sterile hybrid between them, has brightened so many streams and burns (Figure 3.38).

Then there are the introduced species of balsam. *Impatiens parviflora* (small balsam) from central Asia was first recorded from Battersea in London in 1851. It is remarkably polymorphic within its native range but all European populations are uniform, suggesting a single introduction (Coombe, 1957). *Impatiens glandulifera* (Indian balsam) from the Himalayas, *I. parviflora* and *I. capensis* (orange balsam) from North America are more showy. Subtle they aren't but who can deny their beauty?

Figure 3.38 *Mimulus luteus* (blood-drop-emlets) beside a stream in Upper Teesdale.

3.7.4 THE RISE OF FORESTRY

The most important impact of alien plants on our vegetation in the twentieth century has been through the establishment of large forestry plantations of exotic conifers (Figure 3.39).

Modern forestry can perhaps be dated back to the introduction of *Larix decidua* (European larch) in the early 1600s from the Alps. Large plantations were established on the estates of the Duke of Atholl from about 1800. By 1832 there were 10 000 acres of 14 million trees. Unadapted to our mild climate, it suffers from larch canker or die-back, though a variety from Poland and Czechoslovakia performs better.

Figure 3.39 Plantation of *Picea sitchensis* (Sitka spruce).

A high proportion of Britain's forestry plantations have been established on marginal upland sites in the north and west. In these areas the soils are poor, frequently waterlogged or compacted and so root development may be restricted. Windthrow is a serious problem, but can be minimized by careful choice of forestry species and by preparation of the soil before planting. Ploughing may help by breaking up podzols or improving drainage, but certain kinds of ploughing, spaced furrow ploughing, may actually restrict lateral spread of the roots to the ridges.

Pinus contorta (lodgepole pine) is the best pioneer species. It withstands exposure well and is preferred over spruce in frost hollows. *Picea sitchensis* (Sitka spruce), *Tsuga heterophylla* (western hemlock) and *Pseudotsuga menziesii* (Douglas fir) are liable to be checked in their growth by the early competition of *Calluna vulgaris* (heather). *Pinus sylvestris* (pine) and *Larix decidua* (larch) are good nurse trees for these other species. *Larix decidua* (larch) is more resistant to air pollution than most species because it is deciduous.

The history of the Gwydr Forest in North Wales can stand as an example for the establishment of upland forestry plantations throughout Britain (Shaw, 1971). Like many areas exploited for forestry, the area has a high proportion of poor soils; podzols on shale and blanket peat. Rainfall is high. Much of the area is at relatively high altitude, and on the plateau and in sheltered lowland sites plants can suffer from spring frosts. There are frost hollows, and freeze–thaw can lift young trees out of the soil. Most importantly, there is moderate to severe exposure to wind in over 75% of the forest. The effects of wind are magnified due to altitude, shallow peaty soils, inadequate drainage and the topography, which can channel winds.

At first, in the eighteenth century the existing oak of the valley slopes was exploited. Between 1754 and 1760 £50 000 worth had been felled and floated down the

Conway. The plateau lands had long been a moorland and wet heath, supporting about one sheep per 3 acres. First there was introduction of a wide range of conifers, *Larix decidua* (European larch), *Picea sitchensis* (Sitka spruce), *Picea abies* (Norway spruce), *Pinus sylvestris* (Scots pine), *Pinus strobus* (Weymouth pine) and *Abies alba* (silver fir), and there were also broad-leaved introductions such as *Fagus sylvatica* (beech) and *Acer pseudoplatanus* (sycamore). During the First World War these were devastated by felling and clearance. The Forestry Commission interest started in the 1920s. In the first 3 years 909 acres were reafforested: *Pseudotsuga menziesii* (Douglas fir) on the lower sheltered ground, *Larix decidua* (European larch) on intermediate slopes, *Pinus sylvestris* (Scots pine) on thin-soiled ridge tops and *Picea* (spruce) on moist sites. *Pinus nigra* ssp. *laricio* (Corsican pine) was tried at one site but proved susceptible to *Brunchorstia* fungus. *Larix kaempferi* (Japanese larch) was introduced in 1923.

There were early problems. Fences erected to protect saplings from sheep allowed *Ulex europaeus* (gorse) to invade, giving the young trees intense competition. Where not fenced, plants were pulled up as they were chewed; spruce was severely checked or killed, and larch survived but grew spindly. Fencing could not keep rabbits out of the large areas involved. *Abies alba* (European silver fir) and *Larix decidua* (European larch) had to be staked to prevent windthrow. New species were introduced, including *Tsuga heterophylla* (western hemlock). *Pinus contorta* (lodgepole pine) emerged as a potentially successful species of very poor soils, exposure and drought, though it has proved vulnerable to the pine beauty moth in Scotland. *Larix kaempferi* (Japanese larch) gained favour because it was more resistant to competition from *Ulex europaeus* (gorse) and to larch canker. *Larix decidua* (European larch) was not planted after 1937. *Picea abies* (Norway spruce) proved useful in frost hollows on

wetter patches of the plateau lands, but here the root system is shallow and so it is liable to windthrow if the site was exposed.

Successful tree-planting has been achieved on bogs by planting on upturned turves (Hibberd, 1986). Ploughing provides raised turf for planting. Once established, the trees dry out peat and clay by transpiration. The existing vegetation is quickly changed by shade and litter fall after 10–15 years, when the tree canopy closes. The peat surface becomes so dry that even if the plantation subsequently fails, the bog does not re-establish but instead there is a wet heath of *Calluna vulgaris* (heather), *Trichophorum cespitosum* (deergrass), *Molinia caerulea* (purple moor-grass) and *Eriophorum vaginatum* (hare's-tail cotton-grass). Weedy species such as *Chamerion angustifolium* (rosebay willow herb) can invade. Normally this community only has a temporary life because replanting takes place. Now, however, ploughing is difficult because of the old tree roots which are uneconomical to remove. New planting has to be carried out on the raised areas around old trunks.

Picea sitchensis (Sitka spruce), first introduced by David Douglas from north-west America in 1831, has become the main tree of blanket peat throughout the land. By 1965, 600 000 acres had been planted and planting was continuing at a rate of 30 million/year.

The Forestry Commission forests covered 1.75 million acres by 1970, comprising 36% *Picea sitchensis* (Sitka spruce) and 18% *Pinus sylvestris* (Scots pine). There has been a dramatic loss of blanket bog and wet heath. By 1987, 17% (67 000 ha) of the Flow lands of Caithness and Sutherland had been planted.

Much has been written or said about the monotonous forestry plantations, but in fact the Forestry Commission latterly took great pains to ameliorate the countryside in its care and minimize the landscape impact of forestry activities. Trees were planted in great, natural-looking sweeps, rather than straight rows. The demise of the Forestry

Commission will be greatly regretted as the monotonous rows of private forestry are seen more often. Forestry plantations have brought new habitats to marginal areas. They include the digging of drainage channels, cultivation by deep-rip ploughs 10–40 cm deep, to break up the podzol, vegetation control with tools, machinery and chemicals, fertilizers especially to remedy phosphate deficiency, fencing to control wildlife and the construction of roads. Places for a diverse flora have been created at wood margins, at fire breaks and along roads. Part of the poor perception of conifer plantations as lacking diversity arises only because they are young habitats. In time they may gain a more diverse flora. The flip side is that our unique blanket-bog lands have been eroded by forestry.

3.7.5 THE LORDSHIP OF THE EYE

Already marked out in the nineteenth century, there was a richer potential future for the countryside. There has been a shift away from reliance on the countryside for food and for survival, so that the countryside has come to be regarded as 'heritage' and as an amenity resource. And already our flora has been transformed, enriched, by the introduction of exotic species.

The Forestry Commission and the various National Parks and Trusts have been at the forefront in opening up the countryside for amenity use, providing car parks and pathways in areas of great natural beauty. At best they have been the worthy inheritor of the tradition of landscape management, where not just commercial interest but amenity and aesthetic interest have been combined. This tradition has at its heart a view of an idealized countryside. Perhaps this view dates back to the recreational use of the countryside for hunting or for the provision of venison.

In the Middle Ages, parks were constructed to hold deer. The earliest park is dated at 1045 at Ongar in Essex. In the Domesday Book of 1086, 35 parks are recorded. By 1300 Rackham has estimated that there were 3200 parks in England (Rackham, 1976). The average size was about 100 ha. The deer were fenced in with a smooth, rounded or polygonal fence to maximize area and minimize fencing costs. The interior was sometimes compartmented into grazing and woodland areas. Venison was an important supply of meat economically but, particularly, socially. Rather like the grouse moors today, forests provided a setting for social gatherings and meat to be given out of grace or favour. Overlooking Chingford Plain in Epping Forest a grandstand was constructed from which Queen Elizabeth I could watch the kill. In the sixteenth and seventeenth centuries the use of parks declined and some reverted to woodland or were retained only as general pasture.

Nevertheless, a cult of trees was not far from the surface. Trees were associated with national strength and defence. Evelyn, in his best-selling book called *Silva*, a report on tree planting to the Royal Society, noted the loss of yew, 'Since the use of bows is laid aside amongst us the propagation of the yew-tree is quite forborn'. He praised oak because 'ships of oak become our wooden walls'. Evelyn was a champion of native trees, not just oak but also elm, beech, ash, chestnut and walnut, the last two actually being introductions. Evelyn championed the planting of trees for their uses and their aesthetic beauty.

Towards the end of the seventeenth century and in the eighteenth century there was a change of sensibility (Hadfield, 1985). The clipped, geometrically shaped symmetrical gardens of the Restoration with their symmetric parterres gave way to rococo asymmetry, which at its margins blurred the distinction between garden and nature, and then there was the exuberance, the wildness of Baroque. Nature and the countryside were colonized again, but now by the intellect. Just as towns and cities were starting to burgeon like great excrescences, gardens were now to be constructed to recall an Arcadian landscape when life was pure and simple. They were to be artfully constructed to present a picture of an idyllic landscape.

The landscaped parks of William Kent, Charles Bridgeman, Lancelot 'Capability' Brown and Humphry Repton marked the final phase of taking possession of the landscape. Now even the wild or seminatural was to become reformed in an ideal image as part of the estate of the landowner. Landscaping was as much part of the idea of improvement as the agricultural developments of the eighteenth century. Jane Austen applied her caustic wit against the improvers in *Mansfield Park*. 'Improvement' marked just the latest stage in the appropriation of the countryside by a privileged few.

Improvement could be theatrical and painterly as at Stourhead, involving large-scale works like the diverting of rivers and streams, the damming of lakes and the moving of mature trees (Figure 3.40).

Gardening on the grand scale can be seen at Blenheim and many other grand country houses. It might also include more subtle changes of landscape. Views were very important. Through the views the park seemed to include the countryside around. This could include, not just scenes of wilderness, with natural features incorporated or enhanced, but the productive countryside of fields and woods. Ha-has, hidden boundary banks and ditches, were constructed so that the riff-raff were excluded but the lordship of the eye was unimpeded.

In part, agricultural development funded the development of landscaped parks but, in its turn, the timber of the landscape parks was an important source of income.

Figure 3.40 Stourhead, Wiltshire, the idealized countryside.

Humphry Repton railed against the *nouveau riche* who only had a commercial interest, but in time the improvement came to be seen more pragmatically as necessary investment for commercial return. There had always been some element of pragmatism. In 1713 Joseph Addison wrote in *The Spectator* (quoted in Daniels and Seymour, 1990),

> why may not a whole Estate be thrown into a kind of Garden by frequent Plantations, that may turn as much to profit as to the Pleasure of the Owner? A Marsh overgrown with Willows, or a Mountain shaded with Oaks, are not only more beautiful, but more beneficial, than when they lie bare and unadorned. Fields of Corn make a pleasant prospect, and if the walks a little taken care of that lie between them, if the natural Embroidery of the Meadows were helped and improved by some small Additions of Art, and the several Rows of Hedges set off by Trees and Flowers, that the Soil was capable of receiving, a man might make a pretty Landskip of his own Possessions.

Grand landscape design was in decline by the beginning of Victoria's reign, but by this time the Romantic movement had taken psychological possession of the landscape and its flora. The countryside and nature became now as much a mental landscape as a real one. At its most shallow, this was expressed as a search, not just for beauty, but for the 'picturesque', a scene capable of being painted. This was a way in which beauty could be circumscribed, described, possessed. It has provided the vocabulary of our appreciation of the countryside and nature. Jane Austen poked fun at the cult of the 'picturesque' in her novel *Sense and Sensibility*, published in 1797. Edward, describing a walk to Marianne, says

> You must not inquire too far, Marianne – remember, I have no knowledge in the picturesque, and I shall offend you by my ignorance and want of taste, if we come to particulars. I shall call hills steep which ought to be bold; surfaces strange and uncouth, which ought to be irregular and rugged; distant objects out of sight, which ought only to be indistinct through the soft medium of a hazy atmosphere.

However, in the works of the romantic poets something more sophisticated and much more important was created, an emotional relationship with nature. For Wordsworth 'Nature' became a medium through which the most profound, even religious, thoughts were conveyed. In his long autobiographical poem, published after his death, called *Prelude*, he writes of his childhood

> . . . even then I felt
> Gleams like the flashing of a shield; – the earth
> And common face of Nature spake to me
> Rememberable things.

Nature could speak through the medium of a 'rugged' landscape, an individual primrose growing from a rock-face, or the wind blowing through some trees.

> . . . or haply, at noon-day,
> While in a grove I walk, whose lofty trees,
> Laden with summer's thickest foliage, rock
> In a strong wind, some working of the spirit,
> Some inward agitations thence are brought.

The romantic poets were writing in the context of the beginnings of industrialization and a world which seemed more and more mechanistic. Their use of the countryside as inspiration is so much more important to us today, so many of us inhabit a completely built environment, in a godless, mechanistic universe. It is so much more urgent for us to protect this alternative landscape of trees and flowers and rocks and rivers, an other-world which is closer to the infinite and to the mystic. The need for an alternative 'landscape' takes strange forms, like the flowering of the hippie 'New Age' counter-culture, which uses nature to commune with otherness. The need for a countryside which we can access is strong. Rambling is perhaps the biggest leisure activity in Britain. Today the spiritual and recreational value of the countryside rivals its economical and commercial importance.

3.7.6 THE URBAN ENVIRONMENT

The built environment covers perhaps 8% of Britain. It is a very varied environment, with numerous different kinds of habitats. There are walls, and pavements, roadsides and roofs. There are public open spaces and allotments. We have much to thank the Victorian and Edwardian sense of public spiritedness and municipal pride for. It has left us a rich legacy of parks and squares. How astonishing it is that Glasgow can truly call itself a 'Dear Green Place' because of its wealth of parks! Sadly, in so many towns and cities today, parks and public gardens are cared for in the most minimal way, because of cut-backs in public spending. City parks can provide a place for many wild plants. Over 300 species were found in the 250 acres of Hyde Park and Kensington Gardens in central London in the period 1958–62 (Allen, 1965).

Most important in the urban environment is the rich patchwork of gardens. The landscape, nature-loving and collecting traditions of earlier centuries have been converted into the creation of 'gardenesque' artificial environments rich in plant species. Some of the gardens are nearly deserts, clipped and pruned into disciplined shape, but so many are semiwild. They have that kind of mixed management, intense here, less intense there, which has over centuries created such diversity in the countryside; cutting, pruning, planting and weeding keep back the rampant

competitors so that diversity is maintained.

Towns and cities provide many transient habitats. There is a strong comparison between the ruderal communities of the town and of arable fields. In the city they face physical disturbance, pollution and the difficulty of colonizing sites isolated by the desert of concrete, brick and tarmac. The level of pollution in the city has declined markedly since the clean air acts of 1956 and 1968. Fallout of soot and other dust in central Birmingham in the 1960s was over 75 tonnes/km^2. Deciduous trees and evergreens with shiny leaves from which dust washes off survive the best. *Ligustrum ovalifolium* (garden privet) becomes deciduous where fallout is greater than about 45 tonnes/km^2. The fallout of particulate matter has declined but there are still intermittent very high levels of atmospheric pollutants from traffic.

The survey for the *Flora of the London Area*, mostly carried out in the 1960s and 1970s, discovered over 2000 species growing wild in an area circumscribed by a 20 mile radius around Saint Paul's Cathedral (Burton, 1983). Although this includes some rural areas on the fringes of metropolitan London, a glance at the maps in the flora shows how many species have been found right in the centre of London, albeit transiently. Some of the profusion relates to the sudden provision of open sites because of bomb damage in the war. Between 1939 and 1953 the City of London itself gained 269 species of plant. With office development such sites are now rare.

However, the list of regular urban inhabitants is long (Haigh, 1980): *Lobularia maritima* (sweet alison) and *Lobelia erinus* (garden lobelia) in pavement cracks. *Antirrhinum majus* (snapdragon), *Erysimum cheiri* (wallflower), *Parietaria judaica* (pellitory-of-the-wall), *Cymbalaria muralis* (ivy-leaved toadflax), *Campanula poscharskyana* (trailing bellflower) and *Dryopteris filix-mas* (male fern), to name just a few, on walls. *Lycopersicon esculentum* (tomato) seedlings growing up beside the benches where people have had their picnic lunch. *Ficus carica* (fig) growing out of the litter at the edges of car parks. *Senecio vulgaris* (groundsel) and *Poa annua* (annual meadow-grass) in the dirt that collects in any odd corner. *Mercurialis annua* (annual mercury) in the untidy front patch of many terraced-houses. And then there is the glory of the Butterfly-bush, *Buddleia davidii*, growing out of walls and on derelict sites. And the various species of *Oxalis* (pink and yellow sorrel) on open ground and *Diplotaxis* spp. (wall rocket), *Tussilago farfara* (colt's-foot), *Rumex crispus* (curled dock) and *Artemisia* spp. (mugworts), as well as the ubiquitous *Epilobium* spp. (willow herbs) and *Senecio* spp. (ragworts), on bits of waste ground. Then there are the mosses, such as *Tortula muralis*, *Barbula unguiculata*, *Encalypta streptocarpa* on walls, *Bryum argenteum* in pavement cracks and at the junctions between wall and pavement, and the liver-wort, *Marchantia*, in gutters and the lichens on roofs (Seaward, 1979).

3.7.7 THE PUBLIC LANDSCAPE AND ITS FLORA

The development of landscaped parks can be seen as part of the long history of the transfer of ownership of our countryside from the many to the few; over the millennia the public countryside has declined, fenced away from the public. This process started with the marking out of the first fields, and continued with the rise of lordship. Then there was the enclosure of shared open fields and the fencing of common land. The process continues today with the wilful or neglectful decline of that last remnant of a countryside owned by everyone, the public footpath system. The gradual erosion of the network of public footpaths through the countryside is a disgrace. At the same time the public use of surviving common land has changed dramatically in recent decades, to its detri-

ment. Upland commons have declined in quality because of overstocking and a change from cattle grazing to sheep grazing. Coarse grasses, which the sheep dislike, and which out-compete more slow-growing flowers, have been encouraged. Bracken, once trampled by cattle, has become more wide-spread. In lowland commons the lack of grazing has led to a decline in quality. Coarse, tall grasses and scrub or birch wood-land have invaded many commons. Recreational use does not provide a means of maintaining the existing vegetation. Management has been evaded with the cost of a decline in diversity and beauty.

Our island is too small to allow the survival of wilderness. It is possible to drive from one end to the other, east–west or north–south in a day, and it scarcely takes longer than half a day to reach any area off the road by walking. It is essential that there is active management of the countryside to maintain its beauty. By happy accident patterns of land use have created an extraordinarily diverse landscape. We are the lucky inheritors of that landscape. We have not had to pay for the beauty we now enjoy, but we must start to pay now.

How precious and fragile our natural heritage is, has been pointed out in recent summers by the complete loss of river systems, piped away for tap water. This outrage has been perpetrated in our name by the water companies, and barely a squeak has been raised in protest. Every year yet another battle is lost to the developers and road builders. Twyford Down lost this year, and although Oxleas Wood seems saved, where will the next threat be? But we all use these improved roads to visit the country-side. If an ancient part of our countryside is to be damaged out of recognition for our convenience, we must extract another price, be willing to pay a tax on that convenience so that improvement in a sympathetic guise can take place. A convenience tax would probably be a pittance in comparison to what it costs to build a road, but a convenience tax could pay for the purchase of a deciduous woodland, for example, and its 'improvement' by re-establishing coppicing or grazing in perpetuity, to encourage diversity and beauty. Our wildlife trusts must be funded on the same scale as institutions like the Royal Opera House. Our natural heritage is more a part of our culture than any alien opera and it is enjoyed by more people. Beside the countryside, a night at the opera is mere froth. Our countryside bears the marks of our history and it is as fragile as any oil-painting.

We must reverse the trend for the appropriation of the countryside by the few. Paradoxically, the car has done more for that cause than anything else. But the system of public footpaths must be maintained and even extended to spread the public pressure on the countryside. It is wrong that many major waterways and moorlands are off limits. For example the bank of the Thames is one of the very few seminatural habitats in London and yet the public does not own the river bank. It is disgraceful that it is impossible to walk from Westminster to Hammersmith on the bank of the Thames except by detours through built-up areas.

At present (1994) at least 10% of farm income comes from farm subsidies and this is set to rise to 30%, and that is ignoring the costs of a system that maintains artificially high prices. This investment in the countryside by the public surely gives the public a right to make demands of the farming community. Paying for 15% of arable land to be set aside must not be allowed to become a folly, but should be a positive gain for our flora. Less intensive farming systems, with periods of regular fallow, with less fertilizer input and with the maintenance of meadowland, must be financially supported. The amenity value of the countryside must be financially supported. There are great dangers to areas of 'natural' beauty. The uplands and seminatural grasslands in the lowlands were intensively exploited in 1970s

and 1980s. Drainage and reseeding have reduced the floristic value of many areas.

There are many problems in the repossession of the countryside by the public. There are just too many of us. One of the most serious threats to areas of outstanding beauty is the environmental pressure on the amenity areas. Many of our hills and mountains are scarred by the ever-widening paths taken by climbers. It is little realized by many fell-walkers how fragile the peat surface is and how easily irreversible erosion can set in. We need proper paths up our hills so that damage is restricted to the smallest possible area. But we must be careful that we do not do too much. Our coastal flora is under greater threat from efforts to protect the coast from erosion and weathering. The building of visitor centres, tea huts and all, should be resisted.

A few areas may be too precious to allow unimpeded access, but the removal of car parks, making a long walk necessary to visit them will protect many areas. If access to particularly pressured sites has to be rationed, it should not be by cost. Having to apply for a parking ticket before being allowed to enter Lakeland may become necessary, but the cost of a parking ticket must not be so expensive as to restrict access to the rich.

So much has been achieved already. We must be thankful for the establishment of our national parks and the large protected areas owned by the national trusts and wildlife charities. The National Trust now owns 512 km of the coast. Designation as a Site of Special Scientific Interest has its value, but it must not lull us into a false sense of security; 49% of the coast is designated in one way or another as worthy of protection, but how much is actually safe?

What happened to the Broads is a cautionary example (Moss, 1987). From the 1940s more and more people took boating holidays on the Broads. Fringing reed beds, damaged by coypu, did not regenerate and so there was little to protect the banks from erosion due to bow waves of boats. Expensive, ugly piling had to be constructed to protect the banks. This, combined with eutrophication, has reduced the value and the beauty of the Broads. They must not be allowed to degenerate further into a series of boating lakes which might have been constructed anywhere.

Our countryside is a strange paradox. So much of it, like the Broads, is unnatural, and yet so much of it is so beautiful. Our countryside and its flora is a document, a text recording our past, and yet it must change for our future. Our wild lands are a mirage. Our wildernesses are as much psychological as real. One thing is sure, to preserve their beauty they cannot be left alone. Our natural flora must be subtly managed, cultivated, gardened. The countryside is wild but only within boundaries set by man. One of the greatest gifts to the culture of the world from these islands was the development of landscape gardens in the eighteenth century. It is time to renew our acquaintance with the principles of the landscape gardeners, and apply them to our whole countryside and its flora. As Pope put it in his poem on landscape design, 'Of Taste', published in 1731, 'Consult the genius of the place in all'.

REFERENCES

Aaby, B. (1976) Cyclic climatic variations in climate over the past 5,500 years reflected in raised bogs. *Nature*, **263**, 281–4.

Aalen, F.H.A. (1978) *Man and Landscape in Ireland*, Academic Press, London.

Adam, P. (1978) Geographical variation in British saltmarsh vegetation. *Journal of Ecology*, **66**, 339–66.

Allen, J.E. (1965) The flora of Hyde Park and Kensington Gardens, 1958–1962. *Proceedings of the Botanical Society of the British Isles*, **6**, 1–20.

Allison, J., Godwin, H. and Warren, S.H. (1952) Late-Glacial deposits at Nazeing in the Lea Valley, North London. *Philosophical Transactions of the Royal Society of London, B*, **236**, 169.

Alvin, K.L. (1974) Leaf anatomy of *Weichselia* based on fusainized material. *Palaeontology*, **17**, 587–98.

Alvin, K.L. (1982) Cheirolepidaceae: biology, structure and palaeoecology. *Review of Palaeobotany and Palynology*, **37**, 71–98.

Alvin, K.L. (1983) Reconstruction of a Lower Cretaceous conifer. *Botanical Journal of the Linnean Society*, **86**, 169–76.

Andersen, S.T. (1970) The relative pollen productivity and pollen representation of North European trees and correction factors for tree pollen spectra. *Danmarks Geologiske Undersogelse, Serie B*, **96**, 1–99.

Anderton, R., Bridges, P.H., Leeder, M.R. and Sellwood, B.W. (1979) *A Dynamic Stratigraphy of the British Isles*, Chapman & Hall, London.

Andrews, H.N. (1948) Fossil tree ferns of Idaho. *Archaeology*, **1**, 190–5.

Andrews, H.N. (1960) Notes on Belgium specimens of *Sporogonites*. *Palaeobotanist*, **7**, 85–9.

Andrews, H.N. (1961) *Studies in Palaeobotany*, Wiley, New York.

Andrews, J.T. and Tedesco, K. (1992) Detrital carbonate rich sediments, northwestern Labrador-Sea. Implications for ice-sheet dynamics and iceberg rafting (Heinrich) events in the North Atlantic. *Geology*, **20**, 1087–90.

ApSimon, A.M. (1976) Ballynagilly and the beginning and end of the Irish Neolithic, in *Acculturation and Continuity in Atlantic Europe*, (ed. S.J. de Laet), Quaternary Research Association, Bristol, pp. 15–30.

Arber, E.A.N. and Goode, R.H. (1915) On some fossil plants from the Devonian rocks of north Devon. *Proceedings of the Cambridge Philosophical Society for Biological Sciences*, **18**, 89–104.

Ashbee, P., Smith, I.F., and Evans, J.G. (1979) Excavation of three long barrows near Avebury, Wiltshire. *Proceedings of the Prehistory Society*, **45**, 207–300.

Askew, G.P., Payton, R.W. and Shiel, R.S. (1985) Upland soils and land clearance in Britain during the second millenium BC, in *Upland Settlement in Britain. The Second Millenium BC and After*, (eds D. Spratt and C. Burgess), BAR series 143, BAR, Oxford, pp. 5–27.

Avery, B.W. (1990) *Soils of the British Isles*, C.A.B. International, Wallingford, Oxon.

Baker, A.R.H. (1976) Changes in the later Middle Ages, in *A New Historical Geography of England before 1600*, (ed. H.C. Darby), Cambridge University Press, Cambridge, pp. 186–247.

Baker, C.A., Moxey, P.A. and Oxford, P.M. (1978) Woodland continuity and change in Epping Forest. *Field Studies*, **4**, 645–69.

Baker, H. (1937) Alluvial meadows: a comparative study of grazed and mown meadows. *Journal of Ecology*, **25**, 408–20.

Barber, K.E., Dumayne, L. and Stoneman, R. (1993) Climatic change and human impact during the late Holocene in northern Britain, in *Climate Change and Human Impact on the Landscape*, (ed. F.M. Chambers), Chapman & Hall, London, pp. 225–36.

Barnowsky, A.D. (1986) 'Big Game' extinction caused by Late Pleistocene climatic change. Irish Elk (*Megaloceros giganteus*) in Ireland. *Quaternary Research*, **25**, 128–35.

Barnowsky, C. (1988) A Late-glacial and Post-glacial pollen record from the Dingle Peninsula,

County Kerry. *Proceedings of the Royal Irish Academy*, **88B**, 23–37.

Barr, C.J., Bunce, R.G.H., Cummins, R.P. *et al.* (1992) Hedgerow changes in Great Britain, in *Annual Report 1991–2, Institute of Terrestrial Ecology*, HMSO for NERC, London, pp. 21–4.

Bassett, J.A. and Curtis, T.G.F. (1985) The nature and occurrence of sand-dune machair in Ireland. *Proceedings of the Royal Irish Academy*, **85B**, 1–20.

Bassett, M.G. (1984) Lower Palaeozoic of Wales – a review of studies in the past 25 years. *Proceedings of the Geologist's Association*, **95**, 291–311.

Bateman, R.M. and Rothwell, G.W. (1990) A reappraisal of the Dinantian floras at Oxroad Bay, East Lothian, Scotland. 1. Floristics and the development of whole-plant concepts. *Transactions of the Royal Society of Edinburgh: Earth Sciences*, **81**, 127–59.

Bateman, R.M. and Scott, A.C. (1990) A reappraisal of the Dinantian floras at Oxroad Bay, East Lothian, Scotland. 2. Volcanicity, palaeoenvironments and palaeoecology. *Transactions of the Royal Society of Edinburgh: Earth Sciences*, **81**, 161–94.

Bateman, R.M., DiMichele, W.A. and Willard, D.A. (1992) Experimental cladistic analysis of anatomically preserved arborescent lycopsids from the Carboniferous of Euramerica: an essay on palaeobotanical phylogenetics. *Annals of the Missouri Botanical Garden*, **79**, 500–59.

Battarbee, R.W., Appleby, P.G., Odell, K. and Flower, R.J. (1985) ^{210}Pb dating of Scottish lake sediments, afforestation and accelerated soil erosion. *Earth Surface Processes and Landforms*, **10**, 137–42.

Batten, D.J. (1974) Wealden palaeoecology from the distribution of plant fossils. *Proceedings of the Geological Association, London*, **85**, 43–58.

Beck, R.B, Funnell, B.M. and Lord, A.R. (1972) Correlation of Lower Pleistocene Crag at depth in Suffolk. *Geological Magazine*, **109**, 137–9.

Beddows, A.R. (1959) *Dactylis glomerata* L. Biological flora of the British Isles. *Journal of Ecology*, **47**, 223–239.

Beddows, A.R. (1961) *Holcus lanatus* L. Biological flora of the British Isles. *Journal of Ecology*, **49**, 421–30.

Beddows, A.R. (1973) *Lolium multiflorum* Lam. Biological flora of the British Isles. *Journal of Ecology*, **61**, 587–600.

Bell, J.N.B. and Tallis, J.H. (1973) *Empetrum nigrum*. Biological flora of the British Isles. *Journal of Ecology*, **61**, 289–305.

Bennett, K.D. (1983a) Devensian late-glacial and Flandrian vegetational history at Hockham Mere, Norfolk, England. I. Pollen percentages and concentrations. *New Phytologist*, **95**, 457–87.

Bennett, K.D. (1983b) Devensian late-glacial and Flandrian vegetational history at Hockham Mere, Norfolk, England. II. Pollen accumulation rates. *New Phytologist*, **95**, 489–504.

Bennett, K.D. (1986) Competitive interactions among forest tree populations in Norfolk, England, during the last 10,000 years. *New Phytologist*, **103**, 603–20.

Bennett, K.D. (1988) Holocene pollen stratigraphy of central East Anglia, England, and comparison of pollen zones across the British Isles. *New Phytologist*, **109**, 237–53.

Birkbeck, G. (ed.) (1950) *Boswell's Life of Johnson V. The tour to the Hebrides and Journey into North Wales*, (revised and enlarged by L.F. Powell), Clarendon Press, Oxford.

Birks, H.H. (1970) Studies in the vegetational history of Scotland. I A pollen diagram from Abernethy Forest, Invernesshire. *Journal of Ecology*, **58**, 827–46.

Birks. H.H. (1972a) Studies in the vegetational history of Scotland. III A radiocarbon dated pollen diagram from Loch Maree, Ross and Cromarty. *New Phytologist*, **71**, 731–54.

Birks, H.H. (1972b) Studies in the vegetational history of Scotland IV. Pine stumps in Scottish blanket peats. *Philosophical Transactions of the Royal Society of London, B*, **270**, 181–226.

Birks, H.J.B. (1973) *The Past and Present Vegetation of the Isle of Skye – a Palaeoecological Study*, Cambridge University Press, London.

Birks, H.J.B. (1986) Late Quaternary biotic changes in terrestrial and lacustrine environments, with particular reference to north-west Europe, in *Handbook of Holocene Palaeoecology and Palaeohydrology*, (ed. B.E. Berglund), John Wiley and Sons, New York, pp. 3–65.

Birks, H.J.B., Deacon, J. and Peglar, S. (1975) Pollen maps for the British Isles 5000 years ago. *Proceedings of the Royal Society of London, B*, **189**, 87–105.

Blackford, J. (1993) Peat bogs as sources of proxy climate data: past approaches and future research, in *Climate Change and Human Impact on the Landscape*, (ed. F.M. Chambers), Chapman & Hall, London, pp. 47–56.

Blackman, G.E. and Rutter, A.J. (1954). *Endymion non-scriptus* (L.) Garcke. Biological flora of the British Isles. *Journal of Ecology*, **42**, 629–38.

Bold, H.C., Alexopoulos, C.J. and Delevoryas, T.

(1980) *Morphology of Plants and Fungi*, Harper and Row, New York.

Bond, G. , Heinrich, H., Broecker, W. *et al.* (1992) Evidence for massive discharges of icebergs in the North Atlantic Ocean during the last glacial period. *Nature*, **360**, 245–9.

Bond, T.E.T. (1952) *Elymus arenarius* L. Biological flora of the British Isles. *Journal of Ecology*, **40**, 217–27.

Boulter, M.C. (1971) A palynological study of two Neogene plant beds in Derbyshire. *Bulletin of the British Museum of Natural History (Geology)*, **19**, 361–410.

Boulter, M.C. (1980) Irish Tertiary plant fossils in a European context. *Journal of Earth Sciences of the Royal Dublin Society*, **3**, 1–11.

Boulter, M.C. and Chaloner, W.G. (1970) Neogene fossil plants from Derbyshire (England). *Review of Palaeobotany and Palynology*, **10**, 61–78.

Boulter, M.C. and Craig, D.L. (1979) A middle Oligocene pollen and spore assemblage from the Bristol Channel. *Review of Palaeobotany and Palynology*, **28**, 259–72.

Boulter, M.C. and Kvaček, Z. (1990) *The Palaeocene flora of the Isle of Mull*. Special Papers in Palaeontology No. 42, Palaeontological Association, London.

Boulter, M.C. and Manum, S.B. (1989) Brito-Arctic Igneous Province Flora around the Paleocene/ Eocene boundary. *Proceedings of the Ocean Drilling Programme, Scientific Results*, **104**, 663–80.

Bowen, D.Q. and Sykes, G.A. (1988) Correlation of marine events and glaciations on the northeast Atlantic margin. *Philosophical Transactions of the Royal Society of London, B*, **318**, 619–35.

Bowen, D.Q., Rose, J., McCabe, A.M. and Sutherland, D.G. (1986) Correlation of Quaternary glaciations in England, Ireland, Scotland and Wales. *Quaternary Science Reviews*, **5**, 299–341.

Bowen, H.J.M. (1965) Sulphur and the distribution of British Plants. *Watsonia*, **6**, 114–19.

Bradshaw, A.D. and McNeilly, T. (1981) *Evolution and Pollution*, Edward Arnold, London.

Bradshaw, R. (1993) Forest response to Holocene climatic change: equilibrium or non-equilibrium, in *Climate Change and Human Impact on the Landscape*, (ed. F.M. Chambers), Chapman & Hall, London, pp. 57–65.

Bradshaw, R. and Browne, P. (1987) Changing patterns in the post-glacial distribution of *Pinus sylvestris* in Ireland. *Journal of Biogeography*, **14**, 237–48.

British Museum (Natural History) (1975) *British Fossils*, Trustees of the British Museum (Natural History).

Brown, A.P. (1977) Late Devensian and Flandrian vegetational history of Bodmin moor, Cornwall. *Philosophical Transactions of the Royal Society of London, B*, **276**, 251–320.

Browning, E.A. (ed.) (1953) *English Historical Documents*, Eyre Methuen/Oxford University Press, London, Vol. 8, pp. 1660–714.

Bunce, R.G.H. and Jeffers, J.N.R. (1977) *Native Pinewoods of Scotland*, Institute of Terrestrial Ecology, Cambridge.

Burdon, J.J. (1983) *Trifolium repens* L. Biological flora of the British Isles. *Journal of Ecology*, **71**, 307–30.

Burgess, C. (1985) Population, climate and upland settlement, in *Upland Settlement in Britain. The Second Millenium BC and after*, (eds D. Spratt and C. Burgess), BAR series 143, BAR, Oxford, pp. 195–215.

Burgess, N.D. and Edwards, D. (1988) A new Palaeozoic plant closely allied to *Protoaxites* Dawson. *Botanical Journal of the Linnean Society*, **97**, 189–203.

Burrows, C.J. (1990) *Processes of Vegetation Change*, Unwin Hyman, London.

Burton, R.M. (1983) *Flora of the London Area*. London Natural History Society.

Byfield, A. (1991) The Lizard Peninsula. *British Wildlife*, **3**, 92–105.

Cahn, M.A. and Harper, J.L. (1976) The biology of leaf mark polymorphism in *Trifolium repens* L. II. Evidence for the selection of leaf marks by rumen fistulated sheep. *Heredity*, **37**, 327–33.

Campbell, B.M.S. (1990) People and land in the Middle Ages, 1066–500, in *An Historical Geography of England and Wales*, 2nd edn, (eds R.A. Dodgshon and R.A. Butlin), Academic Press, London, pp. 69–113.

Campbell, J.B. (1977) *The Upper Palaeolithic of Britain: A Study of Man and Nature in the Late Ice Age*, Clarendon Press, Oxford.

Caseldine, C. and Hatton, J. (1993) The development of high moorland on Dartmoor: fire and the influence of Mesolithic activity on vegetation change, in *Climate Change and Human Impact on the Landscape*, (ed. F.M. Chambers), Chapman & Hall, London, pp. 119–32.

Caseldine, C.J. and Maguire, D.J. (1986) Late glacial/early Flandrian vegetation change on northern Dartmoor, South West England. *Journal of Biogeography*, **13**, 255–64.

Chaloner, W.G. (1962) Rhaeto-Liassic plants from

the Henfield Borehole. *Bulletin of the Geological Survey Great Britain*, **19**, 16–28.

Chaloner, W.G. (1968) The cone of *Cyclostigma kiltorkense* Haughton, from the Upper Devonian of Ireland. *Journal of the Linnean Society (Botany)*, **61**, 25–36.

Chaloner, W.G. (1970) The rise of the first land plants. *Biological Reviews*, **45**, 353–77.

Chaloner, W.G. (1972) Devonian plants from Fair Isle, Scotland. *Review of Palaeobotany and Palynology*, **14**, 49–61.

Chaloner, W.G. and Collinson, M.E. (1975) An illustrated key to the commoner British Upper Carboniferous plant compression fossils. *Proceedings of the Geologist's Association*, **86**, 1–44.

Chaloner, W.G. and MacDonald, P. (1980) *Plants Invade the Land*, HMSO/Royal Scottish Museum.

Chaloner, W.G., Hill, A.J. and Lacey, W.S. (1976) First Devonian platyspermic seed and its implications for gymnospermous evolution. *Nature*, **265**, 233–5.

Chambers, F.M. (1988) Archaeology and flora of the British Isles: The moorland experience, in *Archaeology and the Flora of the British Isles*, (ed. M. Jones), Oxford University Press, Oxford, pp. 107–15.

Chandler, M.E.J. (1957) The Oligocene flora of the Bovey Tracey Lake Basin, Devonshire. *Bulletin of the British Museum Natural History (Geology)*, **3**, 71–123.

Chandler, M.E.J. (1961) *The Lower Tertiary Floras of Southern England, I. Palaeocene Floras. London Clay Flora (Supplement)*, British Museum (Natural History), London.

Chandler, M.E.J. (1964) *The Lower Tertiary Floras of Southern England. IV. A Summary and Survey of Findings in the Light of Recent Botanical Observations*, British Museum (Natural History), London.

Chandler, T.J. and Gregory, S. (1976) *The Climate of the British Isles*, Longman, London.

Chapman, V.J. (1947a) *Suaeda maritima* (L.) Dum. Biological flora of the British Isles. *Journal of Ecology*, **35**, 293–302.

Chapman, V.J. (1947b) *Suaeda fruticosa* Forsk. *Journal of Ecology*, **35**, 303–11.

Chapman, V.J. (1950) *Halimione portulacoides* (L.) Aell. Biological Flora of the British Isles. *Journal of Ecology*, **38**, 214–22.

Chapman, V.J. (1976) *Coastal vegetation*, Pergamon Press.

Clapham, A.R. (ed.) (1978) *Upper Teesdale, the Area and its Natural History*, Collins, London.

Clark, J.G.D. (1954) *Excavations at Star Carr*, Cambridge University Press, Cambridge.

Clark, J.G.D. and Godwin, H. (1962) The Neolithic in the Cambridgeshire Fens. *Antiquity*, **36**, 10.

Clayton, K. (1981) Explanatory description of the landforms of the Malham area. *Field Studies*, **5**, 389–423.

Cleal, C.J. (1988) British palaeobotanical sites, in *The Use and Conservation of Palaeontological Sites*, Special Papers in Palaeontology no. 40, (ed. P.R. Crowther and W.A. Wimbledon), Palaeontological Association, London, pp. 57–71.

Cloud, P. (1976) Beginnings of biospheric evolution and their biogeochemical consequences. *Paleobiology*, **2**, 351–87.

Clymo, R.S. and Hayward, P.M. (1982) The ecology of *Sphagnum*, in *Bryophyte Ecology*, (ed. A.J.E. Smith), Chapman & Hall, London, pp. 229–89.

Coles, B. and Coles, B. (1986) *Sweet Track to Glastonbury*, Thames and Hudson, London.

Collinson, M.E. (1983) *Fossil Plants of the London Clay*, Palaeontological Association, Field Guide to Fossils, No. 1.

Collinson, M.E., Fowler, K. and Boulter, M.C. (1981) Floristic changes indicate a cooling climate in the Eocene of southern England. *Nature*, **291**, 315–17.

Conway, V.M. (1942) *Cladium* P.Br. Biological flora of the British Isles. *Journal of Ecology*, **30**, 211–16.

Coombe, D.E. (1957) *Impatiens parviflora* DC. Biological flora of the British Isles. *Journal of Ecology*, **45**, 701–13.

Coope, G.R. (1977) Fossil coleopteran assemblages as sensitive indicators of climatic changes during the Devensian (last) cold stage. *Philosophical Transactions of the Royal Society of London*, B, **280**, 313–40.

Coope, G.R., Shotton, F.W. and Strachan, I. (1961) A late Pleistocene fauna and flora from Upton Warren, Worcs. *Philosophical Transactions of the Royal Society of London*, B, **165**, 389.

Cornet, B. and Habib, D. (1992) Angiosperm-like pollen from the ammonite-dated Oxfordian (Upper Jurassic) of France. *Review of Palaeobotany and Palynology*, **71**, 269–94.

Couper, R.A. (1958) Upper Mesozoic and Cainozoic spores and pollen grains from New Zealand. *Palaeontological Bulletin, Wellington*, **32**, 1–77.

Courtney, R. (1992) No global warming. *New Scientist*, **134**, 52.

Cowlishaw, S.J. and Alder, F.E. (1960) The

grazing preferences of cattle and sheep. *Journal of Agricultural Science*, **54**, 157–65.

Cox, H.M.M. (1954) The fossil plants of the Permian beds of England. *Congrèsse International Botanique 8ème, Paris*, Section 5, pp. 172–4.

Coxon, P. (1985) A Hoxnian interglacial site at Athelington, Suffolk. *New Phytologist*, **99**, 611–21.

Coxon, P. and Flegg, A. (1985) A Middle Pleistocene Interglacial deposit from Ballyline Co. Kilkenny. *Proceedings of the Royal Irish Academy*, **85**, 107–20.

Coxon, P. and Flegg, A.M. (1987) A Late Pleistocene deposit at Poulnahallia, near Headford, County Galway. *Proceedings of the Royal Irish Academy*, **87**, 15–42.

Crackles, E. (1990) *Flora of the East Riding of Yorkshire*, Hull University Press/Humberside County Council.

Crane, P.R. (1981) Betulaceous leaves and fruits from the British Upper Palaeocene. *Botanical Journal of the Linnean Society*, **83**, 103–36.

Crane, P.R. (1988) *Abelia*-like fruits from the Palaeogene of Scotland and North America. *Tertiary Research*, **9**, 21–30.

Crane, P.R. and Lidgard, S. (1990) Angiosperm radiation and patterns of Cretaceous palynological diversity, in *Major Evolutionary Radiations*, (eds P.D. Taylor and G.P. Larwood), Systematics Association Special Volume No. 42, Clarendon Press, Oxford, pp. 377–407.

Crane, P.R., Friis, E.M. and Pedersen, K.R. (1986) Lower Cretaceous angiosperm flowers: fossil evidence on early radiation of dicotyledons. *Science*, **232**, 852–4.

Crawford, R.M.M. (1989) *Studies in Plant Survival*, Blackwell Scientific Publications, Oxford.

Crawford, T.J. and Jones, D.A. (1986) Variation in the colour of the keel petals in *Lotus corniculatus* L., 2. Clines in Yorkshire and adjacent counties. *Watsonia*, **16**, 15–19.

Croft, W.N. and Lang, W.H. (1946) The Lower Devonian flora of the Senni Beds of Monmouthshire and Breconshire. *Philosophical Transactions of the Royal Society of London, B*, **231**, 131–63.

Crookall, R.M. (1929) *Coal Measure Plants*, Edward Arnold, London.

Crookall, R.M. (1955) Fossil plants of the Carboniferous rocks of Great Britain. *Memoirs of the Geological Survey of Great Britain, Palaeontology*, **IV**, 1–3.

Cross, J.R. (1975) *Rhododendron ponticum*. Biological flora of the British Isles. *Journal of Ecology*, **63**, 345–64.

Daley, B. (1972) Some problems of the early Tertiary climate of Southern Britain. *Palaeogeography, Palaeoclimatology and Palaeoecology*, **11**, 177–90.

Daniels, R.E. (1978) Floristic analyses of British mires and more communities. *Journal of Ecology*, **66**, 773–802.

Daniels, S. and Seymour, J. (1990) Landscape design and the idea of improvement, 1730–1900, in *An Historical Geography of England and Wales*, 2nd edn, (eds R.A. Dodgshon and R.A. Butlin), Academic Press, London, pp. 487–520.

Dansgaard, W., White, J.W.C. and Johnsen, S.J. (1989) The abrupt termination of the Younger Dryas climate event. *Nature*, **339**, 532–4.

Darby, H.C. (1956) *The Draining of the Fens*. Cambridge University Press, Cambridge.

Darby, H.C. (1976) Domesday England, in *A New Historical Geography of England before 1600*, (ed. H.C. Darby), Cambridge University Press, Cambridge, pp. 39–74.

Day, S.P. (1991) Postglacial vegetational history of the Oxford region. *New Phytologist*, **119**, 445–70.

Deacon, J. (1974) The location of refugia of *Corylus avellana* L. during the Weichselian Glaciation. *New Phytologist*, **73**, 1055–63.

DeBeer, E.S. (1955) *The Diary of John Evelyn*. Clarendon Press, Oxford.

Defoe, D. (1971) *A Tour Through the Whole Island of Great Britain*, (first published 1706), Penguin, London.

Delcourt, H.R. and Delcourt, P.A. (1991) *Quaternary Ecology, A Paleoecological Perspective*, Chapman & Hall, London.

Devoy, R.J. (1980) Postglacial environmental change and man in the Thames estuary: a synopsis, in *Archaeology and Coastal Change*, (ed. F.H. Thompson), Society of Antiquaries, London, Occasional Paper 1, pp. 134–48.

Devoy, R.J. (1985) The problem of late Quaternary land bridges between Britain and Ireland. *Quaternary Science Reviews*, **4**, 43–58.

Dickinson, C.H., Pearson, M.C. and Webb, D.A. (1964) Some microhabitats of the Burren, their micro-environments and vegetation. *Proceedings of the Royal Irish Academy*, **63B**, 291–302.

Dickson, C.A., Dickson, J.H. and Mitchell, G.F. (1970) The Late-Weichselian flora of the Isle of Man. *Philosophical Transactions of the Royal Society of London, B*, **258**, 31–79.

Dimbleby, G.W. (1962) *The Development of British Heathlands and their Soils*, Clarendon Press, Oxford.

Dimbleby, G.W. (1965) Post-Glacial changes in soil

profiles. *Proceedings of the Royal Society of London, B*, **161**, 355–62.

Dimbleby, G.W. and Evans, J.G. (1974) Pollen and land snail analysis of calcareous soils. *Journal of the Archaeological Sciences*, **1**, 117–33.

Donkin, R.A. (1976) Changes in the early Middle Ages, in *A New Historical Geography of England before 1600*, (ed. H.C. Darby), Cambridge University Press, Cambridge, pp. 75–135.

Duffey, E. (1974) *Grassland Ecology and Wildlife Management*, Chapman & Hall, London.

Duigan, S.L. (1963) Pollen analysis of the Cromer Forest Bed series in East Anglia. *Philosophical Transactions of the Royal Society of London, B*, **246**, 149–202.

Edwards, D. (1968) A new plant from the Lower Old Red Sandstone of South Wales. *Palaeontology*, **11**, 683–90.

Edwards, D. (1970) Further observations on the Lower Devonian plant, *Gosslingia breconensis*, Heard. *Philosophical Transactions of the Royal Society of London, B*, **258**, 225–53.

Edwards, D. (1975) Some observations on the fertile parts of *Zosterophyllum myretonianum* Penhallow from the Lower Old Red Sandstone of Scotland. *Transactions of the Royal Society of Edinburgh*, **69**, 251–65.

Edwards, D. (1976) The systematic position of *Hicklingia edwardii* Kidston and Lang. *New Phytologist*, **76**, 173–81.

Edwards, D. (1982) Fragmentary non-vascular plant microfossils from the late Silurian of Wales. *Botanical Journal of the Linnean Society*, **84**, 223–56.

Edwards, D. and Davies, E.C. (1976) Oldest recorded *in situ* tracheids. *Nature*, **263**, 494–5.

Edwards, D. and Davies, M.S. (1990) Interpretations of early land plant radiations: 'facile adaptationist guesswork' or reasoned speculation? in *Major Evolutionary Radiations*, (eds P.D. Taylor and G.P. Larwood), Systematics Association Special Volume No. 42, Clarendon Press, Oxford, pp. 351–76.

Edwards, D. and Fanning, U. (1985) Evolution in the late Silurian–early Devonian: the rise of the pteridophytes. *Philosophical Transactions of the Royal Society of London, B*, **309**, 147–65.

Edwards, D. and Kenrick, P (1986) A new zosterophyll from Lower Devonian of Wales, *Botanical Journal of the Linnean Society*, **92**, 269–83.

Edwards, D. and Richardson, J.B. (1974) Lower Devonian (Dittonian) plants from the Welsh Borderland. *Palaeontology*, **17**, 223–56.

Edwards, K.J. (1993) Models of mid-Holocene forest farming for north-west Europe, in *Climate Change and Human Impact on the Landscape*, (ed. F.M. Chambers), Chapman & Hall, London, pp. 133–45.

Edwards, K.J. and Hirons, K.R. (1984) Cereal pollen grain in pre-elm decline deposits: implications for the earliest agriculture in Britain and Ireland. *Journal of Archaeological Science*, **11**, 71–8.

Edwards, K.J. and McIntosh, C.J. (1988) Improving the detection rate of cereal-type pollen grains in *Ulmus* decline and earlier deposits from Scotland. *Pollen and Spores*, **30**, 179–88.

Edwards, W. and Trotter, F.M. (1954) *British Regional Geography, The Pennines and Adjacent Areas*, Natural Environment Research Council, Institute of Geological Sciences, Geological Survey and Museum, HMSO, London.

Elkington, T.T. (1963) *Gentiana verna* L. Biological flora of the British Isles. *Journal of Ecology*, **51**, 755–67.

Elkington, T.T. (1971) *Dryas octopetala* L. Biological flora of the British Isles. *Journal of Ecology*, **59**, 887–905.

Elkington, T.T. and Woodell, S.R.J. (1963) *Potentilla fruticosa* L. Biological flora of the British Isles. *Journal of Ecology* **51**, 769–81.

Evans, J.G. (1975) *The Environment of Early Man in the British Isles*, Elek Books, London.

Evans, J.G. (1993) The influence of human communities on the English chalklands from the Mesolithic to the Iron Age: molluscan evidence, in *Climate Change and Human Impact on the Landscape*, (ed. F.M. Chambers), Chapman & Hall, London, pp. 147–56.

Evelyn, J. (1664) *Sylva, or a Discourse of Forest Trees, and their Propagation of Timber in His Majesties Dominions.* . . . J. Martyn and J. Allesby, London. (Facsimile reprint 1972, Scolar Press, Menston.)

Fanning, U. and Edwards, D. (1992) A diverse assemblage of early land plants from the Lower Devonian of the Welsh Borderland. *Botanical Journal of the Linnean Society*, **109**, 161–88.

Fanning, U., Edwards, D. and Richardson, J.B. (1990) Further evidence for diversity in Late Silurian land vegetation. *Journal of the Geological Society*, **147**, 725–8.

Ferguson, N.P., Lee, J.A. and Bell, J.N.B. (1978) Effects of sulphur pollutants on the growth of *Sphagnum* species. *Environmental Pollution*, **16**, 151–61.

Ferry, B.W., Waters, S.J.P. and Jury, S.L. (eds) (1989) *Dungeness, the Ecology of a Shingle Beach*, Academic Press/Linnean Society, London.

Firth, F.M. (1984) *The Natural History of Romney Marsh*, Meresborough Books, Rainham, Kent.

Florin, R. (1951) Evolution in cordaites and conifers. *Acta Horticulturae Bergiani*, **15**, 285–388.

Fogg, G.E. (1950) *Sinapis arvensis* L. Biological flora of the British Isles. *Journal of Ecology*, **38**, 415–29.

Francis, J.E. (1983) The dominant conifer of the Jurassic Purbeck formation, England. *Palaeontology*, **26**, 277–94.

Francis, J.E. (1984) The seasonal environment of Purbeck (Upper Jurassic) Fossil Forests. *Palaeogeography, Palaeoclimatology, Palaeoecology*, **48**, 285–307.

Friis, E.M. and Skarby, A. (1981) Structurally preserved angiosperm flowers from the Upper Cretaceous of southern Sweden. *Nature*, **291**, 484–6.

Friis, E.M., Chaloner, W.G. and Crane, P.R. (eds) (1987) *The Origins of Angiosperms and their Biological Consequences*, Cambridge University Press, Cambridge.

Friis, E.M., Crane, P.R. and Pedersen, K.R. (1986) Floral evidence for Cretaceous chloranthoid angiosperms. *Nature*, **320**, 163–4.

Friis-Christensen, E. and Lassen, K. (1991) Length of the solar cycle: an indicator of solar activity closely associated with climate. *Science*, **254**, 698.

Fuller, R.M. (1987) The changing extent and conservation interest of lowland grasslands in England and Wales: a review of grassland surveys 1930–84. *Biological Conservation*, **40**, 281–300.

Gensel, P.G. and Andrews, H.N. (1987) The evolution of early land plants. *American Scientist*, **75**, 478–89.

Ghillam, M.E. (1977) *The Natural History of Gower*, D. Brown and Sons, Crowbridge and Bridgend, South Wales.

Gibbard, P.L. and Aalto, M.M. (1977) A Hoxnian interglacial site at Fisher's Green, Stevenage, Hertfordshire. *New Phytologist*, **78**, 505–23.

Gibbons, S. (1977) *Cold Comfort Farm*. Penguin, Harmondsworth.

Gilbert, O.L. (1980) Juniper in Upper Teesdale. *Journal of Ecology*, **68**, 1013–24.

Gimmingham, C.H., Gemmell, A.R. and Greig-Smith, P. (1948) The vegetation of sand-dune system in the Outer Hebrides. *Transactions of the Botanical Society of Edinburgh*, **35**, 82–96.

Gladfelter, B.G. (1975) Middle Pleistocene sedimentary sequences in East Anglia (United Kingdom) in *After the Australopithecines: Stratigraphy, Ecology and Culture Change in the Middle Pleistocene*, (eds K.W. Butzer and G.L. Isaac), Moouton, The Hague, pp. 225–58.

Godwin, H. (1940) Fenland pollen diagrams IV. Postglacial changes of relative land- and sea-level in the English fenland. *Philosophical Transactions of the Royal Society of London, B*, **230**, 239.

Godwin, H. (1941) Studies on the post-glacial history of British vegetation. IV Post-glacial changes of relative land- and sea-level in the English fenland. *Philosophical Transactions of the Royal Society of London, B*, **230**, 285–304.

Godwin, H. (1955) Vegetational history at Cwm Idwal: a Welsh plant refuge. *Svensk Botanisk Tidskrift*, **49**, 35–43.

Godwin, H. (1975) *The History of the British Flora: A Factual Basis for Phytogeography*, Cambridge University Press, Cambridge.

Godwin, H. (1978) *Fenland: Its Ancient Past and Uncertain Future*, Cambridge University Press, Cambridge.

Godwin, H. (1984) *History of the British Flora*, 2nd edn, Cambridge University Press, London.

Godwin, H. and Newton, L. (1938a) The submerged forest at Borth and Ynyslas, Cardiganshire. *New Phytologist*, **37**, 331–44.

Godwin, H. and Newton, L. (1938b) Stratigraphy and development of two raised bogs near Tregaron, Cardiganshire. *New Phytologist*, **37**, 425–54.

Godwin, H., Clowes, D.R. and Huntley, B. (1974) Studies in the ecology of Wicken Fen. V Development of fen carr. *Journal of Ecology*, **62**, 197–214.

Gordon, W.T. (1909) On the nature and occurrence of the plant-bearing rocks at Pettycur, Fife. *Transactions of the Edinburgh Biogeological Society*, **9**, 355–60.

Gray, A.J. and Scott, R. (1977) *Puccinellia maritima* (Hudson) Parl. *Journal of Ecology*, **65**, 699–716.

Green, B.H. and Pearson, M.C. (1968a) The ecology of Wybunbury Moss, Cheshire 1. *Journal of Ecology*, **56**, 47–59.

Green, B.H. and Pearson, M.C. (1968b) The ecology of Wybunbury Moss, Cheshire 2. *Journal of Ecology*, **56**, 793–814.

Gregor, J.W. (1930) Experiments on the genetics of wild populations 1. *Plantago maritima*. *Journal of Genetics*, **22**, 15–25.

Gregor, J.W. (1938) Experimental taxonomy. 2. Initial population differentiation in *Plantago maritima* in Britain. *New Phytologist*, **37**, 15–49.

Greig, J. (1988) Some evidence of the development of grassland plant communities, in *Archaeology and the Flora of the British Isles*, (ed. M. Jones), Oxford University Committee for Archaeology, pp. 39–54.

Greig-Smith, P. (1948) *Urtica* L. Biological flora of the British Isles. *Journal of Ecology*, **36**, 339–55.

Grime, J.P. (1963) An ecological investigation at a junction between two plant communities in Coombsdale on the Derbyshire limestone. *Journal of Ecology*, **51**, 391–402.

Grime, J.P., Hodgson, J.G. and Hunt, R. (1988) *Comparative Plant Ecology: A Functional Approach to Common British Species*, Unwin Hyman, London.

Grove, J.M. (1988) *The Little Ice Age*, Routledge, London.

Gynn, E.G. and Richards, A.J. (1985) *Acaena novae-zelandiae*. Biological flora of the British Isles. *Journal of Ecology*, **73**, 1055–63.

Hadfield, M. (1985) *A History of British Gardening*, Penguin, London.

Haigh, M.J. (1980) Ruderal communities in English cities. *Urban Ecology*, **4**, 329–38.

Hains, B.A. and Horton, A. (1969) *British Regional Geology, Central England*, HMSO, London.

Handford, S.A. (trans.) (1951) *Julius Caesar, The Conquest of Gaul*, Harmondsworth, London.

Hardy, E.M. (1939) The Shropshire and Flint maelor mosses. *New Phytologist*, **38**, 364.

Harper, J.L. (1957) *Ranunculus acris* L., *Ranunculus repens* L., *Ranunculus bulbosus* L. Biological flora of the British Isles. *Journal of Ecology*, **45**, 289–342.

Harper, J.L. (1977) *Population Biology of Plants*, Academic Press, London.

Harper, J.L. and Wood, W.A. (1957) *Senecio jacobaea* L. Biological flora of the British Isles. *Journal of Ecology*, **45**, 617–37.

Harris, T.M. (1938) *The British Rhaetic flora*, British Museum (Natural History), London.

Harris, T.M. (1939) Naiadata, a fossil bryophyte with reproductive organs. *Annals of Bryology*, **12**, 57–70.

Harris, T.M. (1961) *The Yorkshire Jurassic Flora I*, British Museum (Natural History), London.

Harris, T.M. (1964) *The Yorkshire Jurassic Flora II*, British Museum (Natural History), London.

Harris, T.M. (1969) *The Yorkshire Jurassic Flora III*, British Museum (Natural History), London.

Haskins, L.E. (1978) The vegetational history of south-east Dorset. Ph.D. thesis, University of Southampton.

Haslam, S.M. (1972) *Phragmites communis* L.

Biological flora of the British Isles. *Journal of Ecology*, **60**, 585–610.

Hayward, I.M. and Druce, G.C. (1919) *The Adventive Flora of Tweedside*, Buncle, Arbroath.

Heery, S. (1991) The plant communities of the grazed and mown grasslands of the River Shannon Callows. *Proceedings of the Royal Irish Academy*, **91B**, 1–19.

Heinrich, H. (1988) Origin and consequences of cyclic ice rafting in the North-East Atlantic Ocean during the past 130,000 years. *Quaternary Research*, **29**, 143.

Hepburn, A. (1943) A study of the vegetation of sea cliffs in north Cornwall. *Journal of Ecology*, **31**, 30–9.

Hibberd, B.G. (1986) *Forestry Practice*, Forestry Commission Bulletin 14, HMSO, London.

Hicks, S.P. (1971) Pollen-analytical evidence for the effect of prehistoric agriculture on the vegetation of north Derbyshire. *New Phytologist*, **70**, 647–67.

Hill, C.R., Moore, D.T., Greensmith, J.T. and Williams, R. (1985) Palaeobotany and petrology of a Middle Jurassic ironstone bed at Wrack Hills, North Yorkshire. *Proceedings of the Yorkshire Geological Society*, **45**, 277–92.

Hodder, I. and Millett, M. (1990) The human geography of Roman Britain, in *An Historical Geography of England and Wales*, 2nd edn, (eds R.A. Dodgshon and R.A. Butlin), Academic Press, London, pp. 25–44.

Holgate, R. (1988) *Neolithic settlement of Thames Basin*, BAR series 194, BAR, Oxford.

Holliwell, B. (1981) Lady's slipper orchid – concern for its preservation. *BSBI News*, **28**, 22–3.

Hooker, J.J., Insole, A.N., Moody, R.T.J. *et al.* (1980) The distribution of cartilagenous fish, turtles, birds and mammals in the British Palaeogene. *Tertiary Research*, **3**, 1–45.

Horton, A., Keen, D.H., Field, M.H. *et al.* (1992) The Hoxnian Interglacial deposits at Woodston, Peterborough. *Philosophical Transactions of the Royal Society of London, B*, **338**, 131–64.

Howarth, S.E. and Williams, J.T. (1972) *Chrysanthemum segetum* L. Biological flora of the British Isles. *Journal of Ecology*, **60**, 573–84.

Hubbard, R.N.L.B. and Boulter, M.C. (1983) Reconstruction of Palaeogene climate from palynological evidence. *Nature*, **301**, 147–50.

Hughes, N.F. (1975) Plant succession in the English Wealden strata. *Proceedings of the Geologist's Association*, **86**, 439–55.

Hughes, N.F. and McDougall, A.B. (1990) Barremian–Aptian angiospermid pollen records from southern England. *Review of Palaeobotany and Palynology*, **65**, 145–51.

Hughes, N.R., Drewry, G.E. and Laing, J.F. (1979) Barremian earliest angiosperm pollen. *Palaeontology*, **22**, 513–35.

Huiskes, A.H.L. (1979) *Ammophila arenaria* (L.) Link (*Psamma arenaria* (L.) Roem. et Schult.), Biological flora of the British Isles. *Journal of Ecology*, **67**, 363–82.

Hultén, E. (1971) *The Circumpolar Plants. II Dicotyledons*. Almqvist and Wiksell, Stockholm.

Hunter, R.F. (1962) Hill sheep and their pasture: a study of sheep grazing in south-east Scotland. *Journal of Ecology*, **50**, 65–80.

Huntley, B. (1993) Rapid early-Holocene migration and high abundance of hazel (*Corylus avellana* L.): alternative hypotheses, in *Climate Change and Human Impact on the Landscape*, (ed. F.M. Chambers), Chapman & Hall, London, pp. 205–15.

Huntley, B. and Birks, H.J.B. (1983) *An Atlas of Past and Present Pollen Maps for Europe: 0–3,000 years ago*, Cambridge University Press, Cambridge.

Hutchings, M.J. and Barkham, J.P. (1976) An investigation of shoot interactions in *Mercurialis perennis* L. a rhizomatous perennial herb. *Journal of Ecology*, **64**, 723–43.

Hutchinson, T.C. (1968) *Teucrium scorodonia* L. Biological flora of the British Isles. *Journal of Ecology*, **56**, 901–11.

Imbrie, J. and Imbrie, J. (1979) *Ice Ages: Solving the Mystery*, Enslow Publishers, Short Hills, New Jersey.

Ingrouille, M.J. and Smirnoff, N. (1986) *Thlaspi caerulescens* J. & C. Presl. (*T. alpestre* L.) in Britain. *New Phytologist*, **102**, 219–33.

Ingrouille, M.J. and Pearson, J. (1987) The pattern of morphological variation in the *Salicornia europaea* L. aggregate (Chenopodiaceae). *Watsonia*, **16**, 269–81.

Ingrouille, M.J., Pearson, J. and Havill, D.C. (1990) The pattern of morphological variation in the *Salicornia dolichostachya* Moss group from different sites in southern England. *Acta Botanica Neerlandica*, **39**, 263–73.

Innes, J.B. and Simmons, I.G. (1988) Disturbance and diversity: floristic changes associated with pre-elm decline woodland recession in north-east Yorkshire, in *Archaeology and the Flora of the British Isles*, (ed. M. Jones), Oxford University Committee for Archaeology, Oxford, pp. 7–20.

Institute of Terrestrial and Freshwater Ecology (1993) *Countryside Survey*, Institute of Terrestrial and Freshwater Ecology, HMSO, London.

Iversen, J. (1944) *Viscum, Hedera* and *Ilex* as climatic indicators. *Geologiska Föreningens I Stockholm Förhandlingar*, **66**, 463.

Iversen, J. (1958) The bearing of glacial and interglacial epochs on the formation and extinction of plant taxa. *Uppsala Universiteit Arssk*, **6**, 210–15.

Ivimey-Cook, R.B. (1959) *Agrostis setacea* Curt. Biological flora of the British Isles. *Journal of Ecology*, **47**, 697–706.

Ivimey-Cook, R.B. and Proctor, M.C.F. (1966) The plant communities of the Burren, Co. Clare. *Proceedings of the Royal Irish Academy*, **64**, 211–67.

Jeffrey, D.W. (1987) *Soil–Plant Relationships. An Ecological Approach*, Croom Helm, Beckenham, Kent.

Jeram, A.J., Selden, P.A. and Edwards, D. (1990) Land animals in the Silurian: arachnids and myriapods from Shropshire, England. *Science*, **250**, 658–61.

Jessen, K., Andersen, S.V. and Farrington, A. (1959) The Interglacial deposit near Gort, Co. Galway, Ireland. *Proceedings of the Royal Irish Academy*, **60**, 2–77.

Johnson, N. and Rose, P. (1983) Archaeological Survey and Conservation in West Penwith, Cornwall, Truro, cited in Darvill, T. (1987) *Ancient Monuments in the Countryside: An Archaeological Management Review*, Historic Buildings and Monuments Commission for England, London.

Johnson, S. (1982) *Later Roman Britain*, Granada, London.

Jolley, D.W. (1992) Spore dominated assemblages from the lowermost Reading Beds (Palaeocene) of North Essex. *Proceedings of the Yorkshire Geological Society*, **49**, 149–53.

Jolley, D.W. and Spinner, E. (1991) Spore–pollen associations from the London Clay (Eocene), East Anglia, England. *Tertiary Research*, **13**, 11–25.

Jones, E.W. (1945) *Acer* L. Biological flora of the British Isles. *Journal of Ecology*, **32**, 215–52.

Jones, G.R.J. (1990) Celts, Saxons and Scandinavians, in *An Historical Geography of England and Wales*, 2nd edn, (eds R.A. Dodgshon and R.A. Butlin), Academic Press, London, pp. 45–68.

Jones, L.I. (1967) Studies on hill land in Wales. *Technical Bulletin Welsh Plant Breeding Station*, **2**, 1–179.

Jones, M. (1988) The arable field: a botanical battleground, in *Archaeology and the Flora of the*

British Isles, (ed. M. Jones), Oxford University Committee for Archaeology, pp. 86–91.

Jones, M.G. (1933) Grassland management and its influence on the sward II. The management of a clover sward and its effects. *Empirical Journal of Experimental Agriculture*, **1**, 224–34.

Jones, R.L. and Keen, D.H. (1993) *Pleistocene Environments in the British Isles*, Chapman & Hall, London.

Jones, V. and Richards, P.W. (1957) *Saxifraga oppositifolia* L. Biological flora of the British Isles. *Journal of Ecology*, **44**, 300–16.

Kay, Q.O.N. (1971) *Anthemis cotula* L. Biological flora of the British Isles. *Journal of Ecology*, **59**, 623–36.

Kay, Q.O.N. (1972) Variation in sea mayweed (*Tripleorospermum maritimum* (L.) Koch.) in the British Isles. *Watsonia*, **9**, 81–107.

Kay, Q.O.N. and Harrison, J. (1970) *Draba aizoides* L. Biological flora of the British Isles. *Journal of Ecology*, **58**, 877–88.

Keef, P.A.M., Wymer, J.J. and Dimbleby, G.W. (1965) A mesolithic site on Iping Common, Sussex, England, *Proceedings of the Prehistory Society*, **31**, 85–92.

Kemp, E.M. (1968) Probable angiosperm pollen from British Barremian to Albian strata. *Palaeontology*, **11**, 421–34.

Kenrick, P. and Crane, P.R. (1991) Water conducting cells in early fossil land plants: implications for the early evolution of Tracheophytes. *Botanical Gazette*, **152**, 335–56.

Kidston, R. and Lang, W.H. (1917–21) On Old Red Sandstone Plants showing structure from the Rhynie Chert Bed, Aberdeenshire. Parts I–V. *Transactions of the Royal Society of Edinburgh*, **51**, 761–84; **52**, 603–27, 643–80, 831–54, 855–902.

Kuo, C., Lindberg, C. and Thomson, D.J. (1990) Coherence established between atmospheric carbon dioxide concentration and mean global temperature, *Nature*, **343**, 709–14.

Kurten, B. (1972) *The Ice Age*. Rupert Hart-Davis, London.

Laing, L. (1979) *Celtic Britain*, Granada, London.

Laing, L. and Laing, J. (1982a) *Anglo Saxon England*, Granada, London.

Laing, L. and Laing, J. (1982b) *The Origins of Britain*, Granada, London.

Lamb, H.H. (1981) Climate from 1000 BC to 1000 AD, in *The Environment of Man: the Ice Age to the Anglo-Saxon Period*, (ed M. Jones and G. Dimbleby) BAR British Series 87, BAR, Oxford.

Lambert, D. (1990) *Dinosaur Data Book*, Facts on File, New York.

Lambert, J.M., Jennings, J.N., Smith, C.A. *et al.* (1960) *The Making of the Broads*, Royal Geographical Society Research Series 3, Murray, London.

Lambrick, G. and Robinson, M. (1988) The development of floodplain grassland in the Upper Thames Valley, in *Archaeology and the Flora of the British Isles*, (ed. M. Jones), Oxford University Committee for Archaeology, pp. 55–75.

Lang, W.H. (1926) Contributions to the study of the Old Red Sandstone flora of Scotland. IV. On a specimen of *Protolepidodendron* from the Middle Old Red Sandstone of Caithness. *Transactions of the Royal Society of Edinburgh*, **54**, 785–92.

Lang, W.H. (1927) Contributions to the study of the Old Red Sandstone flora of Scotland. VII. On a specimen of *Pseudosporochnus* from the Stromness Beds. *Transactions of the Royal Society of Edinburgh*, **55**, 450–5.

Lang, W.H. (1932a) Contributions to the study of the Old Red Sandstone flora of Scotland. VI. On *Zosterophyllum myretonianum* Penh., and some other plant remains from the Carmyllie Beds of the Lower Old Red Sandstone. *Transactions of the Royal Society of Edinburgh*, **55**, 443–50.

Lang, W.H. (1932b) Contributions to the study of the Old Red Sandstone flora of Scotland. VIII. On *Arthrostigma, Psilophyton* and some associated plant-remains from Strathmore Beds of the Caledonian Old Red Sandstone. *Transactions of the Royal Society of Edinburgh*, **57**, 491–521.

Lele, K. and Walton, J. (1961) Contributions to the knowledge of *Zosterophyllum myretonianum* Penhallow from the Lower Red Sandstone of Angus. *Transactions of the Royal Society of Edinburgh*, **64**, 469–75.

Lidgard, S. and Crane P.R. (1990) Angiosperm diversification and Cretaceous floristic trends: a comparison of palynofloras and leaf macrofloras. *Paleobiology*, **16**, 77–93.

Lindroth, S. (1983) Two faces of Linnaeus, in *Linnaeus the Man and his Work*, (ed. T. Frängsmyr), University of California Press, Berkeley, pp. 1–62.

Lockwood, W. (1979) Water balance of Britain, 50,000 yr BP to the present day. *Quaternary Research*, **12**, 297–10.

Lodge, R.W. (1959) *Cynosurus cristatus* L. Biological flora of the British Isles. *Journal of Ecology*, **47**, 511–18.

Lousley, J.E. (1960) How sheep influence the travels of plants. *New Scientist*, **8**, 353–5.

Lowe, J.J. (1993) Isolating the climatic factors in early-mid-Holocene paleobotanical records from Scotland, in *Climate Change and Human Impact on the Landscape*, (ed. F.M. Chambers), Chapman & Hall, London, pp. 67–82.

Lowe, J.J. and Lowe, S. (1989) Interpretation of the pollen stratigraphy of Late Devensian lateglacial and early Flandrian sediments at Llyn Gwernan, near Cader Idris, North Wales. *New Phytologist*, 113, 391–408.

Lowe, J.J. and Walker M.J.C. (1984) *Reconstructing Quaternary Environments*, Longman, London.

Lynch, A. (1981) *Man and environment in southwest Ireland, 4000 BC–AD 800: a study of man's impact on the development of soil and vegetation*, BAR British Series 85, BAR, Oxford.

McNeilly, T. and Antonovics, J.A. (1968) Evolution in closely adjacent plant populations. IV. Barriers to gene flow. *Heredity*, 23, 205–18.

McVean, D.N. (1953) *Alnus glutinosa* (L.) Gaertn. Biological flora of the British Isles. *Journal of Ecology*, 41, 447–66.

McVean, D.N. (1956) Ecology of *Alnus glutinosa* (L.) Gaertn. V. Notes on some British alder populations. *Journal of Ecology*, 44, 321–30.

McVean, D.N. and Ratcliffe, D.A. (1962) *Plant Communities of the Scottish Highlands*, HMSO, London.

Mai, D.H. (1991) Palaeaofloristic changes in Europe and the confirmation of the Arctotertiary-Palaeotropical geoflora concept. *Review of Palaeobotany and Palynology*, 68, 29–36.

Malloch, A.J.C. (1971) Vegetation of the maritime cliff-tops of the Lizard and Land's End peninsulas, West Cornwall. *New Phytologist*, 70, 1155–97.

Malloch, A.J.C. (1972) Salt spray deposition on the maritime cliffs of the Lizard Peninsula. *Journal of Ecology*, 60, 103–12.

Marston, J.B., Oppenheimer, M., Fujita, R.M. and Gaffin, S.R. (1991) Carbon dioxide and temperature. *Nature*, 349, 573–4.

Martin, M.H. (1968) Conditions affecting the distribution of *Mercurialis perennis* L. in certain Cambridgeshire woodlands. *Journal of Ecology*, 56, 777–93.

Matten, L.C., Lacey, W.S. and Lucas, W.S. (1980) Studies on the cupulate seed genus *Hydrasperma* Long from Berwickshire and East Lothian in Scotland and County Kerry in Ireland. *Botanical Journal of the Linnean Society*, 81, 249–73.

Matthews, J.R. (1955) *Origin and Distribution of the British Flora*, Hutchinson's University Library, London.

Merryfield, D.L. and Moore, P.D. (1974) Prehistoric human activity and blanket peat initiation on Exmoor. *Nature*, 250, 439–41.

Miller, C.N. (1977) Mesozoic conifers. *Botanical Review*, 43, 217–80.

Miller, G.H., Jull, A.J.T., Linick, T. *et al* (1987) Racemization-derived Late Devensian temperature reduction in Scotland. *Nature*, 325, 593–5.

Mitchell, G.F. (1942) A Late-Glacial flora in Co. Monaghan, Ireland. *Nature*, 149, 502.

Moar, N.T. (1969) A radiocarbon-dated pollen diagram from north-west Scotland. *New Phytologist*, 68, 209–14.

Mohamed, B.F. and Gimingham, C.H. (1970) The morphology of vegetative regeneration in *Calluna vulgaris*. *New Phytologist*, 69, 743–50.

Molloy, K. and O'Connell, M. (1993) Early land use and vegetation history at Derryinver Hill, Renvyle peninsula, Co. Galway, in *Climate Change and Human Impact on the Landscape*, (ed. F.M. Chambers), Chapman & Hall, London, pp. 185–99.

Moore, J.J., Dowding, P. and Healy, B. (1975) Glenamoy, Ireland. *Ecological Bulletin*, 20, 321–43.

Moore, N.W. (1962) The heaths of Dorset and their conservation. *Journal of Ecology*, 50, 369–91.

Moore, P.D. (1975) Origin of blanket mires. *Nature*, 256, 267–9.

Moore, P.D. (1988) The development of moorlands and upland mires, in *Archaeology and the Flora of the British Isles*, (ed. M. Jones), Oxford University Committee for Archaeology, pp. 116–22.

Moore, P.D. and Bellamy, D.J. (1974) *Peatlands*, Elek Science, London.

Moore, P.D., Webb, J.A. and Collinson, M.E. (1991) *Pollen Analysis*, 2nd edn, Blackwell Scientific Publications, London.

Morris, M.G. (1991) Nitrogen fertilisers on old peat grasslands, in *Annual Report 1990–1*, Institute of Terrestrial Ecology, Cambridge, pp. 56–8.

Morton-Boyd, J. (1992) Sycamore – a review of its status in conservation in Great Britain. *Biologist*, 39, 29–31.

Moss, B. (1987) The Broads. *Biologist*, 34, 7–13.

Muir, R. and Muir, N. (1987) *Fields*, Macmillan, London.

Myers, A.R. (ed.) (1969) *English Historical Documents 4. 1327–1485*, Eyre Methuen/Oxford University Press, London.

Myerscough, P.J. (1980) *Epilobium angustifolium* L.

(*Chamaenerion angustifolium* (L.) Scop.). Biological flora of the British Isles. *Journal of Ecology*, **68**, 1047–74.

Nathorst, A.G. (1911) Paläobotanische Mitteilungen. *Kungl Svenska Vetenskapsakademiens handlingar*, **46**, 1–33.

National Trust for Scotland (1986) *Ben Lawers*, 3rd edn, National Trust for Scotland, Edinburgh.

Nature Conservancy Council (1984a) *Nature Conservation in Great Britain*, Nature Conservancy Council, Peterborough.

Nature Conservancy Council (1984b) *The Decline and Present Status of English Lowland Heaths*, Focus on nature conservation No. 11, Nature Conservancy Council, Peterborough.

Nature Conservancy Council (1990) *Nature Conservation and Agricultural Change*, Focus on nature conservation No. 25, Nature Conservancy Council, Peterborough.

Naylor, R.E.L. (1972) *Alopecurus myosuroides* Huds. Biological flora of the British Isles. *Journal of Ecology*, **60**, 611–22.

Nelson, E.C. and Whalsh, W. (1991) *The Burren*, Boethius Press/The Conservancy of the Burren, Aberystwyth, Wales.

Niklas, K.J. (1976) Morphological and ontogenetic reconstruction of *Parka decipiens* Fleming and *Pachytheca* Hooker from the Lower Old Red Sandstone, Scotland. *Transactions of the Royal Society of Edinburgh*, **69**, 483–99.

Niklas, K.J. and Banks, H.P. (1990) A reevaluation of the Zosterophyllophytina with comments on the origin of Lycopods. *American Journal of Botany*, **77**, 274–83.

Noble, J.C. (1982) *Carex arenaria* L. Biological flora of the British Isles. *Journal of Ecology*, **70**, 867–86.

Norris, G. (1969) Miospores from the Purbeck Beds and Marine Upper Jurassic of southern England. *Palaeontology*, **12**, 574–620.

Ockendon, D.J. (1968) *Linum perenne* ssp. *anglicum* (Miller) Ockendon. Biological flora of the British Isles. *Journal of Ecology*, **56**, 871–82.

O'Connell, M., Ryan, J.B. and MacGowran, B.A. (1984) Wetland communities in Ireland: a phytosociological review, in *European Mires*, (ed. P.D. Moore), Academic Press, Dublin, pp. 303–64.

Oldfield, F. (1960) Studies in the post-glacial history of the British vegetation: Lowland Lonsdale. *New Phytologist*, **59**, 192–217.

Oldham, T.C.B. (1976) The plant debris beds of the English Wealden. *Palaeontology*, **19**, 437–502.

O'Reilly, H. (1955) Survey of the Gereagh – an area of wet woodland on the River Lee, near Macroom, Co. Cork. *Irish Naturalists Journal*, **11**, 279–86.

Orr, M.Y. (1912) Kenfig Burrows: an ecological study. *Scottish Botanical Reviews*, **1**, 209.

Packham, J.R. (1978) *Oxalis acetosella* L. Biological flora of the British Isles. *Journal of Ecology*, **66**, 669–93.

Packham, J.R. and Harding, D.J.L. (1982) *Ecology of Woodland Processes*, Edward Arnold, London.

Packham, J.R. and Willis, A.J. (1977) The effects of shading on *Oxalis acetosella*. *Journal of Ecology*, **65**, 619–42.

Parry, M.L. (1985) Upland settlement and climatic change: the medieval evidence, in *Upland Settlement in Britain. The Second Millennium BC and After*, (eds D. Spratt and C. Burgess), BAR Series 143, BAR, Oxford, pp. 35–49.

Pattison, J., Smith, D.B. and Warrington, G. (1973) A review of Late Permian and Early Triassic biostratigraphy in the British Isles, in *The Permian Triassic Systems and their Mutual Boundary*, (eds A. Logan and L.V. Mills), Canadian Society of Petroleum Geologists, Calgary, pp. 220–60.

Pearsall, W.H. (1971) *Mountains and Moorlands*, revised edn, Collins, London.

Pearson, J. and Havill, D.C. (1989) The effect of hypoxia and sulphide on culture-grown wetland and non-wetland plants I: Growth and nutrient uptake. *Journal of Experimental Botany*, **39**, 363–74.

Pearson, M.C. and Rogers, J.A. (1962) *Hippophae rhamnoides* L. Biological flora of the British Isles. *Journal of Ecology*, **50**, 501–13.

Pearson, R. (1964) *Animals and Plants of the Cenozoic Era*, Butterworths, London.

Peglar, S.M. (1979) A radiocarbon-dated pollen diagram from Loch of Winless, Caithness, North-east Scotland. *New Phytologist*, **82**, 245–63.

Peglar, S.M. (1993) The development of the cultural landscape around Diss-mere, Norfolk, UK during the past 7,000 years. *Review of Palaeobotany and Palynology*, **76**, 1–47.

Pennington, W. (1970) Vegetation history in the north-west of England: a regional synthesis, in *Studies in the Vegetational History of the British Isles: Essays in Honour of Harry Godwin*, (eds D. Walker and R.G. West), Cambridge University Press, Cambridge, pp. 41–79.

Pennington, W. (1975) The effect of Neolithic man on the environment in north-west England: the use of absolute pollen diagrams, in *The Effect of Man on the Landscape: the Highland Zone*, (eds

J.G. Evans, S. Limbrey and H. Cleere), CBA Research Report Number 11, pp. 74–86.

Pennington, W. (1977) The late Devensian flora and vegetation of Britain. *Philosophical Transactions of the Royal Society of London, B*, **280**, 247–71.

Pennington, W. (1986) Lags in adjustment of vegetation to climate caused by the pace of soil development. Evidence from Britain. *Vegetatio*, **67**, 105–18.

Perkins, J., Evans, J. and Ghillam, M. (1982) *The Historic Taf Valleys*, Vol. 2 *In the Brecon Beacons National Park*, Merthyr Tydfil and Districts Naturalists' Society/D.Brown & Sons, Bridgend, Mid Glamorgan.

Perring, F.P. and Walters, S.M. (eds) (1982) *Atlas of the British Flora*, 3rd edn, Nelson, London.

Perry, I. and Moore, P.D. (1987) Dutch elm disease as an analogue of the Neolithic elm decline. *Nature*, **326**, 72–3.

Peterken, G.F. (1974) A method for assessing woodland flora for conservation using indicator species. *Biological Conservation*, **6**, 239–45.

Pfefferkorn, H.W. and Thomson, M.C. (1982) Changes in dominance patterns in upper Carboniferous plant-fossil assemblages. *Geology*, **10**, 641–4.

Pfitzenmeyer, C.D.C. (1962) *Arrhenatherum elatius* (L.) J.& C.Presl, Biological flora of the British Isles. *Journal of Ecology*, **50**, 235–45.

Phillips, L. (1974) Vegetational history of the Ipswichian interglacial in Britain and continental Europe. *New Phytologist*, **73**, 589–604.

Phillips, L. (1976) Pleistocene vegetational history and geology in Norfolk. *Philosophical Transactions of the Royal Society of London, B*, **275**, 215–86.

Phillips, M.E. (1954) *Eriophorum angustifolium* Honck. *Journal of Ecology*, **42**, 612–22.

Phillips, T.L. (1979) Reproduction of heteroporous arborescent lycopods in the Mississippian–Pennsylvanian of Euramerica. *Review Palaeobotany and Palynology*, **27**, 239–89.

Phillips, T.L. and DiMichelle, W.A. (1992) Comparative ecology and life-history of arborescent lycopsids in Late Carboniferous swamps of Euramerica. *Annals of the Missouri Botanical Garden*, **79**, 560–88.

Pigott, C.D. (1955) *Thymus* L. Biological flora of the British Isles. *Journal of Ecology*, **43**, 365–87.

Pigott, C.D. (1958) *Polemonium caeruleum* L. Biological flora of the British Isles. *Journal of Ecology*, **46**, 507–25.

Pigott, C.D. (1968) *Cirsium acaulon* (L.) Scop.

Biological flora of the British Isles. *Journal of Ecology*, **56**, 597–612.

Pigott, C.D. and Pigott, M.E. (1963) Late glacial and post-glacial deposits at Malham, Yorkshire. *New Phytologist*, **62**, 317–34.

Pigott, M.E. and Pigott, C.D. (1959) Stratigraphy and pollen analysis of Malham Tarn and Tarn Moss. *Field Studies*, **1**, 84–101.

Pilcher, J.R. (1971) Land clearance in the Irish Neolithic: new evidence and interpretation. *Science*, **172**, 560–2.

Pilcher, J.R. (1991) Radiocarbon dating, in *Quaternary Dating Methods – A User's Guide*, (eds P.L. Smart and P.D. Frances), Quaternary Research Association, Cambridge, pp. 16–36.

Pitt, W. (1809) *A General View of the Agriculture of the County of Leicester*, Board of Agriculture/Richard Phillips, London.

Pollard, E., Hooper, M.D. and Moore, N.W. (1974) *Hedges*, Collins, London.

Polunin, O. and Walters, M. (1985) *A Guide to the Vegetation of Britain and Europe*, Oxford University Press, Oxford.

Poole, I. (1992) Pyritized twigs from the London Clay, Eocene, of Great Britain. *Tertiary Research*, **13**, 71–85.

Poore, M.E.D. (1956) The ecology of Woodwalton Fen. *Journal of Ecology*, **44**, 455–92.

Prentice, I.C. and Lehmans, R. (1990) Pattern and process and the dynamics of forest structure: a simulation approach. *Journal of Ecology*, **78**, 340–55.

Proctor, J. and Woodell, S.R.J. (1971) The plant ecology of serpentine. 1. Serpentine vegetation of England and Scotland. *Journal of Ecology*, **59**, 375–95.

Proctor, M.C.F. (1957) *Helianthemum* Mill. Biological flora of the British Isles. *Journal of Ecology*, **45**, 675–92.

Proctor, M.C.F. (1960) *Tuberaria guttata* (L.) Fourr. Biological flora of the British Isles. *Journal of Ecology*, **48**, 243–53.

Proctor, M.C.F. (1965) The distinguishing characters and geographical distribution of *Ulex minor* and *U. gallii*. *Watsonia*, **6**, 177–87.

Proctor, M.C.F. (1974) The vegetation of the Malham Tarn Fens. *Field Studies*, **4**, 1–38.

Purseglove, J. (1989) *Taming the Flood*. Oxford University Press, London.

Rackham, O. (1975) *Hayley Wood, Its History and Ecology*, Cambridgeshire and Ely Naturalists' Trust, Cambridge.

Rackham, O. (1976) *Trees and Woodland in the British Landscape*, Dent, London.

Rackham, O. (1980) *Ancient Woodland: its History, Vegetation and Uses in England*, Edward Arnold, London.

Rackham, O. (1992) *The Last Forest*, Cambridge University Press, Cambridge.

Raistrick, A. and Marshall, C.E. (1939) *The Nature and Origins of Coal and Coal Seams*, English Universities Press, London.

Ranwell, D. (1960a) Newborough Warren, 1. The dune system and dune slack habitat. *Journal of Ecology*, **47**, 571–602.

Ranwell, D. (1960b) Newborough Warren, 2. Plant associes and succession cycles of the sand-dune and dune-slack vegetation. *Journal of Ecology*, **48**, 117–42.

Ranwell, D. (1974) Machair in relation to the British sand dune series, in *Sand Dune Machair*, NERC/Institute of Terrestrial Ecology, Cambridge.

Ratcliffe, D. (1959) *Hornungia petraea* (L.) Rechb. Biological flora of the British Isles. *Journal of Ecology*, **47**, 241–7.

Ratcliffe, D.A. (1977) *A Nature Conservation Review*, Vol. 1, Cambridge University Press, Cambridge.

Ratcliffe, D.A. (1981) The vegetation, in *The Cairngorms*, (eds D. Nethersole-Thompson and A. Watson), The Melvyn Press, Perth, Scotland, pp. 31–41.

Ratcliffe, D.A. (1991) The Mountain flora of Britain and Ireland. *British Wildlife*, **3**, 10–21.

Ratcliffe, D.A. and Oswald, P.H. (eds) (1988) *The Flow Country, the Peatlands of Caithness and Sutherland*, Nature Conservancy Council, Peterborough.

Raven, J. and Walters, M. (1956) *Mountain Flowers*, Collins, London.

Rayner, R.J. (1983) New observations on *Sawdonia ornata* from Scotland. *Transactions of the Royal Society of Edinburgh, Earth Sciences*, **74**, 79–93.

Reed, M. (1983) *The Making of Britain: The Georgian Triumph, 1700–1830*, Routledge, London.

Reid, E.M. (1920) On two Preglacial floras from Castle Eden, County Durham. *Quarterly Journal of the Geological Society, London*, **76**, 2, 104.

Reid, E.M. and Chandler, M.E.J. (1926) *The Bembridge Flora*, Catalogue of the Cainozoic Plants in the Department of Geology I, British Museum (Natural History), London.

Remy, W. and Remy, R. (1980) Devonian gametophytes with anatomically preserved gametangia. *Science*, **208**, 295–6.

Retallack, G.J. and Dilcher, D.L. (1988) Reconstructions of selected seed ferns. *Annals of the Missouri Botanical Garden*, **75**, 1010–57.

Rex, G.M. and Scott, A.C. (1987) The sedimentology, palaeoecology and preservation of the Lower Carboniferous plant deposits at Pettycur, Fife, Scotland. *Geological Magazine*, **124**, 43–66.

Richards, A.J. (1982) Flowers of the Northumberland Whinstone. *Botanical Society of the British Isles News*, **31**, 6–8.

Richards, P.J. and Clapham A.R. (1941a) *Juncus effusus* L. Biological flora of the British Isles. *Journal of Ecology*, **29**, 375–80.

Richards, P.J. and Clapham A.R. (1941b) *Juncus conglomeratus* L. Biological flora of the British Isles. *Journal of Ecology*, **29**, 381–4.

Richards, P.J. and Clapham A.R. (1941c) *Juncus inflexus* L. Biological flora of the British Isles. *Journal of Ecology*, **29**, 369–74.

Ritchie, J.C. (1954) *Primula scotica* Hook. Biological flora of the British Isles. *Journal of Ecology*, **42**, 623–8.

Rodwell, J.S. (ed.) (1991) *British Plant Communities 1. Woodland and Scrub*, Cambridge University Press, London.

Rodwell, J.S. (ed.) (1992a) *British Plant Communities 2. Mires and Heaths*, Cambridge University Press, London.

Rodwell, J.S. (1992b) *British Plant Communities 3. Grasslands and Montane Communities*, Cambridge University Press, London.

Roebuck, J.F. (1987) Agricultural problems of weeds on the crop headland, in *Field Margins*, (eds J.M. Way and P.W. Greig-Smith), British Crop Protection Council Monograph No. 35, BCPC Publications, Thornton Heath.

Rolfe, W.D.I. (1985) Early terrestrial arthropods: a fragmentary record. *Philosophical Transactions of the Royal Society, London, B*, **309**, 207–18.

Rothwell, H. (ed.) (1975) *English Historical Documents 3. 1189–1327*, Eyre Methuen/Oxford University Press, London.

Rowe, N.P. (1988) New observations on the Lower Carboniferous pteridosperm *Diplopteridium* Walton and an associated synangiate organ. *Botanical Journal of the Linnean Society*, **97**, 125–58.

Rowell, T.A. (1984) Further discoveries of the Fen Violet (*Viola persicifolia* Schreber) at Wicken Fen, Cambridgeshire. *Watsonia*, **15**, 122–3.

Rowley, J.R. and Srivastava, S.K. (1986) Fine-structure of *Classopollis* exines. *Canadian Journal of Botany*, **64**, 3059–74.

Rowley, T. (ed.) (1981) *The Origins of Open-field Agriculture*, Croom Helm, Beckenham, Kent.

Ruddiman, W.F., Sancetta, C.D. and McIntyre, A. (1977) Glacial and Interglacial response rate of subpolar North Atlantic waters to climatic change: the record in oceanic sediments. *Philosophical Transactions of the Royal Society of London, B*, **280**, 119–42.

Ruffel, A.H. and Batten, D.J. (1991) The Barremian–Aptain arid phase in western Europe. *Palaeogeography, Palaeoclimatology and Palaeoecology*, **80**, 197–212.

Sagar, G.R. and Harper, J.L. (1964) *Plantago major* L., *P. media* L. and *P. lanceolata* L. Biological flora of the British Isles. *Journal of Ecology*, **52**, 189–221.

Sahni, B. (1932) A petrified *Williamsonia* (*W. sewardiana*, sp. nov. Rajmahal Hills, India). *Memoirs of the Geological Survey, India, Palaeontology*, **20**, 1–19.

Scaife, R.G. (1988) The elm decline in the pollen record of south-east England and its relationship to early agriculture, in *Archaeology and the Flora of the British Isles*, (ed. M. Jones), Oxford University Committee for Archaeology, pp. 21–33.

Schweitzer, H.-J. (1986) The land flora of the English and German Zechstein sequences, in *The English Zechstein and Related Topics*, (eds G.M. Harwood and D.B. Smith), Geological Society/Blackwell Scientific Publishers, pp. 31–54.

Scott, A.C. (1979) The ecology of coal measure floras from northern Britain. *Proceedings of the Geologist's Association*, **90**, 97–116.

Scott, A.C. (1990) Preservation, evolution, and extinction of plants in Lower Carboniferous volcanic sequences in Scotland. *Geological Society of America, Special Paper*, **244**, 25–38.

Scott, A.C. and Chaloner, W.G. (1983) The earliest fossil conifer from the Westphalian B of Yorkshire. *Proceedings of the Royal Society of London, B*, **220**, 163–82.

Scott, A.C. and Ranwell, R.E. (1976) *Crambe maritima* L. Biological flora of the British Isles. *Journal of Ecology*, **64**, 1077–91.

Scott, A.C. and Taylor, T.N. (1983) Plant/animal interactions during the Upper Carboniferous. *Botanical Review*, **49**, 259–307.

Scott, A.C., Galtier, J. and Clayton, G. (1984) Distribution of anatomically-preserved floras in the Lower Carboniferous in Western Europe. *Transactions of the Royal Society of Edinburgh: Earth Sciences*, **75**, 311–40.

Scott, G.A.M. (1963a) The ecology of shingle beach plants. *Journal of Ecology*, **51**, 517–27.

Scott, G.A.M. (1963b) *Glaucium flavum* Crantz. Biological flora of the British Isles. *Journal of Ecology*, **51**, 743–54.

Scott, G.A.M. (1965) The shingle succession at Dungeness. *Journal of Ecology*, **53**, 21–31.

Scott, N.E. (1985) The updated distribution of maritime species on British roadsides. *Watsonia*, **15**, 381–6.

Scourse, J.D., Allen, J.R.M., Austin, W.E.N. et al. (1992) New evidence on the age and significance of the Gortian temperate stage: a preliminary report on the Cork harbour site. *Proceedings of the Royal Irish Academy*, **92**, 21–43.

Scurfield, G. (1954) *Deschampsia flexuosa* (L.) Trin. Biological flora of the British Isles. *Journal of Ecology*, **42**, 225–33.

Scurfield, G. (1959) The ashwoods of the Derbyshire Carboniferous limestone: Monksdale. *Journal of Ecology*, **47**, 357–69.

Sealy, J.R. and Webb, D.A. (1950) *Arbutus* L. Biological flora of the British Isles. *Journal of Ecology*, **38**, 223–36.

Seaward, M.R.D. (1979) Lower plants in the urban landscape. *Urban Ecology*, **4**, 214–25.

Seaward, M.R.D. and Hitch, C.J.B. (1982) *Atlas of the Lichens of the British Isles*, Institute of Terrestrial Ecology, Cambridge.

Seward, A.C. (1926) The plant bearing beds of Western Greenland. *Philosophical Transactions of the Royal Society of London, B*, **215**, 57–172.

Seward, A.C and Holttum, R.E. (1924) Tertiary plants from Mull, in *Tertiary and Post-Tertiary Geology of Mull*, (eds E.B. Bailey, C.T. Clough, W.B. Wright et al.), Memoirs of the Geological Survey of Scotland, pp. 67–90.

Shaw, D.L. (1971) *Gwydr Forest in Snowdonia*, Forestry Commission Booklet No. 28, HMSO, London.

Sheail, J. (1971) *Rabbits and their History*. David and Charles, Newton Abbot.

Shirreffs, D.A. (1985) *Anemone nemorosa* L. Biological flora of the British Isles. *Journal of Ecology*, **73**, 1005–20.

Shotton, F.W. (1977) Chronology, climate and marine record. The Devensian Stage: its development, limits and substages. *Philosophical Transactions of the Royal Society of London, B*, **280**, 107–18.

Shute, C.H. and Edwards, D. (1989) A new rhyniopsid with novel sporangium organization from the Lower Devonian of South Wales. *Botanical Journal of the Linnean Society*, **100**, 11–37.

Silvertown, J.W. (1983) The distribution of plants in limestone pavement: tests of species interaction and niche separation against null hypotheses. *Journal of Ecology*, **71**, 819–28.

Simmonds, N.W. (1945) *Polygonum* L. em. Gaertn.

Biological flora of the British Isles. *Journal of Ecology*, **33**, 117–43.

Simmons, I.G. and Cundhill, P.R. (1974) Late Quaternary vegetational history of the North York Moors. I Pollen analyses of blanket peats. *Journal of Biogeography*, **1**, 159–69.

Simmons, I.G. and Tooley, M.J. (1981) *The Environment in British Prehistory*, Duckworth, London.

Simpson, D.A. (1986) Taxonomy of *Elodea* Michx. in the British Isles. *Watsonia*, **16**, 1–14.

Simpson, I.M. and West, R.G. (1958) The stratigraphical palaeobotany of a late Pleistocene deposit at Chelford, Cheshire. *New Phytologist*, **57**, 239–50.

Simpson, R.B. (1961) The Tertiary Flora of Mull and Ardamurchan. *Transactions of the Royal Society of Edinburgh*, **64**, 421–68.

Sinker, C.A. (1962) The North Shropshire meres and mosses: a background for ecologists. *Field Studies*, **1**, 101–37.

Sissons, J.B. (1979) The Loch Lomond stadial in the British Isles. *Nature*, **280**, 199–203.

Smart, P.L. and Frances, P.D. (1991) *Quaternary Dating Methods: A User's guide*. Quaternary Research Association, Cambridge.

Smith, A.G. and Cloutman, E.W. (1988) Reconstruction of Holocene vegetation history in three dimensions at Waun Fignen Felen, an upland site in South Wales. *Philosophical Transactions of the Royal Society of London, B*, **322**, 159–219.

Smith, A.G. and Willis, E.H. (1962) Radiocarbon dating of the Fallahogy Landnam phase. *Ulster Journal of Archaeology*, **24–5**, 16–24.

Smith, A.H.V. and Butterworth, M.A. (1967) *Miospores in the coal seams of the Carboniferous of Great Britain*, Special Papers in Palaeontology No. 1, The Palaeontological Association, London.

Smith, D.B., Brunston, R.G.W., Manning, P.I. *et al.* (1974) A correlation of Permian rocks in the British Isles. *Journal of the Geological Society, London*, **130**, 1–45.

Smith, R.W. (1984) The ecology of Neolithic farming systems as exemplified by Avebury region of Wiltshire. *Proceedings of the Prehistory Society*, **50**, 99–120.

Sobey, D.G. (1981) *Stellaria media* (L.) Vill. Biological flora of the British Isles. *Journal of Ecology*, **69**, 311–35.

Spicer, R. A. and Hill, C. (1979) Principal components and correspondence analyses of quantitative data from a Jurassic plant bed. *Review of Palaeobotany and Palynology*, **28**, 273–99.

Stace, C.A. (1975) *Hybridization and the Flora of the British Isles*, Academic Press, London.

Stace, C.A. (1991) *New Flora of the British Isles*, Cambridge Unversity Press, Cambridge.

Steers, J.A. (1937) The Culbin Sands. *Geographical Journal*, **90**, 498–528.

Stevenson, W. (1815) *A General View of the Agriculture of the County of Dorset*. Board of Agriculture. Sherwood, Neely and Jones, London.

Stewart, W.N. and Rothwell, G.W. (1993) *Paleobotany and Evolution of Plants*, 2nd edn, Cambridge University Press, Cambridge.

Stonely, H.M. (1958) The Upper Permian flora of England. *British Museum (Natural History) Bulletin of Geology*, **3**, 295–337.

Strahan, A. (1910) *Guide to the Geological Model of Ingleborough and District*, Memoirs of the Geological Survey, England and Wales, HMSO, London.

Stuart, A.J. (1976) The history of the mammal fauna during the Ipswichian/last interglacial in England. *Philosophical Transactions of the Royal Society of London, B*, **276**, 221–50.

Stuart, A.J. (1982) *Pleistocene Vertebrates in the British Isles*, Longman, London.

Stuart, A.J. and Wijngaarden-Bakker, L.H. (1985) Quaternary vertebrates, in *The Quaternary History of Ireland*, (eds K.J. Edwards and W.P. Warren), Academic Press, London, pp. 221–49.

Summerfield, R.J. (1974) *Narthecium ossifragum* (L.) Huds. Biological flora of the British Isles. *Journal of Ecology*, **62**, 325–39.

Sutcliffe, A.J. (1985) *On The Track of Ice Age Mammals*, Harvard University Press, Cambridge, Massachusetts.

Sydes, C. and Grime, J.P. (1981a) Effects of tree litter on herbaceous vegetation in deciduous woodland I. *Journal of Ecology*, **69**, 237–48.

Sydes, C. and Grime, J.P. (1981b) Effects of tree litter on herbaceous vegetation in deciduous woodland II. *Journal of Ecology*, **69**, 249–62.

Synge, F.M. (1985) Coastal evolution, in *The Quaternary History of Ireland*, (eds K.J. Edwards and W.P. Warren), Academic Press, London, pp. 115–31.

Szafer, W. (1946–47) The Pliocene flora of Kroscienko in Poland. *Rozpr. Wydz. mat. przyr. Akad. Um.*, **72**.

Tallis, J.H. (1965) Studies on southern Pennine peats, IV Evidence of recent erosion. *Journal of Ecology*, **53**, 509–20.

Tansley, A.G. and Adamson, R.S. (1925) Studies

of the vegetation of the English chalk. III. The chalk grasslands of the Hampshire–Sussex border. *Journal of Ecology*, **13**, 177–223.

Taylor, D.W. (1990) Palaeobiogeographic relationships of angiosperms from the Cretaceous and Early Tertiary of the North American area. *Botanical Review*, **56**, 279–417.

Taylor, D.W. and Hickey, L.J. (1992) Phylogenetic evidence for the herbaceous origin of angiosperms. *Plant Systematics and Evolution*, **180**, 137–42.

Taylor, F.J. (1956) *Carex flacca* Schreb. Biological flora of the British Isles. *Journal of Ecology*, **44**, 281–90.

Taylor, K.C., Lamorey, G.W., Doyle, G.A. *et al.* (1993) The 'flickering switch' of late Pleistocene climate change. *Nature*, **361**, 432–6.

Thomas, B.A. and Spicer, R.A. (1987) *The Evolution and Palaeobiology of Land Plants*, Croom Helm, Beckenham, Kent.

Thomas, H.H. (1913) The fossil flora of the Cleveland District. *Quarterly Journal of the Geological Society, London*, **59**, 223–51.

Thomas, H.H. (1925) The Caytoniales, a new group of angiospermous plants from the Jurassic rocks of Yorkshire. *Philosophical Transactions of the Royal Society of London, B*, **213**, 299–363.

Thompson, J.D. (1990) *Spartina anglica*, characteristic feature or invasive weed of coastal saltmarshes. *Biologist*, **37**, 9–12.

Tiffney, B.H. (1985) The Eocene North Atlantic Land Bridge: its importance in Tertiary and modern phytogeography of the Northern Hemisphere. *Journal of the Arnold Arboretum*, **66**, 243–73.

Tooley, M.J. (1982) Sea-level changes in northern England. *Proceedings of the Geologist's Association*, **93**, 43–51.

Trist, P.J.O. (ed.) (1979) *An Ecological Flora of Breckland*, EP Publishing Limited, East Ardley, Yorkshire.

Trueman, A. (ed.) (1954) *The Coalfields of Great Britain*, Edward Arnold, London.

Turner, C. (1970) The Middle Pleistocene deposit at Mark's Tey, Essex. *Philosophical Transactions of the Royal Society of London, B*, **257**, 373–440.

Turner, J. (1978) History of vegetation and flora, in *Upper Teesdale, the Area and its Natural History*, (ed. A.R. Clapham), Collins, London, pp. 88–101.

Tutin, T.G. (1942) *Zostera* L. Biological flora of the British Isles. *Journal of Ecology*, **30**, 217–26.

Tutin, T.G. (1957) *Allium ursinum* L. Biological flora of the British Isles. *Journal of Ecology*, **45**, 1003–10.

Van Konijnenburg-van Cittert, J.H.A. (1989) The flora from the Kimmeridgian (Upper Jurassic) of Culgower, Sutherland, Scotland. *Acta Botanica Neerlandica*, **20**, 1–96.

Wacher, J. (1982) *The Coming of Rome*, Granada, London.

Wagner, R.H. (1983) A lower Rotligende flora from Ayrshire. *Scottish Journal of Geology*, **19**, 135–55.

Walker, D. (1970) Direction and rate of some postglacial British hydroseres, in *Studies in the Vegetational History of the British Isles*, (eds D. Walker and R.G. West), Cambridge University Press, London, pp. 117–40.

Walton, J. (1957) On *Protopitys* (Göppert): with a description of a fertile specimen '*Protopitys scotica*' sp.nov. from the Calciferous Sandstone series of Dunbartonshire. *Transactions of the Royal Society of Edinburgh*, **63**, 333–40.

Walton, J. (1969) On the structure of a silicified stem of *Protopitys* and roots associated with it from the carboniferous (Mississippian) of Yorkshire, England. *American Journal of Botany*, **56**, 808–13.

Walton, J.R. (1990) Agriculture and rural society, 1730–1914, in *An Historical Geography of England and Wales*, 2nd edn, (eds R.A. Dodgshon, and R.A. Butlin), Academic Press, London, pp. 323–50.

Waton, P.V. (1983) The origins and past land-use of south-east Dorset heaths, in *Heathland Management in Amenity Areas*, (ed. J.L. Daniels), Countryside Commission, Cheltenham.

Watson, E. (1977) The Periglacial environment. The periglacial environment of Great Britain during the Devensian. *Philosophical Transactions of the Royal Society of London, B*, **280**, 183–98.

Watson, J. (1988) The Cheirolepidiaceae, in *Origin and Evolution of Gymnosperms*, (ed. C. B. Beck), Columbia University Press, New York, pp. 382–447.

Watson, J. and Batten, D.J. (1990) A revision of the English Wealden Flora, II. Equisetales. *Bulletin of the British Museum Natural History (Geology)*, **46**, 37–60.

Watson, J. and Sincock, C.A. (1992) *Bennettitales of the English Wealden*, Monograph of the Palaeontographical Society No. 588, Palaeontological Society, London.

Watt, A.S. (1955) Bracken versus heather, a study in plant sociology. *Journal of Ecology*, **43**, 490–506.

Watt, A.S. (1962) The effect of excluding rabbits from Grassland A (Xerobrometum) in Breckland, 1936–60. *Journal of Ecology*, **50**, 181–98.

Watt, A.S. and Fraser, G.K. (1933) Tree roots and the field layer. *Journal of Ecology*, **21**, 404–14.

Watt, A.S., Perrin, R.M.S. and West, R.G. (1966) Patterned ground in Breckland. *Journal of Ecology*, **54**, 239–58.

Watts, W.A. (1959) Interglacial deposits at Kilbeg and Newtown. *Proceedings of the Royal Irish Academy*, **60**, 79–134.

Watts, W.A. (1964) Interglacial deposits in Kildromin Td, near Herbertstown, Co. Limerick. *Proceedings of the Royal Irish Academy*, **63**, 167–90.

Watts, W.A. (1967) Interglacial deposits in Kildromin Townland, near Herbertstown, Co. Limerick. *Proceedings of the Royal Irish Academy*, **65**, 339–49.

Watts, W.A. (1977) The late Devensian vegetation of Ireland. *Philosophical Transactions of the Royal Society of London, B*, **280**, 273–93.

Watts, W.A. (1985) Quaternary vegetation cycles, in *The Quaternary History of Ireland*, (K.J. Edwards and W.P. Warren), Academic Press, London, pp. 155–86.

Watts, W.A. (1988) Europe, in *Vegetation History*, (eds B. Huntley and T. Webb), Kluwer, Dordrecht, pp. 155–92.

Webb, D.A. (1955) *Erica mackaiana* Bab. Biological flora of the British Isles. *Journal of Ecology*, **43**, 319–30.

Webb. N. (1986) *Heathlands*, Collins, London.

Webb, N.R. and Haskins, L.E. (1980) An ecological survey of heathland in the Poole Basin, Dorset, England in 1978. *Biological Conservation*, **17**, 281–96.

Wein, R.W. (1973) *Eriophorum vaginatum* L. Biological flora of the British Isles. *Journal of Ecology*, **61**, 601–15.

Wells, T.C.E. (1968) Land use changes affecting *Pulsatilla vulgaris* in England. *Biological Conservation*, **1**, 37–43.

Wells, T.C.E., Sheail, J. and Ball, D.F. (1976) Ecological Studies on the Porton Ranges: relationships between vegetation, soils and land-use history. *Journal of Ecology*, **64**, 589–626.

Wesley, A. (1973) Jurassic plants, in *Atlas of Palaeobiogeography*, (ed. A. Hallam), Elsevier Scientific Publishing Company, Amsterdam, pp. 329–38.

West, R.G. (1956) The Quaternary deposits at Hoxne, Suffolk. *Philosophical Transactions of the Royal Society of London, B*, **239**, 265–356.

West, R.G. (1961) Vegetational history of the Early Pleistocene of the Royal Society borehole at Ludham, Norfolk. *Proceedings of the Royal Society of London*, **155**, 437–53.

West, R.G. (1970) Pollen zones in the Pleistocene of Great Britain and their correlation. *New Phytologist*, **69**, 1179–83.

West, R.G. (1977) Flora and fauna. Early and Middle Devensian flora and vegetation. *Philosophical Transactions of the Royal Society of London, B*, **280**, 229–46.

West, R.G. (1980a) Pleistocene forest history in East Anglia. *New Phytologist*, **85**, 571–622.

West, R.G. (1980b) *The Pre-glacial Pleistocene of the Norfolk and Suffolk Coasts*, Cambridge University Press, Cambridge.

Wheeler, B.D. (1980) Plant communities of rich-fen systems in England and Wales, parts 1–3. *Journal of Ecology*, **68**, 368–95, 405–20, 761–88.

Whitelock, D. (ed.) (1979) *English Historical Documents 1. c.500–1042*, Eyre Methuen/Oxford University Press, London.

Whitlock, R. (1983) *The English Farm*, Dent, London.

Wilkinson, G. (1981) *A History of Britain's Trees*, Hutchinson, London.

Wilkinson, G.C. and Boulter, M.C. (1980) Oligocene pollen and spores from the western part of the British Isles. *Palaeontographica B*, **175**, 27–83.

Wilkinson, G.C., Bailey, R.A.B. and Boulter, M.C. (1980) The geology and palynology of the Oligocene Lough Neagh Clays, Northern Ireland. *Journal of the Geological Society, London*, **137**, 65–75.

Williams, C.H. (ed.) (1971) *English Historical Documents 5. 1485–1558*, Eyre Methuen/Oxford University Press, London.

Williams, D.S.F. (1963) Farming patterns in Craven. *Field Studies*, **1**, 116–39.

Williams, J.T. (1963) *Chenopodium album* L. Biological flora of the British Isles. *Journal of Ecology*, **51**, 711–25.

Williamson, T. (1987) Early co-axial field systems on the East Anglian boulder clays. *Proceedings of the Prehistoric Society*, **53**, 419–31.

Willis, A.J., Folkes, B.F. Hope-Simpson, J.F. and Yemm, E.W. (1959) Braunton Burrows: the dune system and its vegetation, parts 1 and 2. *Journal of Ecology*, **47**, 1–24, 249–88.

Wills, L.J. (1910) On the fossiliferous Lower Keuper Rocks of Worcestershire. *Proceedings of the Geologist's Association*, **21**, 249–332.

Wilson, P. and Farrington, O. (1989) Radiocarbon

dating of the Holocene evolution of Magilligan Foreland, Co. Londonderry. *Proceedings of the Royal Irish Academy*, **89B**, 1–23.

Wiltshire, P.E.J. and Edwards, K.J. (1993) Mesolithic, early Neolithic and later prehistoric impacts on vegetation at a riverine site in Derbyshire, England, in *Climate Change and Human Impact on the Landscape*, (ed. F.M. Chambers), Chapman & Hall, London, pp. 157–68.

Wolfe, W.D.I., Durant, G.P., Fallick, A.E. *et. al.* (1990) An early terrestrial biota preserved by Visean vulcanicity in Scotland. *Geological Society of America, Special Paper*, **244**, 13–24.

Woodell, S.R.J. (1958) *Daboecia cantabrica* K.Koch, Biological flora of the British Isles. *Journal of Ecology*, **46**, 205–16.

Woodell, S.R.J. (1965) Natural hybridization between the cowslip (*Primula veris* L.) and the primrose (*P. vulgaris* Huds.) in Britain. *Watsonia*, **6**, 190–202.

Woodhead, N. (1951) *Lloydia serotina* (L.) Rchb. Biological flora of the British Isles. *Journal of Ecology*, **39**, 98–203.

Woodward, F.I. (1987) Stomatal numbers are sensitive to increases in CO_2 from pre-industrial levels. *Nature*, **327**, 617–18.

Woodward, F.I. (1993) Plant responses to past concentrations of CO_2. *Vegetatio*, **104**, 145–55.

Woodward, M. (1972) *Leaves from Gerard's Herball*, Thorsons, London.

Wu, L., Bradshaw, A.D. and Thurman, D.A. (1975) The potential for evolution of heavy metal tolerance in plants. II The rapid evolution of tolerance in *Agrostis stolonifera*. *Heredity*, **34**, 165–87.

Wymer, J. (1985) *The Palaeolithic sites of East Anglia*, Geo Books, Norwich.

Yelling, J. (1990) Agriculture 1500–1730, in *An Historical Geography of England and Wales*, 2nd edn, (eds R.A. Dodgshon and R.A. Butlin), Academic Press, London. pp. 181–98.

Zagwijn, W.H. (1960) Aspects of the Pliocene and early Pleistocene vegetation in the Netherlands. *Mededeelingen van de Geologische Stichting Serie C, III*, **1**, no. 5.

INDEX

Page numbers appearing in **bold** refer to figures

Fossil sites, quarries, rock exposures and boreholes, mentioned in the text

1. Pen-y Glog
2. Llangammarch Wells, Capel Horeb, Cwm Craig Ddu
3. Perton Lane, Rockhall, Targrove
4. Freshwater East
5. Llanover
6. Rhynie
7. Parkhill
8. Devilsbit Mountain
9. Bay of Skaill
10. Brecon Beacons
11. Stromness
12. Ballanacutor Farm
13. Spital, Halkirk
14. Burren Hill
15. Kiltorcan Beds
16. Plaistow/Sloley
17. Baggy Point
18. Whiteadder River
19. Kingwater
20. Oxroad Bay
21. Pettycur
22. East Kirkton
23. Mauchline Basin
24. Kimberley Ralway Cutting
25. Middridge, Shildon Station
26. Hilton Beck, Vale of Eden
27. Corley
28. Radstock
30. Cnap Twt
31. Stonesfield
32. Brearreraig
33. Roseberry Topping
34. Hayburn Wyke
35. Lulworth Cove
36. Culgower
37. Weald
38. Luccombe Chine, Hanover Point
39. Arsliignish Arda, Shiaba, Bremanour
40. Antrim
41. Herne Bay, Sheppey
42. Bognor
43. Bembridge
44. Bovey Tracey
45. Lough Neagh
46. Tremadoc
47. Bee's Nest, Grassington
48. Hollymount
49. Corallline Crag
50. Castle Eden
51. Red Crag, Walton on the Naze
52. Ludham
53. Easton Bavents
54. Paston, West Runton
55. Corton Cliffs
56. Beeston, Roosting Hill (Beetley)
57. Hoxne
58. Bobbitshole, Ipswich
59. Woodston
60. Mark's Tey
61. Gort
62. Kilbeg
63. Baggotstown
64. Wretton
65. Chelford
66. Upton Warren
67. Ballynahgilly
68. Fallahogy
69. Barfield Tarn
70. Tregaron
71. Whixall Moss
72. Bloak Moss
73. Iping Common
74. Kennet Valley
75. Star Carr

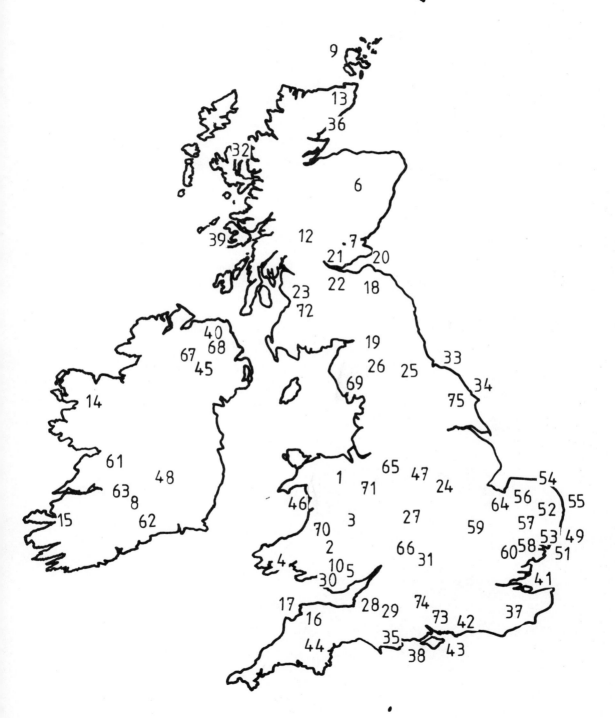